Advanced Computer-aided Fixture Design

Advanced Computer-aided Fixture Design

Dr. Yiming (Kevin) Rong
Worcester Polytechnic Institute
Worcester, MA

Dr. Samuel H. Huang
University of Cincinnati
Cincinnati, OH

Dr. Zhikun Hou
Worcester Polytechnic Institute
Worcester, MA

ELSEVIER
ACADEMIC
PRESS

Amsterdam • Boston • Heidelberg • London
New York • Oxford • Paris • San Diego
San Francisco • Singapore • Sydney • Tokyo

Elsevier Academic Press
30 Corporate Drive, Suite 400, Burlington, MA 01803, USA
525 B Street, Suite 1900, San Diego, California 92101-4495, USA
84 Theobald's Road, London WC1X 8RR, UK

This book is printed on acid-free paper. ⊚

Library of Congress Cataloging-in-Publication Data
Rong, Yiming, 1958-
  Advanced computer-aided fixture design / Yiming (Kevin) Rong, Samuel H. Huang,
Zhikun Hou.
    p. cm.
  Includes bibliographical references and index.
  ISBN 0-12-594751-8 (casebound : alk. paper) 1. Jigs and fixtures--Computer-aided
design. I. Huang, Samuel H. II. Hou, Zhikun. III. Title.
  TJ1187.R62 2005
  621.9'92--dc22

2004026784

British Library Cataloguing in Publication Data
A catalogue record for this book is available from the British Library

ISBN: 0-12-594751-8

For all information on all Elsevier Academic Press publications
visit our Web site at www.books.elsevier.com

Transferred to Digital Printing 2008

PREFACE

Manufacturing industry involves tooling intensive operations. Fixturing is an important manufacturing operation which contribute greatly to the production quality, cycle time, and cost. As computer technology, especially computer-aided design and manufacturing (CAD/CAM), developed, more and more attention has turned to design and verify fixtures digitally. For more than 15 years, computer-aided fixture design (CAFD) techniques have been developed and gradually applied in industry. The motivation of CAFD is to generate conceptual and detail fixture designs rapidly even in product and production design stages, to provide tools for fixture design and process verification, and to implement the CAD/CAM integration.

A first book introducing the CAFD technique comprehensively was published in 1999. *Computer-aided Fixture Design* (published by Marcel Dekker, New York, 1999, ISBN: 0-8247-9961-5) summarized the author research work of CAFD from 1992 -1999. Since the book published, a lot of advanced research on CAFD has been carried out. Especially the CAFD has been expanded to and integrated with machining systems planning and fixture design analysis for process verification. The aim of this book is to provide a comprehensive knowledge of CAFD and to introduce the recent research advance on CAFD as well as in relevant fields. This book is mainly based on the authors' long time research work on CAFD. The content of the book is uniquely designed for a thorough understanding of the CAFD and related topics, from fixture planning to tolerance analysis both inter-setups and intra-setup, and from fixture structural design to fixture design verification. In the new era of supplier-based manufacturing, CAFD techniques and systems are particular important in technical specifications and business quoting as well as process verifications between OEM and their suppliers.

This book can be used as a text book for engineering graduate students in class study or an engineering reference book for manufacturing engineers in workshop practice.

The CAFD problem has been increasely studied. Development of a practical techniques and applications system in CAFD is a major task in the study. This will be the second comprehensive-technical reference book in the area. The related research has been supported by National Science Foundation (NSF) and major manufacturing companies, such as Delphi Corporation, Caterpillar, Ford Motor Company, Pratt & Whitney, and GE Aircraft Engines.

AUTHOR BIOGRAPHIES

Yiming (Kevin) Rong is the Higgins Professor of Mechanical Engineering and the Director of Computer-aided Manufacturing Laboratory (CAM Lab.) at Worcester Polytechnic Institute (WPI). He received a B.S. Degree in Mechanical Engineering from Harbin University of Science and Technology, China, in 1981; an M.S. Degree in Manufacturing Engineering from Tsinghua University, Beijing, China, in 1984; an M.S. Degree in Industrial Engineering from University of Wisconsin, Madison, WI in 1987; and a Ph.D. in Mechanical Engineering from University of Kentucky, Lexington, KY, in 1989. Dr. Rong worked as a faculty member at Southern Illinois University at Carbondale for eight years before joining WPI in 1998.

Dr. Rong's research area is Computer-aided Manufacturing, including Manufacturing Systems, Manufacturing Processes, and Computer-aided Fixture Design (CAFD). His research on CAFD has been recognized both nationally and internationally. He is the principal investigator of many research projects funded by NSF, DOE, the Air Force, SME, and several major manufacturing companies. Dr. Rong is a fellow of ASME and a member of SME and ASEE. He has published many technical papers in professional journals and conference proceedings.

Samuel H. Huang is currently Associate Professor and Director of Intelligent CAM Systems Laboratory at the Department of Mechanical, Industrial and Nuclear Engineering (MINE), University of Cincinnati. He has over ten years of machining-related research experience and published over 80 technical papers. He also worked for UGS/PLM Solutions in 1996 and 1997 and managed several projects on CAM functionality improvement for the Unigraphics software package. In the summer of 1993, he worked as a guest researcher at NIST's Manufacturing Systems Integration Division. Dr. Huang received a B.S. Degree in Instrument Engineering from Zhejiang University (P. R. China) in 1991, and M.S. and Ph.D Degrees in Industrial Engineering from Texas Tech University in 1992 and 1995, respectively. He is a member of SME, IEEE, IIE, ASEE, and Alpha Pi Mu.

Zhikun Hou is currently Professor of Mechanical Department at Worcester Polytechnic Institute. He received his Masters of Engineering Degree from Tongji University, China, in 1981; and Masters of Science and Ph.D. degrees from California Institute of Technology in 1985 and 1990, respectively. He has served on the ASCE Dynamic Committee and ASME Committee of Uncertainty and Probability and is an Associate Editor of *The Journal of Shock and Vibration*. He has authored over 100 technical papers in the area of dynamics and vibration, including random vibration and reliability analysis, earthquake engineering and structural dynamics, smart materials and intelligent structure, and, most recently, structural health monitoring.

TABLE OF CONTENTS

CHAPTER 1

Introduction

In a report on manufacturing technology, it was estimated that during the latter half of the 1990s the innovations in manufacturing technologies contributed nearly $1 trillion to the U.S. economy. Manufacturing is described as a cornerstone of the American economy. Manufacturing is critical to national security, defense, advanced technologies, precision machineries, and even our standard of living and creation of wealth. However, U.S. manufacturing is in decline. Between July of 2000 and December of 2003, 2.8 million manufacturing jobs disappeared. Since 1997, machine tool orders have decreased by 68%, from $8 billion to $3.35 billion, forcing 10% of the industry to close its doors (AMT, 2004).

U.S. manufacturers need to be able to compete in the global marketplace and have maximum agility in customizing their products for regional and personal preferences. The key is to develop enabling technologies to produce low- to medium-volume customized products with mass-production efficiency. Recent advances in flexible automation technologies, such as computer numerical controlled (CNC) machines, high-speed networking, and e-business, have made such quantum productivity jumps possible.

Computer-aided design and computer-aided manufacturing (CAD/CAM) tools have been used for product and process design for decades. As the power of computing continues to increase, the use of CAD/CAM has been ubiquitous, extending to the smallest shop and the most remote countries in the world. More recently, major CAD/CAM companies have been promoting product life-cycle management (PLM) in an attempt to integrate all product life-cycle functions, including engineering, purchasing, and manufacturing, into CAD-based systems. These PLM products have increased engineers' productivity noticeably. However, the true potential of product life-cycle utilities is far from being fully realized. The problem is in part due to the complexity of product life-cycle functions. For example, in an internal survey Delphi found that a typical product drawing has 300 corresponding manufacturing documents. Manufacturing of a single component requires detailed consideration of candidate processes, tooling, fixtures, machines, and process parameters with complex interactions and constraints among them. Current PLM products are general-purpose tools and often involve many discrete modules or independent software that require extensive training and a great deal of manual input. The almost endless possibilities in tooling, work holding, and machines make general-purpose tools cumbersome and difficult to use.

To maximize overall manufacturing efficiency, the automotive industry, under extreme cost pressure and global competition, has developed elaborate processes to bring innovative products to market at minimum cost. These processes typically start with voice-of-customer (VOC), which defines product requirements. From these requirements, the product design is derived based on engineering principles and technical innovations. The product design dictates dimensions, surface finish, and tolerance requirements for each component. These requirements are subsequently used to determine appropriate processes, manufacturing systems, equipment design, build and runoff, and finally production launch and continuous improvement. Manufacturing system design (MSD) comes after capable processes are determined, but before hardware design and build are committed. MSD is the most opportunistic stage for optimization. It can help resolve design-for-manufacturing issues in making a product more manufacturable, and it is the right time to put all manufacturing options on the table and to determine the best solution before committing to a costly hardware build.

In facilitating the MSD process, it is important to take advantage of CAD technologies that help engineers visualize complex 3D geometric relationships of the workpiece, tooling, fixtures, machine components, tool paths, and so on toward detecting design errors, misfits, or interferences. Starting with a part family model, the manufacturing features are first recognized automatically and assigned with appropriate process parameters. Setup planning and machine tool/fixture selection and design can then be done based on the best-practice knowledge captured and represented in bill of processes (BOP). A series of verification functions is executed to validate the manufacturing plan. If all of these functions can be done within a few minutes or even hours, the MSD system can be a very effective tool for evaluating many process and manufacturing system concepts in an MSD workshop. Modeling complexity can be an issue. The key is to reduce system complexity by taking advantage of the similarities within a part family. By the same token, similarities in machines, fixtures, tooling, and machining features can be utilized to provide a structure to a solution to the problem. Within a well-defined problem or subproblem, verification functions can be used to provide feedback and toward identifying optimal solutions to a problem. The advanced optimization techniques, such as genetic algorithms, fuzzy logic, and neural network, can be applied to provide comprehensive optimization functions for fixture layout, locator and clamping layout, process sequencing, manufacturing system design, tool path, cycle time, cost, and so on. These optimization functions are especially useful in new product families for which little history or production experience exists to help derive an optimal solution by heuristic rules.

Manufacturing systems consist of manufacturing equipment laid out in sequence as a production line. Manufacturing equipment mainly includes machine tools, fixtures, and processing tools, which may be provided by different vendors. The manufacturing equipment contains the capabilities to generate geometric forms with the combination of the primary/feed motions provided by a machine tool and the cutting edge provided by a cutting tool; to process features in a certain position and orientations in one setup through the synthesis of machining table, spindle, and fixture; and to ensure production precision. Requirements regarding production time, cost, and quality; the process capabilities of available common equipments and, more complicatedly, customized equipments place major constraints on process planning and optimization.

A model for manufacturing equipment capability is very important in rapid MSD. Having established model of manufacturing equipment elements (e.g., machine, cutter, and fixtures), the overall capability of the subsystem can be established. Together with process information, the performance of the subsystem can be estimated with measures of quality, cycle time, and flexibility. Further studies are required to integrate manufacturing knowledge into the analysis of kinematics, stiffness, and accuracy information of machine tools, fixtures, and cutting tools so that mapping between manufacturing equipment capability data and manufacturing requirements identified at the part design stage can be performed.

Strategies regarding common manufacturing equipment exist for rapid MSD and optimization. Common manufacturing equipment is defined as manufacturing equipment qualified by current best practice for a family of parts in particular operations. Such equipment might be a machine, machine module, fixture, or combination of thereof. The commonality of equipment in processing similar parts simplifies the development of new equipment enhancing optimal performance. Via flexible combinations of common equipment, a variety of parts and processes can be dealt with based on similarities among those parts and processes. It has been proved that adapting common equipment to new production requirements can result in optimal solutions quickly.

Fixture design is also part of MSD. The objective of fixture design is to generate fixture configurations to hold parts firmly and accurately during manufacturing processes. Therefore, in rapid MSD, fixture design should be conducted at an early stage, in particular because conceptual fixture design contributes to the feasibility validation of MSD. Fixture design should also be verified to avoid later modification of the production system and processes, which can increase costs significantly. In other words, an intelligent fixture design is desired in MSD. An intelligent fixture design is necessary in adapting product and process designs while maintaining an optimal design for function and structural performance. An intelligent fixture design involves

- automated generation of a fixture design,
- use of best-practice knowledge in the design,
- reuse of best proven structural designs for specific functions,
- parametric design based on its correlation with required performance, and
- self-verification capability to ensure the design quality.

Although fixtures can be designed using CAD functions, a lack of scientific tools and a systematic approach in evaluating design performance reduces to a process of trial-and-errors, resulting in several problems. Such problems include the over-design of functions, which is very common and sometimes degrades performance; compromised quality of design before production; and the long lead time of fixture design, fabrication, and testing, which may take weeks if not months. Therefore, computer-aided fixture design (CAFD) has been motivated.

CAFD incorporates automated modular fixture design, where standard fixture components are used to construct desired fixture configurations (Rong 1999; Kow 1998; Brost 1996); dedicated fixture design with predefined fixture component types (An 1999; Chou 1993); rule-based and case-based reasoning (CBR) fixture design (Nee 1991; Kumar 1995; Pham 1990; Sun 1995; Boyle 2003); variation fixture design for part families (Rong 2003); and fixture design verification (Fuh 1994; Kang 2003). CAFD research has provided a methodology and concept-proven prototypes. How to make use of best-practice knowledge in fixture design and how to verify fixture design quality under various conditions remain areas of study. The applications of CAFD are still very limited because many operational constraints need to be considered.

CAFD has been a research focus at the Computer-aided Manufacturing Laboratory (CAM-Lab) at Worcester Polytechnic Institute (WPI), including association with Southern Illinois University at Carbondale, in collaboration with other research teams, for almost 15 years. Early work was presented in the first comprehensive book in the area of CAFD (Rong 1999). This book summarizes recent CAFD research work.

During such research it became clear that fixture design is part of manufacturing systems planning, in that fixtures are part of a manufacturing system. Fixture design is also part of manufacturing process verification, in that fixture performance contributes to the performance of manufacturing processes significantly, in both quality assurance and process stability, as well as to ease of process operation, which contributes to the efficiency of production cycle time and operation ergonomics. Therefore, this book takes the perspective that CAFD research is integrated with computerized manufacturing planning and computerized verification of fixture design and manufacturing processes.

In a supplier-based manufacturing environment, fixture design activities become very "agile" in terms of by whom and where they may be produced. Although most major manufacturing companies (sometime referred to as OEMs, original equipment manufacturers) are not performing a significant amount of fixture design, not many people doubt the importance of the quality and efficient delivery time of fixture design contributes to an "agile" production environment. CAFD is particularly favorable to fixture designers (usually in the tier 2 or 3 suppliers), but also has much to offer major manufacturing companies, although the application of CAFD in production is at present quite limited. Fixture design involves a lot of science and knowledge, both analytical and empirical. In many cases, fixture design technology is domain dependent and different practice may require different applications. Although CAFD-related techniques and systems have been studied in a comprehensive way regarding individual aspects of the problem, it is difficult to formulate a comprehensive "system" that will address a majority of problems. Therefore, in this book each chapter represents a topic within the problem that has been studied in detail and is relatively independent of others. Each chapter offers solutions to the problem at hand, and each chapter includes introductory materials and a list of references cited in the text. Readers have the option of selecting chapters that address specific applications.

The book begins with a system view of manufacturing system planning, including fixture design integration. It continues with several generations of CAFD studies and design verification. Finally, more advanced fixture analysis is presented in Chapters 5 and 6. The book content includes both theoretical and practical studies of CAFD. The book serves as a valuable reference for professional personnel in problem solving and as a source of study for the academic side. Research in CAFD continues, which will produce further applicable results in the future.

The authors would like to show their appreciation and acknowledgements to many research associates and graduate research assistants contributed to the work presented in the book. Ac-knowledgement also goes to the National Science Foundation (NSF) and several major manufacturing companies, such as Delphi Corporation and Ford Motor Company, for funding the research projects and providing production data and knowledge.

References

AMT. "Why Do We Need Manufacturing Technology Made in America?," Association of Manufacturing Technology (AMT) Annual Report 2004.

An, Z., S. Huang, J. Li, Y. Rong, and S. Jayaram. "Development of Automated Fixture Design Systems with Predefined Fixture Component Types: Part 1, Basic Design," *International Journal of Flexible Automation and Integrated Manufacturing* 7:3/4, pp. 321–341, 1999.

Boyle, I., D. Brown, and Y. Rong. "Case-based Reasoning in Fixture Design," Photonics East, Oct. 27–30, 2003, Providence, RI. Proceedings of SPIE 5263, *Intelligent Manufacturing*, pp. 85–96

Brost, R. C., and K. Y. Goldberg. "A Complete Algorithm for Synthesizing Modular Fixtures for Polygonal Parts," *IEEE Transactions on Robotics and Automation* 12:1, pp. 31–46, 1996.

Chou, Y. C. "Automated Fixture Design for Concurrent Manufacturing Planning," *Concurrent Engineering: Research and Applications* 1, pp. 219–229, 1993.

Fuh, J.Y.H., and A.Y.C. Nee. "Verification and Optimization of Workholding Schemes for Fixture Design," *Journal of Design and Manufacturing* 4, pp. 307–318, 1994.

Kang, Y., Y. Rong, and J.-C. Yang. "Computer-aided Fixture Design Verification: Part 1, The Framework and Modeling; Part 2, Tolerance Analysis; and Part 3, Stability Analysis," *International Journal of Advanced Manufacturing Technology* 21, pp. 827–849, 2003.

Kow, T. S., A. S. Kumar, and J.Y.H. Fuh. "An Integrated Computer-aided Modular Fixture Design System for Interference-free Design," MED-8, pp. 909–916, ASME IMECE, Anaheim, CA, Nov. 15–20, 1998.

Kumar, A. S., and A.Y.C. Nee. "A Framework for a Variant Fixture Design System Using Case-based Reasoning Technique," MED-2:1, pp. 763–775, ASME WAM, 1995.

Nee, A.Y.C., and A. S. Kumar. "A Framework for an Object/Rule-based Automated Fixture Design System," *Annals of the CIRP* 40:1, pp. 147–151, 1991.

Pham, D. T., and A. de Sam Lazaro. "AUTOFIX: An Expert CAD System for Fixtures," *International Journal of Machine Tools and Manufacture* 30:3, pp. 403–411, 1990.

Rong, Y., and X. Han. "Computer-aided Reconfigurable Fixture Design," CIRP Second International Conference on Reconfigurable Manufacturing, Ann Arbor, MI, Aug. 20–22, 2003.

Rong, Y., and Y. Zhu. *Computer-aided Fixture Design*. New York: Marcel Dekker, 1999.

Sun, S. H., and J. L. Chen. "A Modular Fixture Design System Based on Case-based Reasoning," *International Journal of Advanced Manufacturing Technology* 10, pp. 389–395, 1995.

CHAPTER 2

Computerized Manufacturing Setup Planning

2.1 AN OVERVIEW

2.1.1   Setup Planning

2.1.1.1 A Quick Look at Setup Planning

Manufacturing setup planning involves developing detailed work instructions for setting up a part for machining. The purpose is to ensure the stability of the part during machining, and more importantly the precision of the machining process. Note that our discussion focuses on material removal processes such as milling and turning, as these processes constitute the vast majority of manufacturing activities and represent $60 billion per annum in U.S. industrial activities (DeGarmo 1997). Figure 2.1 shows a setup plan for a shaft support. Four surfaces (shaded) are machined in four setups. A, B, and C within the diagrams denote the primary, secondary, and tertiary datums, respectively.

2.1.1.2 The Important Role of Setup Planning

To understand the importance of setup planning, it is necessary to review current state-of-the-art manufacturing. The U.S. economy enjoyed a prolonged expansion in the 1990s, which is attributed in general to the rapid advance in information technology (IT) and in particular to the advent of the Internet. It may be a surprise to many that manufacturing actually played a significant role in this economic expansion. A study showed that real output of durable goods manufacturer increased at twice the rate of the overall economy from 1992 to 1997 (SME 2000). This performance is remarkable, considering the fact that the U.S. manufacturing industry has faced fierce competition from the global market for many years and has lost millions of manufacturing jobs to developing countries. Whereas developing countries take advantage of their low labor costs to manufacture inexpensive goods, U.S. manufacturers are able to stay on top of the competition through improved productivity. Manufacturing productivity improvement is fueled by the adoption of high technology, including computer-aided design (CAD) and computer-aided manufacturing (CAM). CAD/CAM systems are now standard engineering tools used to reduce the time and costs of product design and manufacture. The present problem, however, is the inability to completely realize the potential efficiency of CAD/CAM systems due to the tedious and costly human interaction required.

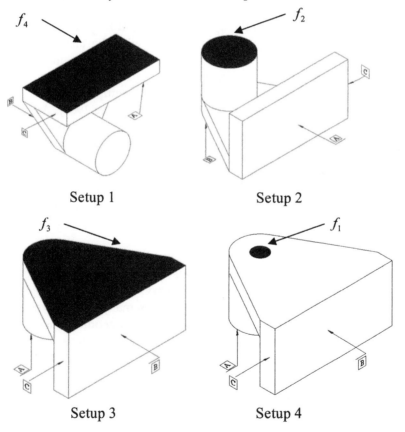

Figure 2.1. A setup plan for a shaft support.

It is well known that there is a functional gap between CAD and CAM, and thus the need for tedious human interaction. Computer-aided process planning (CAPP) has been proposed as the link to achieve complete CAD/CAM integration (Kiritsis 1995; Alting 1989; Ham 1988; Chang 1985). Although basic research in CAPP has advanced significantly during the past three decades, very few CAPP systems have reached the stage of commercialization. Because very large industrial proprietary R&D efforts have not been successful in achieving CAD/CAM integration using CAPP systems, the few standalone commercial CAPP systems, e.g., PART by Tecnomatix, have had a very limited impact on the manufacturing industry. Compared with the widespread use of CAD/CAM systems, the development of CAPP systems is hardly a success story.

The barrier that hinders the application of CAPP systems is mainly due to the enormous complexities associated with the task of process planning. Whereas there have been great successes in automating portions of the process planning task, such as operation sequencing and cutter path optimization, other areas such as setup generation and automatic fixturing remain elusive (Sarma 1996; Hayes 1989). A variety of commercial software tools can automatically generate NC (numerically controlled) codes for a machining operation once required parameters (machining geometry, feeds, speeds, and so on) are specified. However, none of them can provide solutions to problems such as setup formation and sequencing, workpiece locating, and workholding. As a result, the task of setup planning and fixture design is still performed manually, which constitutes the need for the "missing link" in CAD/CAM integration.

The difficulty in setup planning is further complicated by advances in NC technology, such as the high-performance machining centers utilized widely in the metal-cutting industry. Present technology has allowed machine motions to be controlled in as many as five axes, with high accuracy in producing products with complex geometry using both rough and finishing processes. As a result, multi operations in a single setup is a common and desired configuration for ensuring product quality and production flexibility. As pointed out by Trappey (1992), under current condition, fixture design has become a major restriction in multi operation setup planning, which influences operation planning and the determination of a production system. As shown in Figure 2.2, a successful fixture design strategy includes three steps; namely, setup planning, fixture planning, and fixture configuration design. The objective of setup planning is to determine the number of setups needed, the orientation of the workpiece, and the machining surfaces in each setup. Fixture planning is used to determine the locating, supporting, and clamping points on workpiece surfaces. The task of fixture configuration design is to select fixture elements and place them into a final configuration to locate and clamp the workpiece. It is recognized that setup planning is the key to CAD/CAM integration because it takes product design and manufacturing requirement information from CAD models and provides information to CAM for NC programming and fixture design.

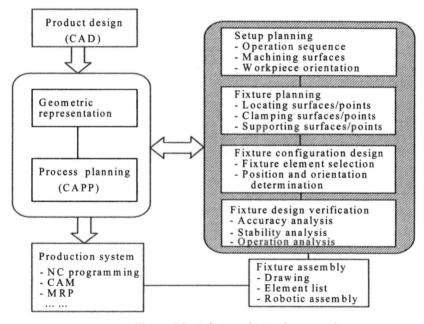

Figure 2.2. A fixture design framework.

2.1.1.3 A Closer Look at Setup Planning

It is now clear that setup planning is an important part of process planning. Process planning is "the act of preparing detailed work instructions to produce a part" (Chang 1985). Generally, a part design defines its specifications and functionalities, yet provides no information on how the part is to be manufactured. Confined by a certain production environment, process planning maps a design to a number of manufacturing steps and provides detailed instructions for each step, following which a piece of raw material is converted to a finished part. To gain an in-depth understanding of process planning, see Halevi (1995).

Setup planning evolved from the practice of decomposing the complex task of process planning into a number of manageable parts. Figure 2.3 illustrates a typical scheme of partitioning a material removal process planning task into several sequential phases. Each phase makes decisions in a relatively independent manner and issues information down to the subsequent phase. When a phase involves more than one problem domain, subtasks are further developed, which are normally highly interrelated. Setup planning refers to phase III in the illustrated scheme. However, there is no well-acknowledged standard regarding its scope. Setup planning was considered as the selections of setups, fixturing elements, and access direction, which along with operation sequencing constitute the macro level of process planning (Sarma 1996). Rong (1997) treats setup planning as the determination of number of setups needed, the orientation of workpieces in each setup, and the manufacturing surface in each setup. Huang (1998) defines setup planning as the act of preparing the instruction for setting up parts for machining, which consists of setup formation, datum selection, and setup sequencing. Joneja and Chang (1999) incorporate into setup planning, fixture, and cutting tool selection.

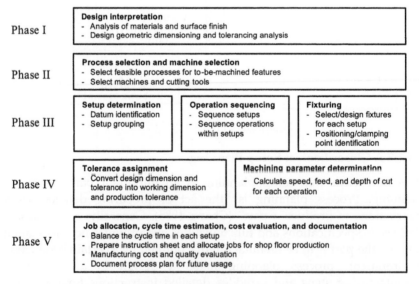

Figure 2.3. Decomposition of the process planning task.

The vagueness of setup planning boundaries in the literature reflects the diversity of industrial practices. Guided by the same "divide and conquer" strategy, different production entities having different equipment, factory layouts, manpower, and targeted markets often take completely different dividing methods in developing a process plan to manufacture a similar part. Nevertheless, a formal definition of setup planning is warranted as a systematic solution cannot exist without a formal problem definition. Furthermore, the standardization of process planning has been an inevitable trend in the industry. Here, we define setup planning as an intermediate phase of process planning that takes to-be-machined features and selected machines as inputs and performs the following subtasks:

- Identify tool-approach directions and machining datums
- Group and sequence setups
- Sequence operations within each setup
- Configure fixturing components and/or provide guidelines to fixture design
- Output the previous information as a plan to a subsequent process planning phase

This definition is graphically shown in Figure 2.4. It breaks down setup planning into a number of subtasks. To accomplish these subtasks, certain manufacturing constraints have to be respected. Before elucidating these constraints, we will look at the inputs to setup planning first. The to-be-machined features contain geometry and tolerance information of those features that require machining. The former includes feature shape and dimension, whereas the latter provides dimensional and geometrical tolerances. The selected machines are a collection of machines needed to produce the part. Take the shaft support shown in Figure 2.1 for example. Industrial production of the shaft support begins with a die-casting process in which the part is made into shape first (usually without the hole f_1), followed by material removal processes that produce the final part. Therefore, only features f_1, f_2, f_3, and f_4 require machining (to meet tolerance specification required for assembly). The machining processes can be carried out using a vertical machining center, whereby features f_2, f_3, and f_4 are milled and feature f_1 is first drilled and then bored.

Given the input information, setup planning starts by identifying the tool-approach directions and machining datums for the to-be-machined features. The tool-approach direction of a feature is an unobstructed path a tool can take to access the feature in the workpiece. Some features have multiple tool-approach directions such as f_1, which is a through hole that has two tool-approach directions +Y and –Y. Features f_2, f_3, and f_4 all have one tool-approach direction. They are +Y, –Y, and +Z, respectively.

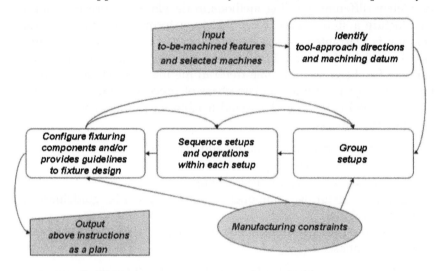

Figure 2.4. A formal definition of setup planning.

Machining datum identification refers to identifying potential main locating surfaces and their corresponding supporting and clamping points for each to-be-machined feature. A main locating surface is a surface in the workpiece, upon which its corresponding to-be-machined feature sits for undergoing the machining process. To have a workpiece secured in a desired position, supporting and clamping forces have to be exerted in certain points, pushing the workpiece against the main locating surface and counteracting the cutting forces from machining. Several factors have to be taken into consideration while choosing the main locating surface, including the accuracy associated with it, the stability it provides, and the availability of its corresponding supporting and clamping points. Usually, the main locating surface of a to-be-machined feature is chosen to be the largest surface on the opposite side of the feature. For example, feature f_3 is the main locating surface to machine feature f_2.

Those to-be-machined features that share common tool-approach-directions and machining datums would naturally be grouped into the same setup. In this context, a setup refers to a set of to-be-machined features that will be machined consecutively without repositioning the workpiece. For those to-be-machined features that have multiple tool-approach directions and machining datums, decisions have to be made to group them into a particular setup. Similarly, decisions have to be made to sequence setups and operations within each setup. The decision-making process is constrained by the physical manufacturing environment, as previously mentioned. Manufacturing constraints include feature interaction, datum and reference requirements, good manufacturing practice, and tolerance requirements.

Feature Interaction. Feature interaction occurs when machining a feature may destroy the requirements of machining another feature. The main concern of feature interaction is clamping. An example is shown in Figure 2.5, where the holes must be drilled before the angle is cut, because otherwise the part could not be clamped firmly on an angled surface (Hayes 1989). Another concern of feature interaction comes from the mechanics of machining processes. For example, for the slot and offset-hole interaction shown in Figure 2.6, the maximum deflection caused by drilling is calculated by the mechanistic models of milling and drilling processes. If the deflection exceeds the allowable cylindricity of the hole, drilling of the hole should be performed prior to milling the slot.

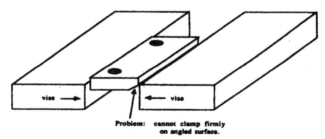

Figure 2.5. Feature interaction due to clamping (Hayes 1989).

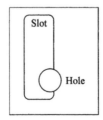

Figure 2.6. Feature interaction due to machining mechanics (Lin 1998).

Datum and Reference Requirements. Datum and reference requirements mandate that a surface must be present before it can be used as a machining datum or reference. For example, in Figure 2.7 the hole is located using two dimensions dist1 and dist2. The dimension dist2 is referenced from the bottom surface of the slot. Therefore, the slot should be machined before the hole is drilled.

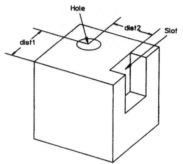

Figure 2.7. Datum and reference requirements (Zhang 1995).

Good Manufacturing Practice. Good manufacturing practice refers to experience that comes from industry practice. For example, it is a rule of thumb to machine a hole having the smallest diameter or the longest depth before another hole of a larger diameter or a lesser depth. This is because the straightness of a thinner hole or of a more precise hole can be impaired by the larger or less precise hole when they intersect.

Tolerance Requirements. As previously mentioned, an important aspect of setup planning is to ensure the precision of the machining process. In other words, a good setup plan must ensure that design tolerance requirements are met after machining. We view tolerance analysis as a driving factor in obtaining a good setup plan.

Once the manufacturing constraints are taken into consideration to determine setup groups and sequences, a preliminary fixturing scheme for each setup has to be developed. A broad fixturing method has to be determined first, such as vise fixturing, modular fixturing, or special fixture design. Then, a fixture configuration for the chosen method has to be generated. Specifically, available fixturing elements or special design fixtures are applied to build a fixturing assembly for each setup, such that each to-be-machined feature can be machined properly. This preliminary fixturing scheme development process is influenced by a number of considerations simultaneously, including tool-path/fixture interference, fixturing stability and stiffness, locating accuracy, fixture mounting difficulty, and so on.

For the sake of elucidation, the subtasks of setup planning are introduced in a sequential manner. However, they are by no means conducted sequentially in practice. To come up with a feasible setup plan, these subtasks are often handled in an iterative manner. Two common loops exist, as shown in Figure 2.4. A sequencing scheme incapable of meeting all manufacturing constraints will result in a setup regrouping. A fixturing scheme incapable of eliminating tool-path/ fixture interferences or not being able to achieve required locating accuracy will lead to a new sequencing scheme or even a regrouping of setups. These subtasks appear even more intertwined when the optimality of a setup plan is desired. Once all of the issues are resolved, setup planning outputs the following information:

- Setup grouping scheme: how to-be-machined features are grouped, and the tool-approach direction for each setup
- Sequencing scheme: how setups and operations are sequenced
- Fixturing scheme: the machining datum, including locating surface, supporting and clamping points, fixturing configuration, and instructions for special fixture design if necessary

2.1.2   Current State of the Art

Setup planning has traditionally been carried out by skilled process planners, who use different rules of thumb accumulated throughout years of hands-on experience. Because these rules are often implicit and biased toward one's own heuristic, setup plans thus developed are prone to flaws and are difficultly standardized. More severely, a huge amount of knowledge will be lost with one experienced process planner's retirement. Concerned by this situation, researchers have been working diligently in the last two decades trying to decode process planners' heuristics and bring systematical methodology into setup planning. The inputs of setup planning have been widely assumed available from upstream CAD stages by researchers. A feature-based design or a feature recognition system is intended to output design information regarding feature geometry and tolerances and raw material specifications. A manufacturing resource database is further assumed by researchers to provide specifications of available machines, cutting tools, and fixturing elements.

Many researchers regard setup planning as a discrete constrained optimization problem associated with multiple competitive objectives. Manufacturing constraints are commonly represented by matrices or graphs. Experience and heuristics are converted to various objectives in directing the search for an optimal solution. The most commonly used objective is probably the minimization of the number of setups required to machine a part. Fewer setups result in higher part accuracy by avoiding errors associated with workpiece locating and clamping being introduced into the machining process, which cuts down refixturing costs. Another objective along similar lines of logic is to minimize the number of tool changes required to machine a part. Some other objectives are maximizing the utilization of available fixturing elements, maximizing the number of features having critical tolerance relationships in the same setup, and minimizing total machining time.

A considerable amount of setup planning literature stresses fixturing planning. One method is based on feedback from fixture design (Sakurai 1992). It starts with generating a cutting-direction table for each to-be-machined feature, based on which setup direction is selected. For each setup, heuristics are applied to find the best locating surfaces. If no such surface is available, some features in the setup must be dropped. The locating and clamping positions are also checked against sets of physical requirements and heuristics to see if that specific setup is fixturable. The feedback is then used to decide whether setup regrouping is needed. Similar approaches are documented by Öztürk (1996) and Zhang (2001). Related research can also be found in Joneja (1999) and Young (1991).

Tolerance analysis has been considered by a stream of researchers a main driving factor in setup planning. The influence of tolerance relations among features on setup planning has been viewed in the following way (Boerma 1988). Errors in aligning machining parts on the machine tool can be equal to or larger than the accuracy requirements of small-tolerance relations. As a result, the position accuracy of a feature, which has already been machined in a previous setup, can be insufficient to realize the required accuracy in the relations between that feature and those that have to be machined in the present setup. Therefore, closely related features have to be machined in one setup, whereas less accurately related features can be machined in different setups. The effect of setup planning has been analyzed in regard to two surfaces that have a tolerance relationship (Zhang 1996). Zhang identifies three setup methods: (1) machining the two surfaces in the same setup, (2) using one surface as the setup datum and machining the other, and (3) using an intermediate setup datum to machine the two surfaces in different setups. Zhang argued that dimensions obtained using setup method (1) entail the least manufacturing errors and hence features with tight tolerance relationships should be arranged into the same setup whenever possible. Machining error stack-up in multi-setups has also been studied (Rong 1996). Rong categorizes the machining error between two features into five models based on the locating methods in setups, and provides recommendation for minimal error stack-up. A graph-based method is used to model and estimate the resultant machining error in multi-setup production. Graphs represent the tolerance and datum relationships among features, which can transform the problem of setup planning into a graph search problem (Huang 1997). The rough setup formation is made following the analysis of tool-approaching direction constraints. Then graphs are used to refine the setup formation and datum selection. The degree of tolerance tightness in a setup with others is used as an indication of sequencing setups. A concept of normalized tolerance has also been proposed to rigorously measure the tightness of tolerance, which is an angle representing the maximum permissible rotation error when locating a part. The smaller normalized tolerance, the tighter underlying tolerance (Huang 2003).

Some researchers chose to take a global search approach. Various algorithms are applied to generate setup plans, and multiple constraints and objectives are then employed simultaneously in screening setup plan alternatives. Generating feasible setup plan alternatives is undoubtedly the most challenging task in this approach. If all possible combinations must be tested with constraints, a search explosion will be inevitably encountered with increases in the complexity and number of features of the machined part. Two methods have been applied to reduce the size of the problem (Zhang 1995). One is merging those features sharing the same geometric and technological attributes into a compound feature. The other is searching for promising combinations only. Zhang (1999) incorporates a gradient descent algorithm with simulated annealing to search for the optimal plan starting with a random and feasible solution. Zhang also proposes a comprehensive cost-based objective function, which embraces five cost factors: machine cost, tool cost, machine change cost, setup change cost, and tool change cost. These cost factors could be used individually, or collectively as a cost compound. Kohonen's self-organizing neural network was applied to generate feasible setups in terms of the constraints of fixture, approach directions, and tolerance relationships. The Hopfield neural network was then used to solve the sequencing problem (Ming 2000). An unsupervised learning algorithm was proposed for setup generation that incorporates multiple objective functions into setup generation (Chen 1993). Related work can be found in Kim (1995), Ong (1994), and Delbressine (1993). It is argued that to deliver a robust setup planning solution applicable to practical problems the following issues must be addressed simultaneously (Huang 2003):

- Geometry analysis, to recognize features and ensure features of the part can be produced by the cutting tools
- Kinematic analysis, to ensure the workpiece can be located in a definite position
- Force analysis, to ensure the stability of the workpiece during machining
- Precedence constraint analysis, to ensure the feasibility as well as optimality of machining, locating, and clamping
- Tolerance analysis, to ensure the precision of the machined part.

Table 2.1 summarizes literature on setup planning according to whether these issues are addressed. It can be seen that although these issues have been separately addressed no integrated solution is available. To solve this problem, Huang (2003) has proposed an integrated modular systems framework to incorporate all the required analyses in a setup planning system. The framework, including the functionality of each module, is shown in Figure 2.8.

2.1.3   Closure

With traditional planning methods, manufacturing companies have devised setup plans that successfully produce all sorts of complicated parts. Hence, the purpose of setup planning research is not to solve a new problem. Rather, it aims to automate all or part of the planning process in order to achieve the efficiency promised by CAD/CAM integration. To a certain extent, this objective has been achieved. There are software tools (including commercial tools such as Cimskil by Technology Answers) that can automatically generate setup plans. These setup plans usually need to be checked/modified by an experienced process planner before they are released to production. In other words, these systems can reduce but not totally eliminate human interaction.

This may sound a little depressing, especially for those researchers that advocate the "unmanned factory of the future." However, we believe in factories where human beings are in charge, while computer systems serve as their digital assistants. Therefore, our challenge is to make a setup planning system smarter, rather than completely automatic. Specifically, a system that can generate optimal or nearly optimal setup plans with limited human interaction is preferred over one that automatically generates mediocre plans.

Table 2.1. Summary of literature on setup planning.

Addressed Issue / Reference	Geometry Analysis	Precedence Constraint	Kinematic Analysis	Force Analysis	Tolerance Analysis
Huang and Liu 2003					✓
Zhang 2001		✓			✓
Ming and Mak 2000	✓	✓			✓
Joneja and Chang 1999	✓	✓		✓	
Zhang and Lin 1999a, 1999b	✓	✓			✓
Zhang 1999	✓	✓			
Huang 1998	✓	✓			✓
Chen 1998	✓	✓			
Wu and Chang 1998	✓	✓			✓
Huang 1997	✓				✓
Lin 1997	✓	✓			
Rong 1997	✓				✓
Sarma and Wright 1996	✓	✓			
Zhang 1996	✓				✓
Öztürk 1996	✓	✓			
Demey 1996	✓	✓			✓
Huang and Zhang 1996					✓
Rong and Bai 1996					✓
Kim 1995	✓	✓			
Zhang 1995	✓	✓			
Huang and Gu 1994					✓
Ong and Nee 1994	✓	✓			✓
Delbressine 1993	✓	✓			
Chen 1993	✓				
Sakurai 1992	✓	✓	✓	✓	
Young and Bell 1991			✓	✓	
Chang 1990	✓	✓			
Hayes and Wright 1989	✓	✓			
Boerma and Kals 1988	✓	✓			✓
van Houten 1986	✓			✓	

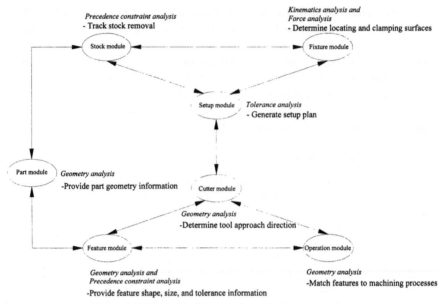

Figure 2.8. Integrated setup planning framework.

Setup planning can be viewed as a multiobjective optimization problem. General objectives are reduced cost, improved quality, and shortened cycle time. Multiobjective optimization is inherently much more difficult than single-objective optimization. Some researchers prefer to convert quality and cycle time into cost, and thus transform setup planning into a cost minimization problem. This may be theoretically sound but its practical applicability could be very poor. In practice, quality and cycle time cannot be accurately converted into monetary terms. Depending on the production environment, a certain objective may dominate monetary cost considerations. An extreme example is the construction of NASA's Martian Rover, for which the primary concern was that it be ready prior to the scheduled launch time. It is our view that the optimality of a setup plan should not be defined universally. Rather, it should be determined by designers and process planners based on a particular production environment. This is consistent with our philosophy of "computer-assisted human-centered" manufacturing.

2.2 Tolerance Analysis in Setup Planning

As previously mentioned, we view tolerance analysis as a driving factor in deriving good setup plans. The reason is that tolerance analysis can be used to devise a setup plan that improves part quality without increasing cost and prolonging cycle time. This is accomplished by minimizing tolerance stack-up while using the same manufacturing resources (machines, cutting tools, and fixtures). To understand why this is possible, we need to study the influence of manufacturing errors on part quality.

2.2.1 Manufacturing Errors and Tolerance Stack-up

In a manufacturing process, due to inevitable cutter-fixture alignment errors, tool wear, motion errors of a machine table, force and thermal effects, vibration, and so on, the dimension generated cannot be exactly equal to theoretically desired dimensions (Bai 1993). The deviation of a generated dimension from its theoretically desired dimension is caused by various types of manufacturing errors and should be controlled in order to meet the design tolerance requirement. In practice, it is common to classify manufacturing errors into two categories (namely, deterministic errors and random errors) for the purpose of analysis and control (Wang 1991; Bai 1993). Deterministic errors are manufacturing errors that arise with evident regularity, whereas random errors arise in a batch of successively machined parts without apparent regularity in the variations of the directions and values (Wang 1991). Usually, the direction and value of a deterministic error can be predicted because its regularity can be found. However, because the regularity of a random error does not exist, its direction and value cannot be predicted. When performing a worst-case manufacturing error analysis, random errors are always summed, whereas deterministic errors might be summed or canceled out.

Manufacturing errors can also be classified based on error sources. To manufacture a part, the workpiece has to be located on the machine tool in a definite manner so that features of the part can be machined. Usually, a fixture is used to provide some type of a clamping mechanism to maintain the workpiece in a desired position. The fixture has to be located on the machine table first. The fixturing datum (e.g., the bottom of a vise) is placed against the machine table and the fixture is then fixed on the machine tool. Next, the setup datum (which may or may not coincide with the design datum) of the workpiece is placed against the clamping surface of the fixture (e.g., the stationary jaw of a vise) and the workpiece is then fixed on the fixture. In this way, the workpiece is located on the machine tool in a definite manner and hence the machining can take place. The features of the part are formed by the cutting tool path. Manufacturing errors involved in such a machining process are shown schematically in Figure 2.9, in which the surfaces connected by a double line are coinciding surfaces.

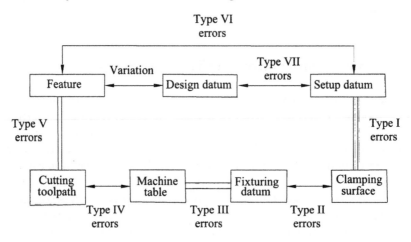

Figure 2.9. Machining errors involved in a machining process.

From Figure 2.9 one can see that there are seven types of manu-facturing errors. Type I errors are errors that occur when locating and clamping the workpiece on the fixture. These errors include workpiece-locating error, workpiece-clamping distortion, and opera-tional error. Type II errors are related to the fixture itself. The major por-tion of errors in this case are caused by geometrical and dimensional inaccuracy of the fixture. Other errors are caused by the deformation of the fixture when clamping force is applied. When the rigidity of the fixture is low, the fixture will deform due to the effect of gravity, clamping force, and cutting force. Type III errors occur when locating and fixing the fixture on the machine table. They are similar to Type I errors except that Type I errors are related to the workpiece and the fixture, whereas Type III errors are related to the fixture and machine table. Type IV errors are related to the machine tool itself. These errors are caused by geometrical and dimensional inaccuracy of the machine tool and its rigidity, stiffness, and thermal deformation. Type V errors are related to the cutting tool itself. These errors are caused by geometrical and dimensional inaccuracy of the cutting tool and by tool wear. Type VI errors are related to the workpiece itself. These errors are mainly caused by deformation of the workpiece due to internal stresses after machining.

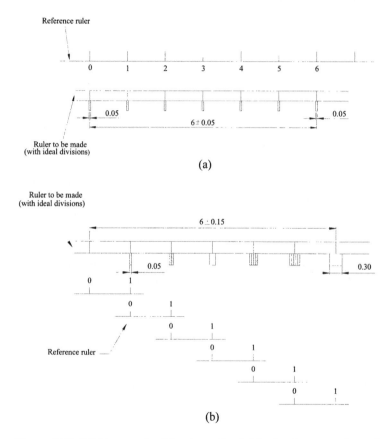

Figure 2.10. Making a ruler (a) using a fixed 6-inch reference ruler and (b) using a 1-inch reference ruler.

Type VII errors occur when the setup datum does not coincide with the design datum. In production planning for economic runs, parts cannot always be machined dimensionally as shown on the blueprint; the setup datum may not coincide with the design datum. Hence, certain blueprint dimensions can be obtained only indirectly. As a result, manufacturing variations will stack up. This phenomenon is commonly known as tolerance stack-up. Although the other types of errors cannot be eliminated, Type VII errors can sometimes be reduced, if not eliminated, via setup planning. To understand why this is possible, examine the example shown in Figure 2.10.

Suppose one would like to make a 6-inch ruler with 1-inch divisions. It is known that the error of making a mark is ±0.025 inch. There are two methods to make the ruler. The first method is to use an accurate 6-inch master ruler as a reference. The ruler to be made is placed against the reference ruler and seven marks are made. In this way, the deviation of each mark from its ideal position will be ±0.025 inch and the total length of the ruler made will be 6±0.05 inch (Figure 2.10a). The second method is to use an accurate 1-inch reference ruler. The first mark is made without any references. The reference ruler is then aligned with the mark so that the second mark can be made. To make the third mark, the reference ruler has to be aligned with the second mark. The rest of the marks are made the same way. In this way, the deviation of the position of a mark will be carried to the making of the next mark. Assuming there is no alignment error, the total length of the rule made will be 6±0.15 inch (Figure 2.10b). This example clearly shows that the key to controlling tolerance stack-up is to reduce datum changes, which can be achieved through setup planning. Note that the analysis is also valid for other tolerance specifications, although not all tolerances are datum related and hence only certain types of tolerances can be controlled through setup planning.

2.2.2   Setup Planning and Tolerance Control

Manufacturing tolerances can be classified into two basic categories; namely, dimensional tolerances and geometric tolerances. Geometric tolerances can be further classified into 14 types per ANSI Y14.5-1994 (ANSI 1995). Dimensional tolerances can also be further classified into two types; namely, those that belong to limit-of-size (e.g., the diameter of a cylinder) and those that do not (e.g., the distance between two parallel surfaces). Certain types of tolerances are assigned to a single feature (the toleranced feature), such as limit-of-size dimensional tolerances and certain geometric tolerances including straightness, flatness, circularity, cylindricity, and profile tolerances when specified without datums. These types of tolerances are mainly determined by machine/process capabilities and do not need to be considered in setup planning. Other types of tolerances involve not only the toleranced feature but one or more reference features (datums), and hence are referred to as relative tolerances. Relative tolerances include non-limit-of-size dimensional tolerances and the following geometric tolerances: parallelism, perpendicularity, angularity, position, concentricity, symmetry, runout (circular and total), and profile tolerances (profile of a line and profile of a surface) when specified with datums. They are influenced not only by machine/process capabilities but by the setup methods applied, and therefore need to be considered in setup planning.

In machining, three setup methods are used so that the dimension between two features can be obtained: (1) machining the two features in the same setup, (2) using one feature as the setup datum and machining the other, and (3) using intermediate setup datums to machine the two features in different setups. The three methods are denoted as setup method I, setup method II, and setup method III, respectively. The dimensions obtained using different setup methods are affected by different manufacturing error sources and should be treated differently. To investigate the manufacturing errors involved in the dimensions obtained using different setup methods, the following assumptions are made:

- The part to be machined is located and clamped on the machine table using a fixture.
- The fixture is perfectly setup on the machine table; that is, Type III errors are negligible.
- The cutting tool is perfectly accurate and the effect of tool wear is trivial; that is, Type V errors are negligible.
- The effect of workpiece deformation due to clamping force, cutting force, gravity, and internal stress is trivial; that is, Type VI errors are negligible.

When locating and clamping (setting up) the workpiece on the fixture, it is desirable that the workpiece coordinate system (WCS) coincide with the machine coordinate system (MCS). However, due to Type I and Type II errors the WCS will vary from the MCS. The variation is called a *setup error*. After a workpiece has been located and clamped, the setup error remains constant unless the workpiece is removed from the fixture. Therefore, a setup error is a deterministic error within each setup. In addition to the setup error, there exists a *machine motion error*, which is the deviation of a machine movement from its ideal position due to Type IV errors. The direction and value of a machine motion error usually cannot be predicted, and hence it is a random error. A worst-case error analysis is depicted in Figure 2.11. The following notations are used: Δ_i, $i = 1, 2, \ldots$, is the setup error of each setup (could be negative); $\pm\frac{1}{2}\delta_j$, $j = 0$, $1, 2, \ldots$, is the machine motion error for each machine movement; x_j, $j = 0$, $1, 2, \ldots$, is the nominal dimension for each machine movement; A, B, $\ldots$, are the part surfaces; A', B', $\ldots$, are the machined surfaces; and L_{PQ} is the length between surfaces P and Q.

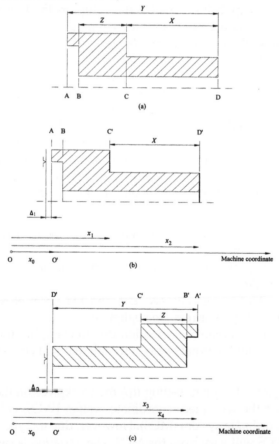

Figure 2.11. Machining error analysis for NC (numerical control) machining (a) part design, (b) the first setup, and (c) the second setup.

Figure 2.11 shows a part machined on an NC turning center. The part design is shown in Figure 2.11a. The part is machined in two setups, as shown in Figures 2.11b and 2.11c, respectively. Prior to machining, the operator would pick a location on the fixture to be the program zero. In this example, the program zero is O'. When setting the program zero O', the cutting tool needs to be brought to O' and the distance between the machine origin O and the program zero O' is recorded. Assume the recorded distance is x_0. Due to the machine motion error, the actual distance would be $x_0 \pm \frac{1}{2}\delta_0$.

X is the length between machined surfaces C' and D'. Surfaces C' and D' are both machined in the first setup, in which surface A is used as a setup datum. Therefore, X is obtained using setup method I. From Eq. 2.1, it can be seen that X is influenced only by two machine motion error components.

$$
\begin{aligned}
X &= L_{C'D'} \\
&= \left(x_2 \pm \frac{1}{2}\delta_2\right) - \left(x_1 \pm \frac{1}{2}\delta_1\right) \\
&= \left(x_2 - x_1\right) \pm \frac{1}{2}\left(\delta_1 + \delta_2\right)
\end{aligned} \tag{2.1}
$$

Y is the length between machined surfaces A' and D'. Surface D' is the setup datum when machining surface A'. Therefore, Y is obtained using setup method II. From Eq. 2.2, it can be seen that Y is influenced by one setup error component and two machine motion error components.

$$
\begin{aligned}
Y &= L_{A'D'} \\
&= \left(x_4 \pm \frac{1}{2}\delta_4\right) - \left(x_0 \pm \frac{1}{2}\delta_0\right) - \Delta_2 \\
&= \left(x_4 - x_0\right) - \Delta_2 \pm \frac{1}{2}\left(\delta_0 + \delta_4\right)
\end{aligned} \tag{2.2}
$$

Z is the length between machined surfaces B' and C'. Surface C' is machined using surface A as a setup datum in the first setup; surface B' is machined using surface D' as a setup datum in the second setup. Therefore, Z is obtained using setup method III. From Eq. 2.3, it can be seen that Z is influenced by one setup error component and four machine error components under the assumption.

$$
\begin{aligned}
Z &= L_{B'C'} = L_{D'B'} - L_{C'D'} \\
&= \left[\left(x_3 \pm \frac{1}{2}\delta_3\right) - \left(x_0 \pm \frac{1}{2}\delta_0\right) - \Delta_2\right] - \left[\left(x_2 - x_1\right) \pm \frac{1}{2}\left(\delta_1 + \delta_2\right)\right] \\
&= \left(x_3 - x_2 + x_1 - x_0\right) - \Delta_2 \pm \frac{1}{2}\left(\delta_0 + \delta_1 + \delta_2 + \delta_3\right)
\end{aligned} \tag{2.3}
$$

From this analysis it can be seen that when using setup method I, the setup error is not included in the dimensions obtained. In general, the geometric relationship of the features machined in the same setup mainly depends on the geometry built into the machine tool. The dimensional relationship, such as the distance between two parallel features, is determined mainly by the accuracy of the control unit, which is a built-in capability of the machine tool. Dimensions obtained using setup method I entail the fewest manufacturing errors. Whenever possible this setup method should be used to facilitate tolerance control.

Setup method II is the most frequently recommended method in the literature (Doyle 1953; Buckingham 1954; Wilson 1963). When setup method II is used, the setup error is included in the dimensions obtained. To control the tolerance of the dimensions, the accuracy of setting up the part has to be considered, which becomes a major part of the tolerance. Dimensions obtained using setup methods II are less accurate than those obtained using setup method I in NC machining. However, setup method II is regarded as a good method when two features cannot be machined in the same setup.

Setup method III is the least desired setup method. In this case, a dimension chain is formed for the dimension obtained. As a result, the tolerances obtained will stack up. When the tolerance of a dimension is tight, this setup method should be avoided. The guidelines for using setup planning to achieve proactive tolerance control can be summarized as follows:

- Features with a tight tolerance relationship should be arranged into the same setup whenever possible.
- When two features with a tight tolerance relationship cannot be machined in the same setup, they should be manually datumed.
- Only when the tolerance relationship between two features is not important can intermediate datums be used.

The key to correctly applying these guidelines is the development of an accurate means of comparing the tightness of different types of tolerances. This is a challenging task because there are no common measures defined in ANSI standards for different types of tolerances. To solve this problem, the use of tolerance factors was proposed to convert the values of geometric tolerances into non-type-specific values that can be compared (Boerma 1988). This approach provides a new direction in developing advanced tolerance analysis techniques. However, the approach has several drawbacks.

- The procedure for converting geometric tolerances to tolerance factors is not well defined. Specifically, the approach provides only certain examples by intuition from part drawings rather than generalized and accurate formulae derived with mathematical rigor. In addition, the conversion scheme is ad hoc in nature and does not accurately reflect the tightness of certain geometric tolerances (e.g., the perpendicularity tolerance).
- The approach cannot be directly applied to convert certain types of geometric tolerances (e.g., concentricity and runout tolerances).
- The approach only deals with geometric tolerances and is thus not able to provide a means of comparing a geometric tolerance to a dimensional tolerance.
- The case of multiple tolerance requirements associated with the same set of features is not addressed.

These drawbacks prevent tolerance factors from becoming a general tolerance analysis tool in setup planning. To overcome these drawbacks, we need to investigate how the tightness of a tolerance relates to setup planning. The tightness of a relative tolerance refers to the difficulty in assuring the accuracy of the toleranced feature with respect to its datum during machining. Machined features cannot be perfect due to a variety of error sources, such as improper locating of the workpiece, tool wear, motion errors of the cutter and machine table, vibration, and thermal effects. Among these error sources, the one related to setup planning is improper locating of the workpiece (locating errors). Locating errors consist of a rotation error component and a translation error component. The translation error influences the final position of the machined feature with respect to its datum. However, it does not influence the size of the tolerance zone, whereas the rotation error does. Therefore, the maximum permissible rotation error in part locating alone can be used as an accurate measure for the tightness of a tolerance.

2.2.3 Tolerance Normalization

A mathematically rigorous approach is presented here to convert relative tolerances into maximum permissible rotation deviations called normalized tolerances, which are measured using the degree of the angle between the ideal datum and the deviated datum. The smaller normalized tolerance, the tighter underlying tolerance. In the following subsections, generalized mathematical formulae for normalizing orientation tolerances (parallelism, perpendicularity, and angularity), location tolerances (position, concentricity, and symmetry), runout tolerances (circular and total), and non-limit-of-size dimensional tolerances (relative dimensional tolerances) are provided. No generalized formulae are presented in normalizing profile tolerances (when specified with datums) because the underlying profile can take a very complicated shape. Generalized formulae will not help in normalizing these profile tolerances. Instead, the use of constrained optimization methods is needed, as discussed in material to follow.

2.2.3.1 Orientation Tolerances

Parallelism. In defining a parallelism tolerance, the two features involved could be two planes, one plane and one axis, or two axes. For a parallelism tolerance defined by the zone between two parallel planes, suppose the equation for the ideal position of the toleranced feature is

$$S_0 : \begin{cases} z = h \\ x, y \in \{(x, y) \mid G(x, y) = 0\}, \end{cases}$$

where h is the distance between the toleranced feature and the reference feature (datum) and $G(x, y) = 0$ is the equation for the contour of the toleranced feature and is either continuous (described by a single function, such as an ellipse or a circle) or segmental continuous (described by a set of functions, such as a polygon). Due to the rotation error in locating, points on the toleranced feature surface will deviate from their ideal positions.

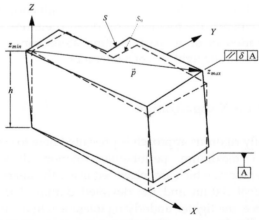

Figure 2.12. Deviation of machined feature due to locating errors.

Figure 2.12 shows an example part the top surface of which has a parallelism tolerance δ with respect to the bottom surface. Assume the part is in its ideal form but located imperfectly. As a result, the toleranced feature deviates from S_0 to a new plane S described by the equation $Ax + By + Cz + D = 0$. Due to the presence of rotation error, points on plane S will have different z coordinate values. Because the bottom surface is parallel to plane S, points on the bottom surface will also have different z coordinate values. Ignoring machining errors, the machined top surface will be parallel to the ideal datum but not the bottom surface. If the part is located such that the difference between the points having the maximum and minimum z coordinate values (on plane S) is less than δ, the parallelism tolerance can be assured. The problem can be formulated as

$$\text{Max } P_z = -\frac{Ax + By + D}{C} \text{ and Min } P_z = -\frac{Ax + By + D}{C}, \tag{2.4}$$

subject to

$$G(x, y) \leq 0, \tag{2.5}$$

where $G(x,y) = 0$ is the equation for the contour of the machined feature and is either continuous (such as a circle) or piecewisely continuous (such as a polygon).

Let z_{max} and z_{min} be the solutions for Eq. 2.4, respectively. To assure the parallelism tolerance, the following equation should hold.

$$\left| z_{max} - z_{min} \right| \leq \delta$$

Constraint given by Eq. 2.5 gives a feasible region in which z_{max} and z_{min} can be found. Because the objective functions in Eq. 2.5 are linear, it is easy to show that z_{max} and z_{min} can be found only on the boundary of $G(x, y)$. Let $\vec{p}$ denote the vector established by a point whose z coordinate equals to z_{min} and a point whose z coordinate equals z_{max}. The value of $\left| z_{max} - z_{min} \right|$ can be calculated as

$$\left| z_{max} - z_{min} \right| = \left\| \vec{p} \right\| \times \sin\theta,$$

where θ is the angle between the vector $\vec{p}$ and the plane S_0.

Let φ denote the angle between the deviated plane S and the ideal plane S_0. Referring to Figure 2.12, because plane S_0 is parallel to the XY plane, the ideal setup datum, and plane S is parallel to the bottom surface of the part, the deviated setup datum, it is obvious that φ is also the angle between the ideal setup datum and the deviated setup datum, which represents the rotation error in part location. Since φ is the angle between the two planes S and S_0 and θ is the angle between a line on plane S and the plane S_0, it can be shown that $\theta \leq \varphi$. Therefore, the maximum value of $|z_{max} - z_{min}|$, denoted Δ_{max}, occurs when the datum is rotated to a certain place when $\theta = \varphi$, i.e., $\Delta_{max} = \|\vec{p}\| \times \sin \varphi$. To ensure the parallelism tolerance, the value of Δ_{max} (maximum deviation) cannot exceed δ. Therefore, the following is derived:

$$\|\vec{p}\| \times \sin \varphi \leq \delta \Rightarrow \sin \varphi \leq \frac{\delta}{\|\vec{p}\|}.$$

The maximum permissible deviating angle can be obtained when the denominator, $\|\vec{p}\|$, arrives at a maximum. Therefore, the normalized tolerance for parallelism defined by two parallel features is

$$\varphi_{//} = \sin^{-1}\left(\frac{\delta}{\|\vec{p}_{max}\|}\right) = \sin^{-1}\left(\frac{\delta}{L_r}\right),$$

where L_r is the *representative length*, which is the line segment of maximum length whose two end points are on the boundary of $G(x, y)$.

The angle $\varphi_{//}$ represents the maximum permissible rotation error in part locating. It is worth noting that deviated locating datums may be different even though they have the same deviating angle. In addition, the part may be rotated around the normal vector of the deviated locating datum plane. Therefore, the part may have different points on the boundary that reach maximum deviation. However, this maximum deviation will never exceed the tolerance zone as long as the rotation deviation of the locating datum is controlled within the normalized tolerance range. In other words, the normalized tolerance is obtained based on a worst-case analysis.

There is another type of parallelism tolerance in which the axis of the toleranced feature is constrained within the tolerance zone (an axis parallel to a plane or an axis parallel to another axis). In this case, the angle used to define normalized tolerance should be the angle between the deviated axis and the ideal axis. Figure 2.13a shows an example of a parallelism tolerance defined as a cylindrical zone in which the axis of the toleranced feature must lie. The size of the tolerance zone is δ.

Consider a coordinate system as shown in Figure 2.13b, in which the ideal position of the toleranced axis lies on the X axis of the coordinate system. The equations for the two end surfaces of the tolerance zone are

$$E_1: \begin{cases} y^2 + z^2 \le (\delta/2)^2 \\ x = 0 \end{cases} \qquad (2.6)$$

and

$$E_2: \begin{cases} y^2 + z^2 \le (\delta/2)^2 \\ x = L \end{cases}, \qquad (2.7)$$

where L is the length of the axis.

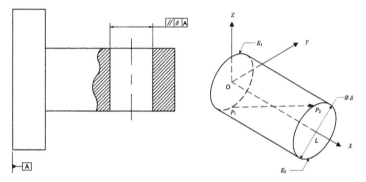

Figure 2.13. Analysis of parallelism tolerance with a cylindrical tolerance zone: (a) tolerance specification and (b) normalization analysis.

Assume the toleranced axis deviates from its ideal position to a new position at P_1P_2. The end points are $P_1(0, y_1, z_1)$ and $P_2(L, y_2, z_2)$, respectively. Then the vector P_1P_2 is

$$P_1P_2 = Li + (y_2 - y_1)j + (z_2 - z_1)k.$$

Let φ denote the angle between P_1P_2 and the XY plane. The following can be derived:

$$\cos\varphi = \frac{L}{\sqrt{L^2 + (y_2 - y_1)^2 + (z_2 - z_1)^2}} \qquad (2.8)$$

To ensure the parallelism tolerance, points P_1 and P_2 must lie within the tolerance zone; that is $(y_2 - y_1)^2 + (z_2 - z_1)^2 \le \delta^2$. Therefore,

$$\cos\varphi \le \frac{L}{\sqrt{\delta^2 + L^2}}.$$

The worst case occurs when the end points of the inclined axis are situated on the exterior circle of the tolerance zone and the angle φ reaches maximum. Therefore, the normalized tolerance can be defined as

$$\varphi_{//} = \cos^{-1}\left(\frac{L}{\sqrt{\delta^2 + L^2}}\right),$$

which can be further simplified as

$$\varphi_{//} = \tan^{-1}\left(\frac{\delta}{L}\right), \tag{2.9}$$

where L is the length of the toleranced axis.

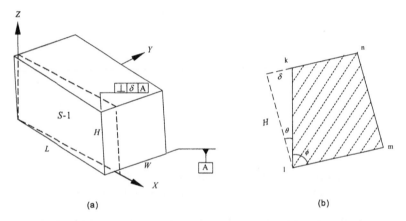

(a)   (b)

Figure 2.14. Analysis of perpendicularity tolerance: (a) part incorrectly located (due to rotation error) and (b) cross-section view of the machined part.

Perpendicularity. The perpendicularity tolerance can also be normalized based on rotation error analysis. Suppose a prismatic part to be machined has a length of L, a width of W, and a height of H (Figure 2.14a). In Figure 2.14a, surface S-1 has a perpendicularity tolerance requirement δ with the bottom surface. To facilitate the analysis, assume that surface S-1 is machined using a horizontal machining center with the bottom surface as the primary setup datum. Ideally, the bottom surface of the part should coincide with the ideal datum plane (machine table) in order to guarantee the perpendicularity between surface S-1 and the bottom surface. Due to the rotation error in locating, the bottom surface of the part deviates from the datum plane. Without loss of generality, let this deviated bottom surface be described as

$$S : \begin{cases} Ax + By + Cz = 0 \\ 0 \leq x \leq L \\ 0 \leq y \leq W \end{cases}$$

in which

$$\sqrt{A^2 + B^2 + C^2} = 1 . \tag{2.10}$$

The rotation error in part locating is defined as the angle between plane S and the ideal datum plane (the XY plane, which can be described using equation $z = 0$). This angle, denoted φ, can be obtained from the following equation:

$$\cos \varphi = C . \tag{2.11}$$

Assuming machining errors can be ignored, the machined surface S-1 will be perpendicular to the ideal datum plane but not to the bottom surface of the part. The cross-section view (using a plane parallel to the plane $x = 0$ that intersects the part) of the machined part is shown in Figure 2.14b. The angle ϕ is the angle between the deviated bottom plane and the plane $y = 0$. The value of ϕ can be obtained from the following equation:

$$\cos \phi = B . \tag{2.12}$$

To assure that the perpendicularity tolerance δ is met, the value of $H \times \tan \theta$ cannot exceed the value of δ [i.e., $H \times \tan \theta = H \times \tan(90° - \phi) \leq \delta$]. To normalize the perpendicularity tolerance δ, we need to convert it into $\varphi_\perp$, which defines the maximum permissible rotation error of the bottom surface. From Eqs. 2.11 and 2.12, we have

$$(\cos \varphi)^2 + (\cos \phi)^2 = C^2 + B^2 . \tag{2.13}$$

From Eqs. 2.10 and 2.13, we have

$$(\cos \varphi)^2 + (\cos \phi)^2 \leq 1 .$$

Therefore, $\varphi + \phi \geq 90° \Rightarrow \phi \geq 90° - \varphi$.

Revisiting the equation $H \times \tan(90° - \phi) \leq \delta$ that assures the perpendicularity tolerance, we can see that the maximum deviation, denoted d_{max}, occurs when ϕ is minimized. Because $\phi \geq 90° - \varphi$, we have $d_{max} = H \times \tan[90° - (90° - \varphi)] = H \times \tan \varphi$. To ensure the perpendicularity tolerance, d_{max} cannot exceed δ (i.e., $d_{max} = H \times \tan \varphi \leq \delta$). Therefore, the maximum permissible rotation error of the bottom surface $\varphi_\perp$ can be calculated as $\varphi_\perp = \tan^{-1}\dfrac{\delta}{H}$, which is the normalized perpendicularity tolerance. This conclusion is applicable to features with different shapes when H is defined as the maximal height of the toleranced feature along the direction perpendicular to the reference feature (datum).

There is another perpendicular tolerance for which the tolerance zone is defined as a cylinder. The tolerance value δ of the perpendicularity control defines the diameter of the cylindrical tolerance zone. The axis of the diameter must lie within the tolerance zone. In this case, Eqs. 2.6 and 2.7 can be used to describe the two end surfaces of the cylindrical tolerance zone, and Eq. 2.8 can be used to describe the angle between the inclined axis and the ideal datum. Thus, the normalized tolerance can be derived in the same form as Eq. 2.9 [i.e., $\varphi_\perp = \tan^{-1}(\dfrac{\delta}{L})$], where L is the length of the axis.

Angularity. In manufacturing, angularity is processed and inspected through the guarantee of parallelism, as shown in Figure 2.15. For example, if a 30-degree angle is needed, a wedge block of 30 degrees is put onto the machine table and the part is mounted on the wedge. If the machined surface is parallel to the machine table, the angularity of 30 degrees between the machined surface S and the datum R is obtained. Thus, the case of angularity in imperfect locating is similar to parallelism and the normalized tolerance for angularity applied to a surface can be defined as

$$\varphi_\angle = \sin^{-1}\left(\frac{\delta}{L_r}\right),$$

where L_r is the largest line segment in the enclosed contour of the machined feature.

For the example shown in Figure 2.15, the representative length L_r is the distance of AC.

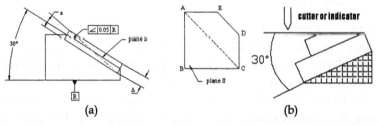

(a)   (b)

Figure 2.15. Angularity tolerance: (a) tolerance specification with auxiliary view of the machined feature, (b) processing and inspection of an angularity tolerance.

When an angularity tolerance is applied to an axis, the normalized tolerance is

$$\varphi_{\angle} = \tan^{-1}\left(\frac{\delta}{L}\right),$$

where L is the length of the axis.

2.2.3.2 Location Tolerances

There are three types of location tolerances: concentricity, symmetry, and position. These tolerances all deal with the control of features that normally have a centerline or center plane. Because of imperfect locating, the expected ideal centerline or center plane will deviate to an imperfect position.

Position. A tolerance of position control is the total permissible variation in the location of a feature about its true position. For position tolerance specified on a cylindrical feature, Eqs. 2.6 to 2.9 can be applied to normalize the tolerance. For positional tolerance specified on other features, the normalization procedure is similar to the case of perpendicularity tolerance (see Figure 2.16 for an illustration). In both cases, the normalized tolerance takes the same form. Therefore, normalized position tolerance can be defined as

$$\varphi_{\oplus} = \tan^{-1}\left(\frac{\delta}{L}\right),$$

where L is the length of the axis, the centerline, or the height of the center plane of the toleranced feature.

Concentricity. Concentricity tolerance is applied to rotational features to control the location of the median points of diametrically opposed surface elements. The tolerance zone for a concentricity control is a cylinder that is coaxial with the datum axis. The diameter of the cylinder is equal to the concentricity tolerance value. This is a situation similar to that when one axis is parallel to another. Therefore, Eqs. 2.6 to 2.9 can be used and hence the normalized concentricity tolerance is defined as

$$\varphi_{\copyright} = \tan^{-1}\left(\frac{\delta}{L}\right),$$

where L is the length of the axis of the toleranced feature.

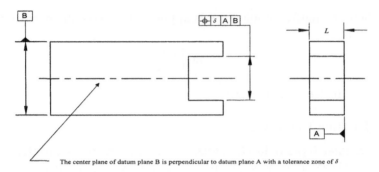

The center plane of datum plane B is perpendicular to datum plane A with a tolerance zone of δ

Figure 2.16. Position tolerance specified on a center plane.

Symmetry. A symmetry tolerance zone is established by two parallel planes in which the median points of the toleranced feature must lie. Symmetry and concentricity are the same concept, except as applied to different part configurations (ANSI 1995). Therefore, the normalized tolerance for symmetry has the same form as that for concentricity. It is defined as

$$\varphi_{\equiv} = \tan^{-1}\left(\frac{\delta}{L}\right),$$

where L is the height of the center plane.

2.2.3.3 Runout Tolerances

Runout tolerances are composite because form, orientation, and location errors are involved in runout. There are two types of runout tolerances: circular runout and total runout. Circular runout provides control of the runout of a single random element in the surface being measured. The tolerance is applied independently at each circular measuring surface as the part is rotated 360 degrees. Suppose only locating errors contribute to the circular runout tolerance (i.e., the circle being verified is perfectly round), whereas the axis deviates from its ideal position (datum). Consider the worst case, in which the biggest deviation, $\Delta_{\max}$, occurs at the end surface. As shown in Figure 2.17, assume a runout tolerance δ is specified, and thus

$$
\begin{aligned}
\Delta_{\max} \quad &= v_u - v_l \\
&= (R + L \times \tan\varphi) \times \cos\varphi - (R - L \times \tan\varphi) \times \cos\varphi \\
&= 2L \times \tan\varphi \times \cos\varphi \\
&= 2L \times \sin\varphi.
\end{aligned}
$$

To ensure the runout tolerance, we have $\Delta_{max} = 2L \times \sin\varphi \le \delta$. Therefore, the normalized runout tolerance is

$$\varphi_\uparrow = \sin^{-1}\left(\frac{\delta}{2L}\right),$$

where L is the length of the axis of the toleranced feature.

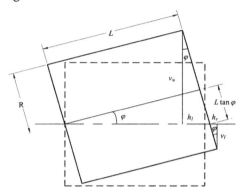

Figure 2.17. Analysis of runout tolerance.

Total runout is inspected while the part is chucked or colleted to the ideal datum axis, and a dial indicator is placed perpendicular to the machined surface. The part is rotated while the dial indicator is continuously traversing the surface axially, perpendicular or parallel to the datum. The tolerance zone for axis total runout is the space between two cylinders with a radius of δ (tolerance value), which the total indicator readings cannot exceed. Ignoring form tolerances, the worst case that may occur during locating is the same as that for circular runout. Therefore, the normalized tolerance for total runout is

$$\varphi_\upharpoonright = \sin^{-1}\left(\frac{\delta}{2L}\right),$$

where L is the length of the axis of the toleranced feature.

Runout also applies to a surface that is perpendicular to the datum axis. Again, if we consider the worst case, circular runout is the same as total runout. Refer to Figure 2.17. The largest deviation, Δ_{max}, can be calculated as

$$\begin{aligned}
\Delta_{max} &= h_l + h_r \\
&= (R + L \times \tan\varphi) \times \sin\varphi + (R - L \times \tan\varphi) \times \sin\varphi \\
&= 2R \times \sin\varphi.
\end{aligned}$$

To ensure the runout tolerance, we have $\Delta_{max} = 2L \times \sin\varphi \le \delta$. Therefore, the normalized runout tolerance is

$$\varphi_\uparrow = \varphi_\Uparrow = \sin^{-1}\left(\frac{\delta}{2R}\right),$$

where R is the radius of the toleranced feature.

2.2.3.4 Relative Dimensional Tolerances and the Multiple Tolerances

A relative dimensional tolerance is specified on two features, usually two parallel planes (relative dimensional tolerance can also be specified between a plane and an axis, or between two axes, just as the case of parallelism tolerance). Again, we assume that the machined feature deviates from its ideal position only because of the rotation error in part locating. As shown in Figure 2.18a, there is a dimensional tolerance $H \pm \frac{1}{2}\delta$ specified on the height of the prismatic part. The two parallel planes (under ideal condition) involved are the top and bottom surfaces. Due to the rotation error in locating, the machined surface (top) is not parallel to the locating surface (bottom). Figure 2.18b shows the left view of the machined part. To ensure that the machined part meets the dimensional tolerance requirement, the vertical distance between the lowest and the highest points on the top surface must not exceed δ, the size of the tolerance (the other necessary condition is that the nominal dimension H is assured, which has nothing to do with the tolerance zone). It is obvious that this condition is the same as specifying a parallelism tolerance δ on the top surface while the bottom surface is used as a datum. Therefore, when normalizing a relative dimensional tolerance for setup planning it can be treated as a parallelism tolerance and the formulae presented in previous sections can be used.

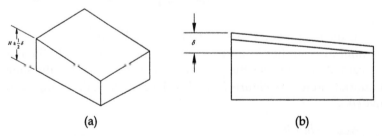

        (a)   (b)

Figure 2.18. Analysis of relative dimensional tolerance: (a) tolerance specification and (b) left view of the machined part.

Sometimes there are multiple tolerances specified on the same set of features. This usually happens when the features involved are two parallel planes. In addition to a dimensional tolerance, a parallelism tolerance is also specified. Based on previous discussion, it is obvious that if the tolerance with the smaller tolerance zone is ensured (usually the parallelism tolerance) the other tolerance (usually the dimensional tolerance) will also be ensured. This conclusion can be easily generalized. Therefore, when there are multiple relative tolerances associated with the same set of features these tolerances should be first normalized independently. Then, the one with the smallest value is used as the normalized tolerance associated with the set of features. Table 2.2 summarizes the formulae for tolerance normalization (except for profile tolerances).

2.2.3.5 Profile Tolerances

A profile control is a geometric tolerance that specifies a uniform boundary along the true profile that elements of the feature must lie within. It is a composite control affecting feature size, form, orientation, and location. When a reference datum is used to specify a profile tolerance, the profile tolerance is a relative tolerance and needs to be normalized.

Table 2.2. Tolerance normalization quick reference table.

Tolerance Type	Toleranced Feature	Normalized Tolerance	Note
• Parallelism • Angularity	Surface	$\sin^{-1}\left(\dfrac{\delta}{L_r}\right)$	L_r: length of the largest line segment whose two end points are on the surface boundary
	Axis	$\tan^{-1}\left(\dfrac{\delta}{L}\right)$	L: length of the axis
• Perpendicularity	All	$\tan^{-1}\left(\dfrac{\delta}{L}\right)$	L: height of the surface or length of the axis
• Position • Concentricity • Symmetry	All	$\tan^{-1}\left(\dfrac{\delta}{L}\right)$	L: length of the axis or centerline, or height of the center plane
• Circular Runout • Total Runout	Axis	$\sin^{-1}\left(\dfrac{\delta}{2L}\right)$	L: length of the axis
	Surface	$\sin^{-1}\left(\dfrac{\delta}{2R}\right)$	R: radius of the circular surface

1. A relative dimensional tolerance is treated as a parallelism tolerance.
2. When multiple tolerances are specified on the same set of features, use the one with minimum value.

Profile of a Line. When a profile of a line control is specified, the tolerance zone is two uniform lines (a 2D tolerance zone). Suppose the equation for the profile is

$$\begin{cases} x_0 = r(\varphi)\cos\varphi \\ y_0 = r(\varphi)\sin\varphi. \end{cases}$$

Assume $P(x_0, y_0)$ is a point on the ideal profile and its polar coordinate is (r, φ). The tolerance zone for the profile is the region constricted by the two enveloping lines consisting of circles with the diameter d equal to the tolerance value δ (Figure 2.19a). Suppose due to errors in locating, the profile feature is machined at a deviated position. Assume the coordinate system established by the imperfect locating datum is $X'O'Y'$ (Figure 2.19b). The equation for the machined profile in the XOY coordinate plane is

$$\begin{bmatrix} x' \\ y' \end{bmatrix} = \begin{bmatrix} \cos\theta & -\sin\theta \\ \sin\theta & \cos\theta \end{bmatrix}\begin{bmatrix} x_0 \\ y_0 \end{bmatrix} + \begin{bmatrix} a \\ b \end{bmatrix} = \begin{bmatrix} x_0\cos\theta - y_0\sin\theta + a \\ x_0\sin\theta + y_0\cos\theta + b \end{bmatrix} = \begin{bmatrix} r(\varphi)\cos(\theta+\varphi)+a \\ r(\varphi)\sin(\theta+\varphi)+b \end{bmatrix}, \quad (2.14)$$

where θ is the rotation angle and (a, b) is the coordinate of O' in the XOY coordinate system.

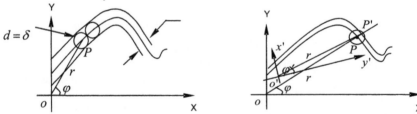

Figure 2.19. Analysis of profile of a line tolerance: (a) ideal condition and (b) deviated profile due to imperfect locating.

Let point P' be a random point on the machined imperfect profile. Connecting the origin O and P', there is a intersecting point (P) between line OP' and the ideal profile. To ensure the profile tolerance, P' should be within the enveloping circles. In other words, the distance between P and P' should be less than $\delta/2$ (i.e., $\|OP\| - \|OP'\| \le \delta/2$), which can be written as

$$\left|\sqrt{x'^2 + y'^2} - r(\varphi)\right| - \delta/2 \le 0. \quad (2.15)$$

Let the following equation denote the result by substituting Eq. 2.14 into Eq. 2.15.

$$F(\varphi, \theta, r) \le 0 \tag{2.16}$$

If Eq. 2.16 holds, then all points on the profile can be controlled within the range of the tolerance zone. In general, solving the equation $\dfrac{\partial F}{\partial \varphi} = 0$ will allow one to find the point that has the maximum deviation, and hence compute the value of normalized tolerance.

Profile of a Surface. When a profile of a surface control is specified, the tolerance zone is a uniform 3D boundary. The surface being controlled must reside within the profile tolerance zone to be accepted. The tolerance zone for the profile of a surface is the region constricted by the two enveloping surfaces consisting of spheres whose diameters equal to the tolerance value δ. The profile of a surface can be defined in various ways, according to the shape and size of the part. Common applications for profile of a surface control including the control of the size, orientation, and form of the following (Krulikowski 1998):

- Planar, curved, or irregular surfaces
- Polygons
- Cylinders, surfaces of revolution, or cones

Whereas only one rotational angle contributes to the locating errors in the control of the profile of a line, two orthogonal rotational angles contribute to the locating errors in the control of the profile of a surface. The normalization of the profile of a surface tolerance is the most complex task compared with that for other types of geometric tolerances. The problem can be generalized as follows:

1. A point $P(x_0, y_0, z_0)$ lies on the ideal surface S_0 and satisfies $F(x_0, y_0, z_0) = 0$.
2. Due to imperfect locating, P moves to a new position P'. The three coordinate values of P' are $x' = g_1(x_0, y_0, z_0, \varphi, \theta)$, $y' = g_2(x_0, y_0, z_0, \varphi, \theta)$, and $z' = g_3(x_0, y_0, z_0, \varphi, \theta)$, where φ is the angle between two datums (plane XOY and $X'OY'$) and θ is the angle between axis Z and Z'.
3. Draw a normal line of the surface $F(x, y, z) = 0$ from point P', let the intersection point of the normal line and $F(x, y, z) = 0$ be $N(x'', y'', z'')$.
4. The points lying between the two enveloping surfaces formed by the tolerance zone should satisfy $(x'-x'')^2 + (y'-y'')^2 + (z'-z'')^2 \le (\delta/2)^2$.

The maximum allowable rotational angle for a datum to which a profile of a surface is specified can be computed via looking for a point closest to the upper or lower boundary formed by the two enveloping surfaces. Due to the involvement of multiple design variables, a generalized analytical model for finding the maximum allowable angle is complicated and cannot provide insight when dealing with a specific profile of a surface tolerance control. Therefore, a mathematical software tool, such as MATLAB, is a better alternative for solving this problem. Given the definition of the geometric shape of a machined part, one can use MATLAB's constrained optimization toolbox to compute the normalized tolerance for the profile of a surface.

2.3 SYSTEM DEVELOPMENT

2.3.1   Information Modeling

As discussed in Section 1.1, the task of setup planning requires part and machine information as input. Part information consists of feature geometry and feature tolerance relationships. Feature geometry determines the machine to be used. Therefore, this section first explores feature and machine information models and then tolerance information models.

2.3.1.1 Feature and Machine Information Models

The term *feature* usually refers to geometry that can be produced with a machining process. For example, a hole is produced by drilling and a slot is produced by milling. In industry, it is a common practice to group parts that have geometric similarity and that provide the same functions within part families. To represent part family information, the definition of *feature* is extended to combined features consisting of primary surfaces, including flat surfaces, cylindrical surfaces, and cone surfaces. Figure 2.20 shows the combined features of a simplified automotive part.

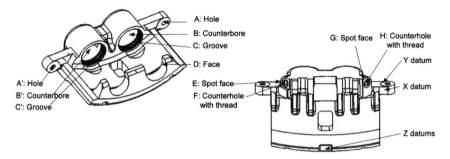

Figure 2.20. A simplified automotive part.

A combined feature is defined as linked geometric entities that can be machined in one or more manufacturing processes. Each combined feature corresponds to a sequence of predefined manufacturing processes in which combination cutters are used to reduce manufacturing cycle time. Figure 2.21 shows the information structure of a combined manufacturing feature. Surfaces are considered primary features and are mathematically represented by operational data sets. Then an object-oriented programming technique is applied for necessary reasoning. Main surfaces are those that determine the feature type, main parameters, position, and orientation. Auxiliary surfaces are those attached to main surfaces. The feature information can be further linked to a tool-path representation. Figure 2.22 shows the definition of the hole feature of the simplified automotive part and its corresponding manufacturing processes.

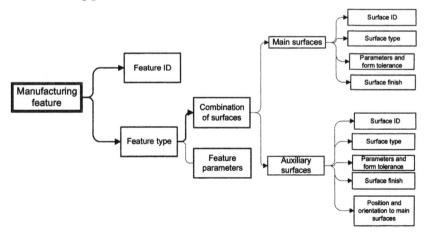

Figure 2.21. Information model for a combined manufacturing feature.

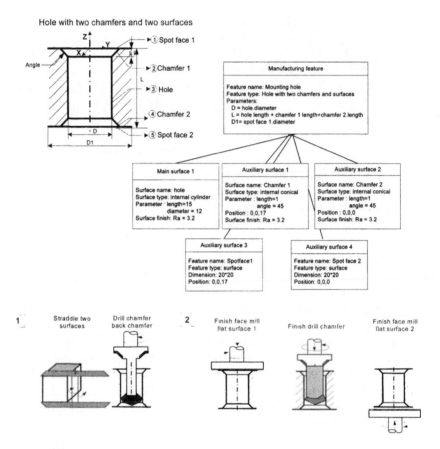

Figure 2.22. A combined hole feature and its manufacturing processes.

A combined feature may be produced by different manufacturing processes, depending on available resources. Figure 2.23 shows the relationship between part features and manufacturing resources. The manufacturing resource capability is described as three levels; namely, station (machine tool), part, and feature. Mapping from resource capability to part feature is driven by cutter capability at the feature level. A feature may have several alternative manufacturing processes. Each process may have specific requirements for cutter design and motions. The information structure of a process model is shown in Figure 2.24.

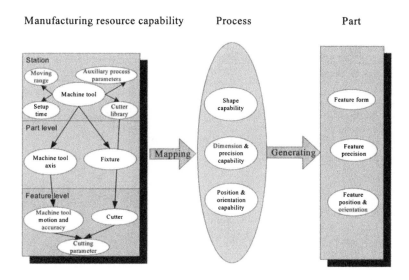

Figure 2.23. Relationship of manufacturing features and resources.

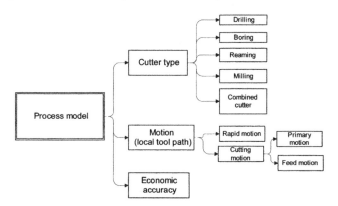

Figure 2.24. Information structure of a process model.

In the process model, the cutter type determines the basic motion types that are divided into primary and feed motions. Both of these motions can be represented mathematically. Cutter parameters and feature parameters determine the machine motion parameters. Figure 2.25 shows several examples of machine motion. Figure 2.26 shows the parameter-driven relationship between the hole feature and the cutter and tool path used to machine this feature. The cutter template and tool-path template are set up based on industry best practice.

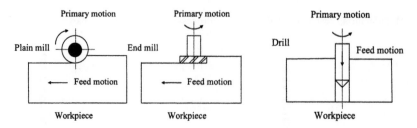

Figure 2.25. Examples of machine motion.

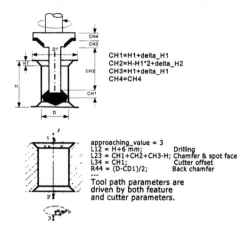

Figure 2.26. An example of cutter design and tool-path design.

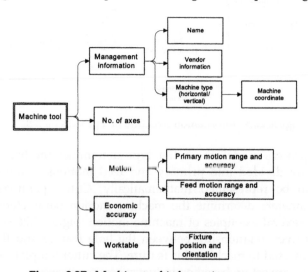

Figure 2.27. Machine tool information structure.

In the mapping from manufacturing resources to features, shape capability is determined by the combination of cutter and machine tool motions. Therefore, the motions in the process model need to be transformed into machine tool kinematic motions with accuracy. Within the industry, a variety of machine tools with similar functions are used. Therefore, an information model is needed to describe the similarities and differences among all possible machine tool capabilities so that manufacturing engineers can make a comparison. A machine tool information model is shown in Figure 2.27.

2.3.1.2 Tolerance Information Model

Tolerance relationships between features are represented using a modified version of a directed graph. A directed graph is a powerful tool used to represent discrete nodes and their relationships (Parker 1993). Mathematically, the definition of a conventional directed graph is a collection of vertices (V) and associated edges (E) given by the pair $G = (V, E)$. A directed graph is a graph with vertices and edges wherein each edge has a specific direction relative to each of the vertices:

$$G = (V, E)$$
$$V = \{v_1, v_2, \ldots, v_n\}$$
$$E = \{e_1, e_2, \ldots, e_m\}$$
$$e_i = (v_j, v_k), \quad i \in m; j, k \in n$$

where G is the graph, V is the set of vertices, and E is the set of edges. A directed graph is most commonly represented as an adjacency matrix $M = [m_{ij}]$, which is an $n \times n$ table where

$$m_{ij} = \begin{cases} 1, & \text{if } (v_i, v_j) \in E \\ 0, & \text{otherwise}. \end{cases}$$

An adjacency-matrix-based directed graph has some drawbacks in computer realization, such as awkward removal and insertion operations, the need to estimate the maximum size to which it might grow, wasted space, and inflexibility (Model 1994). Moreover, this type of graph cannot handle a very common situation in geometric dimensioning and tolerancing; namely, two or more tolerances between two features. Figure 2.28 shows such a situation, which cannot be simply expressed in an adjacency matrix as $m_{13} = 1$.

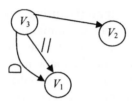

Figure 2.28. Multiple tolerance relations between two features.

```
class Label
{     private:
            char* name;
            int id;
            float value;
      public:
            //constructor, destructor, set and get operations are omitted
};
class Node
{     private:
            char* name;
            int id;
            Link_List<Label> fromEdges;      //Link_List is a container;
            Link_List<Label> toEdges;
      public:
            //constructor, destructor, set and get operations are omitted
};
class Graph
{     private:
            Link_List<Node>        nodes;      //Link_List is a container
      public:
            // constructor, destructor, copy and some overload operations are omitted
            void add(Node);                    //add a node to a graph, shown in Figure3(a)
            void remove(Node);     //remove a node from the current graph, shown in Figure 3(b)
            void join(Node, Label, Node);      //join two nodes with a label, shown in Figure 3(c)
            void unite(Graph g);   //unite declared graph g with current graph, Figure 3(d)
            void subtract(Graph g);            //subtract declared graph g from current graph, Figure 3(e)
            void intersect(Graph g);           //intersect declared graph g with current graph, Figure 3(f)
            int traverseTo(Node nd1, int level, Node nd2);
                              //given a start node nd1 and level, traverse to node nd2 -
                              //output, see Figure 3(g)
            Link_List<Graph> getIndirectedGraph(Node nd1, Node nd2);
                              //given a start node nd1 and end node nd2, get all the routines
                              //expressed as graph, see Figure 3(h)
};
```

Figure 2.29. Object-oriented representation of the extended directed graph

To overcome the shortcomings of the conventional directed graph and make it amenable to model tolerance information, an extended directed graph is proposed. It is marked as G^E:

$$G^E = (V, E^E)$$
$$V = \{v_1, v_2, ..., v_n\}$$

$$E^E = \{e_1, e_2, ..., e_m\}$$
$$e_j = \{v_k, v_l, l_j\} \qquad j \in m;\ k, l \in n,$$

where V is the set of vertices, E is the set of edges, and l_j is a lable describing edge e_j.

An extended directed graph is implemented using an object-oriented model, an example of which is shown in Figure 2.29. Some important operations of the model are shown in Figure 2.30.

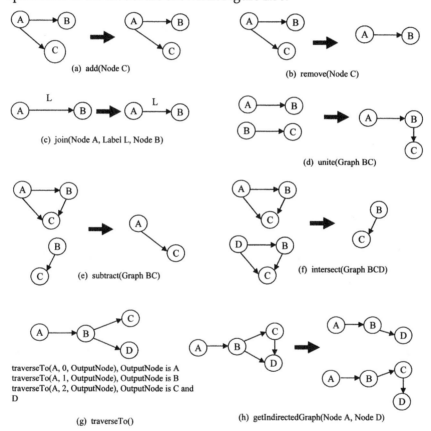

(a) add(Node C)

(b) remove(Node C)

(c) join(Node A, Label L, Node B)

(d) unite(Graph BC)

(e) subtract(Graph BC)

(f) intersect(Graph BCD)

traverseTo(A, 0, OutputNode), OutputNode is A
traverseTo(A, 1, OutputNode), OutputNode is B
traverseTo(A, 2, OutputNode), OutputNode is C and
D

(g) traverseTo()

(h) getIndirectedGraph(Node A, Node D)

Figure 2.30. Main operations in the object-oriented extended directed graph.

An example best illustrates how to model feature tolerance relationships using an extended directed graph. Figure 2.31 shows a simplified front knuckle for a generic automotive chassis system. A front knuckle is the joint at the confluence of the braking system (caliper mounting pads), suspension system (strut), and steering system (features are not shown in the drawing for concise representation). Therefore, a front knuckle requires parallelism between caliper mounting pad surfaces N and N′ and perpendicularity between strut mounting surface D and datum reference A. A diagrammatic representation of the tolerance relationships is shown in Figure 2.32.

2.3.2   Generative Setup Planning Algorithm

Generative setup planning refers to fully automated setup planning. That is, a setup plan is generated without human intervention whenever a part design specification is given. Currently, generative algorithms can gracefully handle parts with moderate complexity. For more complicated parts, certain human interaction is needed to ensure a good solution. This is referred to as an integrated setup planning, explored in the next section.

2.3.2.1 Mathematical Formulation of Setup Planning

Rigorous problem formulation is a prerequisite in ensuring the robustness of solution methodology. To facilitate the development of a robust methodology, the setup planning problem is formulated mathematically using the following notations:

N	Number of surface features of the part
M	Number of surface features to be machined
f_i	Surface feature i of the part, where $i = 1, 2, ..., N$
F	Set of surface features to be machined, where $F = \{f_1, f_2, ..., f_M\}$
$C(f_i)$	Set of tool-approach directions of surface feature i, where $i = 1, 2, ..., M$
R	Number of setups in the setup plan
S_r	Set of surface features to be machined in setup r, where $r = 1, 2, ..., R$
$c(S_r)$	Tool-approach direction of setup r, where $r = 1, 2, ..., R$
M_r	Number of surface features to be machined in setup r, where $r = 1, 2, ..., R$
g_j^r	Surface feature j in setup r, where $j = 1, 2, ... M_r, r = 1, 2, ..., R$
L_r	Set of surface features used to locate setup r (setup datums), where $r = 1, 2, ..., R$

K_r Set of surface features that exist after the workpiece is machined in the rth setup, where K_0 denotes the set of features that exist on the stock (raw material) and $K_r = K_{r-1} \cup S_r$

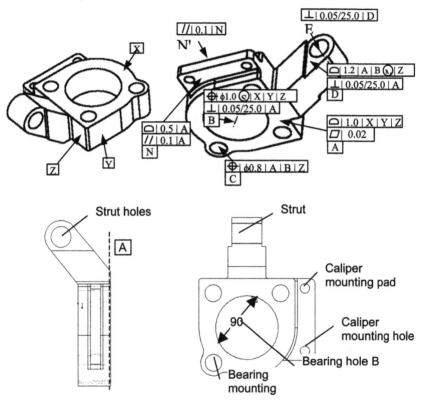

Figure 2.31. Front knuckle example part.

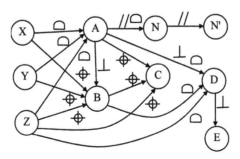

Figure 2.32. Extended directed graph representation of tolerance relationships.

The setup planning problem is equivalent to finding the sets of surface features S_1 (with surface features in L_1 as setup datums), S_2 (with surface features in L_2 as setup datums), and so on through S_R (with surface features in L_R as setup datums) to be machined in that sequence, with the following constraints:

$$S_1 \cup S_2 \cup \ldots \cup S_R = F \tag{2.17}$$

$$C(g_1^r) \cap C(g_2^r) \cap \ldots \cap C(g_{M_r}^r) = \{c(S_r)\} \neq \varnothing, \forall\, r \in \{1, 2, \ldots, R\} \tag{2.18}$$

$$L_r \cap S_r = \varnothing, \forall\, r \in \{1, 2, \ldots, R\} \tag{2.19}$$

$$L_r \subseteq K_{r\text{-}1}, \forall\, r \in \{1, 2, \ldots, R\} \tag{2.20}$$

Constraint Eq. 2.17 guarantees that all of the surface features of the part that require machining are machined. Constraint Eq. 2.18 ensures that all the surface features within the same setup have a common tool-approach direction. Constraint Eq. 2.19 reflects the fact that the setup datums cannot be machined in that particular setup. Constraint Eq. 2.20 warrants that the selected datums exist on the workpiece. All solutions that satisfy these constraints are feasible solutions. To find good solutions, objective functions need to be formulated. However, the formulation of objective functions is not straightforward as different objectives might arise in different production environments. The algorithms presented here are developed based on two common objectives: (1) minimize the number of setups, and (2) minimize machining errors to ensure that design tolerance requirements are met.

2.3.2.2 Preparatory Steps

As previously mentioned, the input of setup planning includes part and machine information. The values of N, M, f_i, and F are extracted from part information. Note that here we define f_i as a *surface feature*, which is different from the conventional feature defined in the literature. A conventional feature is a *machining feature* such as a hole or a slot, which is produced by a certain machining process. In some cases, a surface feature is a machining feature. For example, the flat face of a block (f_1 in Figure 2.33) and the cylindrical surface of a hole (f_5 in Figure 2.33). In other cases, a machining feature consists of several surface features as in, for example, the open slot shown in Figure 2.32 has three surfaces, and thus has three surface features: f_2, f_3, and f_4. The use of surface features in setup planning is necessary because (1) surface features are used for locating and clamping and (2) tolerances are specified on surface features.

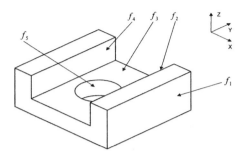

Figure 2.33. Surface features of a part.

Now we need to determine $C(f_i)$, the tool-approach direction of a surface feature. The tool-approach direction of a feature is the orientation of the workpiece when the feature is machined. Therefore, determining the tool-approach direction of a surface feature involves three steps if the surface feature is not a machining feature. First, the machining feature to which the surface feature belongs is identified. Then, the possible workpiece orientation (which might be more than one) where the machining feature can be produced is identified. Finally, the orientation is assigned to the surface feature as its tool-approach direction. In other words, the determination of tool-approach direction is based on machining features. For example, the slot in Figure 2.33 is a machining feature (consists of surface features f_2, f_3, and f_4). It can be produced when the workpiece is oriented as shown in Figure 2.33 (machined from the +Z direction). Therefore, the tool-approach direction of f_2, f_3, and f_4 are the same; namely, +Z.

For features machined on a machining center, six tool-approach directions are possible: +X, –X, +Y, –Y, +Z, and –Z. For example, in Figure 2.33, the tool-approach direction of f_1 is +X; whereas f_5 has two possible tool approach directions (+Z and –Z). For features machined on a lathe, only two tool approach directions are possible; namely, +X and -X (Figure 2.34). Tool-approach directions are determined either via feature recognition capability provided by certain CAPP systems or through the analysis of feature information provided by CAD systems. For example, the popular Pro/Engineer CAD software tool provides feature information via its neutral file. Automatic determination of tool-approach direction for rotational parts from a neutral file can be found in (Zhou, 2002).

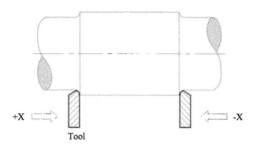

Figure 2.34. Tool-approach directions for a rotational part.

The next step is to determine L_r, the setup datums, which are based on the 6-point locating principle (Halevi 1991). For rotational parts, the principle takes the form of 4-1-1 locating, where the last point is applied with clamping to create friction. As shown in Figure 2.35, the part is first located on its surface using four points (1, 2, 3, and 4). Four degrees of freedom (two translations along and two rotations around the y and z axes) are canceled. A fifth point (5) is added to cancel the translation along the x axis. Finally, a sixth point (6) is applied to cancel the rotation around the x axis. Note that point 6 acts by friction and requires clamping.

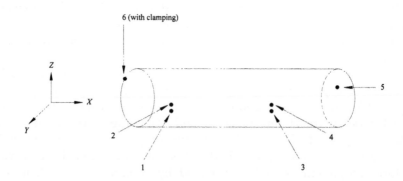

Figure 2.35. Application of six-point locating principle to a rotational part.

For prismatic parts, the six-point locating principle takes the form of 3-2-1 locating, where three orthogonal surfaces are needed to restrict all six degrees of freedom of a workpiece. The normal direction of a surface feature f_i, denoted $d(f_i)$ (perpendicular to the surface pointing outward), can be determined from the CAD model of a part. Because slanted surfaces (such as f_9 of the shaft support shown in Figure 2.36) and curved surfaces (such as f_{11} in Figure 2.36) are generally not used for locating and clamping purposes, only flat surface features (whose normal direction is either +X, –X, +Y, –Y, +Z, or –Z) and holes (when dowel pin locating is used) need to be considered in determining L_r. For clarity, the use of only flat surfaces for locating and clamping is discussed first.

Note that L_r depends on r, and when r is determined S_r is also determined. Therefore, $c(S_r)$ can be obtained. To determine the setup datums, a flat surface feature (say f_i) is examined. If $d(f_i) = -c(S_r)$, then f_i can be used as a primary locating surface for setup S_r. Assume f_i is selected as the primary locating surface (see datum selection algorithm in next section). A candidate for the secondary locating surface, f_j, should satisfy $|d(f_j)| \neq |d(f_i)|$. If f_j is selected as the secondary locating surface, then the tertiary locating surface f_k should satisfy the condition $|d(f_k)| \neq |d(f_j)| \neq |d(f_i)|$. For example, if surface feature f_{10} in Figure 2.36 is to be machined in the first setup (i.e., $r =1$, $S_r = S_1 = \{f_{10}\}$), then $c(S_1) = +Y$. There is only one feature (f_3) whose normal direction is –Y [i.e., $d(f_3) = -Y = -c(S_1)$]. Therefore, f_3 is selected as the primary locating surface. The secondary locating surface is selected from features whose normal direction is neither +Y nor –Y. Say f_4 is selected. Since $d(f_4) = -Z$, only surface features with +X or –X normal direction can be used as the tertiary locating surface.

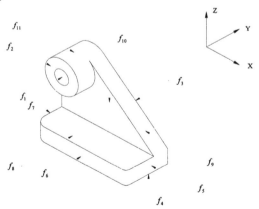

Figure 2.36. A shaft support.

Although holes are not used as locating surfaces in universal fixtures such as a vise, they can be used in modular fixtures when dowel pin locating (Rong 1999) is applied. In this case, a hole can be used as a secondary locating surface. Instead of its normal direction, the axial direction of a hole f_j is determined, denoted $a(f_j)$. The axial direction of a through hole has no sign. For example, the through hole f_1 shown in Figure 2.36 has an axial direction of Y [i.e., $a(f_1) = Y$]. For a blind hole, its axial direction is defined as its tool-approach direction. For example, if f_1 in Figure 2.36 is a blind hole, then $a(f_1) = -Y$. For a through hole f_j, if $|c(\mathbf{S}_r)| = a(f_j)$, then it can be used as a secondary locating surface in L_r; whereas for a blind hole f_j, $a(f_j) = -c(\mathbf{S}_r)$ is required. Once the hole f_j is used as a secondary locating surface, the restriction for the tertiary locating surface f_k is relaxed to $|d(f_k)| \neq |d(f_i)|$ (f_i is the primary locating surface). For example, if dowel pin locating is used, to machine f_{10} in Figure 2.35 f_3 and f_1 can be used as the primary and secondary locating surfaces, respectively, whereas f_4, f_5, f_6, f_7, and f_8 are all candidates for the tertiary locating surface.

2.3.2.3 Setup Planning Algorithms

In this section, algorithms are developed to accomplish three subtasks in setup planning: (1) setup formation, (2) datum selection, and (3) setup sequencing. Objectives incorporated into the setup formation algorithm are (1) minimizing the number of setups and (2) ensuring features with tight tolerance relations are machined in the same setup if possible. The steps, applied to surface features that need to be machined (surface features in the set F), are as follows:

Step 1.   Find all surface features that have multiple tool-approach directions. Let n^+ denote the number of such surface features. If $n^+ = 0$, let $\Omega = \varnothing$, and go to step 5; otherwise, let f_i^+ denote one such surface feature [the number of elements in $C(f_i^+)$ is greater than 1], $i = 1, \dots, n^+$. Let $\Omega = \{f_1^+, \dots, f_{n^+}^+\}$.

Step 2.   Determine whether any f_i^+ has tolerance relationship with a surface feature f_j, in which $f_j \notin \Omega$. If not, go to step 5; otherwise, go to step 3.

Step 3.   Without loss of generality, let f_1^+ denote the surface feature that has the tightest tolerance relationship (based on comparison of normalized tolerance, smaller normalized tolerance value means the underlying tolerance is tighter) with surface feature f_j. Check whether $C(f_1^+) \cap C(f_j) \neq \varnothing$. If affirmative, let $C(f_1^+) = C(f_1^+) \cap C(f_j)$

and remove f_1^+ from Ω (f_1^+ is grouped with f_j in the same setup because they have a tight tolerance relationship); otherwise, ignore the tolerance relationship between f_1^+ and f_j and go to step 2.

Step 4. If $\Omega = \varnothing$, go to step 5; otherwise go to step 2.

Step 5. If $\Omega = \varnothing$, all setups are formed (surface features with the same tool-approach direction are in the same setup); otherwise, utilize the Quine-McClusky algorithm (Lee 1987) or an exhaustive search algorithm (in that there are only two and six tool-approach directions for rotational and prismatic parts, respectively) to determine the minimal number of setups.

Now that all setups are formed, the next task is datum selection for each setup. The objective incorporated into the algorithm is to ensure that surface features with tight tolerance relations are used as datums whenever possible. The steps carried out for a setup S_r, applied to all surface features are as follows:

Step 1. Find surface feature $f_i \notin S_r$ and $d(f_i) = -c(S_r)$ (normal direction is opposite to the tool-approach direction of the setup), if there exist more than one such surface feature, find the one that has the tightest tolerance relationship with surface features in S_r. If none has tolerance relationship, pick the one that has the largest area. Let f_i be the primary locating surface.

Step 2. The selection of the secondary locating surface, f_j, is similar to step 1 except that $|d(f_j)| \neq |d(f_i)|$ should be satisfied instead of $d(f_i) = -c(S_r)$. For rotational parts, stop. For prismatic parts, continue to step 3.

Step 3. The selection of tertiary locating surface f_k is also similar to step 1 except that $|d(f_k)| \neq |d(f_j)| \neq |d(f_i)|$ should be satisfied instead of $d(f_i) = -c(S_r)$. Let $L_r = \{f_i, f_j, f_k\}$.

For dowel pin locating in modular fixture applications, the determination of secondary and tertiary datums should be slightly modified based on discussions in the preparatory steps. The final task is setup sequencing. The objective is to ensure more tolerance relationships in finishing operations. The steps are given as follows:

Step 1. Determine t_r, which is the number of tolerance relationships between elements in S_r and elements in L_r, where $r = 1, \ldots, R$.

Step 2. Sort the setups based on increasing order of t_r.

Step 3. Let $\Omega = \varnothing$. Loop through the setups (r = 1 to R) and perform the following:

> a. Check whether $L_r \subseteq K_{r-1}$.
> b. If not, then Let $\Omega = \Omega \cup (L_r - (L_r \cap K_{r-1}))$.
> c. Let $K_r = K_{r-1} \cup S_r$.

Step 4. If $\Omega = \varnothing$, the setup plan is valid; otherwise, surface features in Ω need to be machined first.

2.3.3   Integrated Setup Planning

Integrated setup planning refers to the use of limited human interaction to ensure that setup plans generated are practically useful even for very complicated parts. The human interaction usually takes the form of grouping features and identifying setup datum frames based on best practice. A graph-based approach can then be used for reasoning and manipulation, converting a feature tolerance graph (FTG) into a datum-machining feature relationship graph (DMG), from which a setup plan is extracted.

Take the automotive part shown in Figure 2.20 as an example. The features that have close position, orientation, or profile tolerance requirements are grouped in the same datum frame, as shown in Table 2.3. The FTG is shown in Figure 2.37.

Table 2.3. Feature grouping and datum identification.

Feature Group	Manufacturing Features	Datum
Group 1	A, A'	X, Y, Z
Group 2	D, B, C, B', C'	X, Y, Z
Group 3	E, F	B, B', Z
Group 4	G, H	B, B', Z

Although it is desirable to machine features with tight tolerance relationships in one setup, the machine may not have the necessary capabilities. Therefore, we need to first attach selected manufacturing processes to the features in the FTG (as shown in Figure 2.38) and then conduct a tool-approach direction analysis. The result is shown in Table 2.4. Each group of features in Table 2.4 may be machined in one setup using standard manufacturing resources such as general machine tools and fixtures. To achieve a high production rate under mass-customization conditions, however, multi-part fixtures and multi-axis NC machines are widely used. This allows the number of setups to be greatly reduced, with a DMG generated as shown in Figure 2.39.

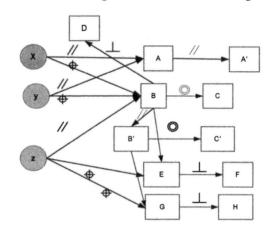

Figure 2.37. FTG of the automotive part.

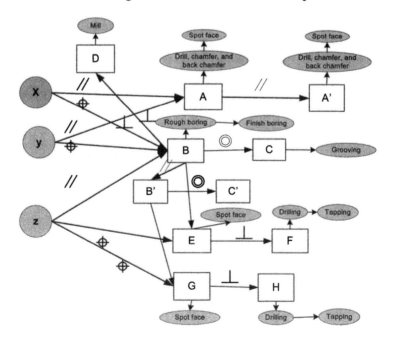

Figure 2.38. FTG with processes attached.

Table 2.4. Process and tool-approach direction analysis.

Feature Group	Manufacturing Features	Datum	Tool-approach Direction
Group 1	A (Drill, chamfer, and back chamfer) A′ (Drill, chamfer, and back chamfer) B, B′ (Rough boring, finish boring) C, C′ (Grooving)	X, Y, Z	+X
Group 2	A (Spot face) A′ (Spot face) D (milling)	X, Y, Z	-X
Group 3	E (Spot face) F (Drilling, tapping)	B, B′, Z	-0.6Y +0.8Z
Group 4	G (Spot face) H (Drilling, tapping)	B, B′, Z	0.6Y +0.8Z

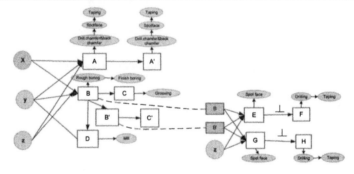

Figure 2-39. DMG generation.

Finally, a process sequence for the setups is generated. The problem of process sequencing is transformed mathematically into searching for an optimal path of traversal for each vertex in the DMG under specified constraints. The time required to traverse each vertex is determined by the number of processes linked to each feature. The constraints are divided into strong and weak constraints. Strong constraints, such as following, cannot be violated:

- Maintaining the manufacturing process sequence of each feature.
- Maintaining the feature-creation sequence of the graph. For example, planes are processed prior to holes, and holes prior to grooves.
- Doing rough cuts first, semi and finish cuts in a prescribed order.

Weak constraints are determined by best practice. For example, the cutter for milling the outboard flange could be combined with the cutter for drilling and chamfering and back-chamfering the mounting holes, which would reduce tool change time. Figure 2.40 shows one solution for the process sequence of the example part.

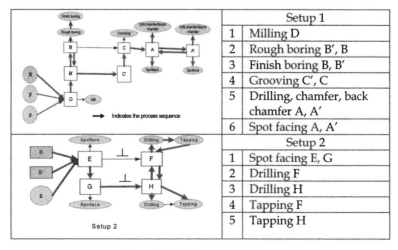

	Setup 1
1	Milling D
2	Rough boring B′, B
3	Finish boring B, B′
4	Grooving C′, C
5	Drilling, chamfer, back chamfer A, A′
6	Spot facing A, A′
	Setup 2
1	Spot facing E, G
2	Drilling F
3	Drilling H
4	Tapping F
5	Tapping H

Figure 2.40. Process sequence.

Once the process sequence is determined at the part level, the setup plan can be generated at the station level. The various solutions to a conceptual fixture design are then analyzed with the following in mind toward generating a feasible setup plan:

- Determine the best process sequence to machine the manufacturing features of all parts with the fixture.
- Determine the best tool-path for machining all features with the fixture based on the process sequence.
- Reduce cycle time and optimize manufacturing processes. Cycle time is the basic criterion applied in choosing a manufacturing plan. Cycle time consists of cutting time, rapid travel time, tool-change time, and machine tool table index time. Cutting time is determined by cutting parameters, and the other time is related to machine tool performance. Therefore, through the adjustment of cutting parameters, the changing of machine tool parameters, and even the changing of machine tools can help reduce cycle time and help identify a good manufacturing solution.
- Generate manufacturing documents. Documents are generated according to a company-specific format. Manufacturing planning

information – including setups, process sequences, machines, tooling, and process parameters – are incorporated in such documents. This helps users understand what the system does and what type of information is used in the decision-making process.

Three solutions of conceptual fixture design for setup 1 of Figure 2.40 are outlined in Table 2.5, including the simplified part and fixture base information. Even with the use of the same machine tool and process parameters, a different cycle time can be achieved. It can be figured out that the bridge is the best solution for setup 1, which has the least cycle time. Corresponding requirements to machine tool are generated and evaluated. Table 2.6 shows the different fixture bases used in the two setups. Figure 2.41 shows part of the process sequence and parameters in the documentation.

Table 2.5. Alternative fixture solutions for the setup 1.

Conceptual Fixture Design	**Bridge**	**Rectangle Plate**	**Round Plate**
Part Layout			
Machine Tool	DMV- 500 provided by Daewoo		
Cutting Time per Part (Sec.)	66.48	66.48	66.48
Non-cutting Time per Part (Sec.)	31.48	40.06	34.08
Cycle Time per Part (Sec.)	97.96	106.54	100.56

In summary, thus far an overall framework of a computerized setup planning system has been developed. The objective is to provide a computerized tool for rapid design and for the simulation of manufacturing systems, with an emphasis on the utilization of best practice and analysis of production planning. In the system, a feature-based part information model is used to represent part information. Combined features are parametrically represented and subsequently used in determining manufacturing methods and processes based on available manufacturing resources and capabilities. Graph-based automated setup planning has been extended to consider flexible manufacturing resources, multi part fixture configuration, and process sequence optimization. Finally the standard manufacturing documentation is automatically generated to a company-specific format by the system.

Table 2.6. Station-level setup planning.

		Setup 1	Setup 2
Fixture Base Type		Bridge	Tombstone
Part Layout on Fixture Base  			
Machine Tool Requirements	No. of Axes	31/2	21/2
	Moving Range X	800 mm	500 mm
	Y	363 mm	500 mm
	Z	765 mm	700 mm

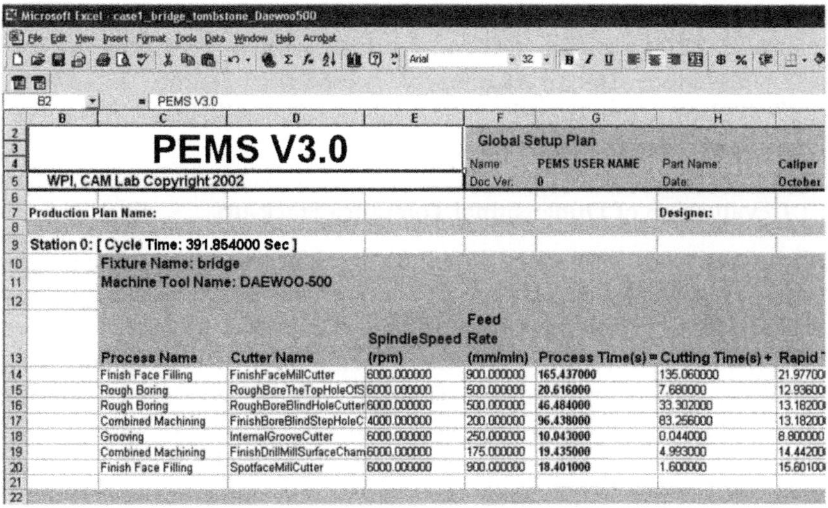

Figure 2.41. Standard document output.

This section focuses on setup planning and station-level planning. Process-level optimization, such as tool-path or process parameter optimization, is not emphasized. Nonetheless, results from process-level optimization can be easily integrated into the system. Two other important setup planning tasks; namely, fixture design and tolerance analysis, are discussed in separate sections.

2.4 ADVANCED TOPICS

2.4.1 Setup Plan Evaluation

Common objectives in setup planning are to (1) minimize the number of setups and (2) minimize tolerance stack-up. Once a setup plan is generated, the number of setups is immediately available. However, the plan's effect on tolerance stack-up cannot be readily quantified. Therefore, how to evaluate the effectiveness of a setup plan, in terms of its ability to minimize tolerance stack-up, becomes an important research issue. As previously mentioned, there are two types of tolerances; namely, dimensional tolerance and geometric tolerance. The evaluation of dimensional tolerance stack-up is relatively easy, as it is a one-dimensional (1D) problem and can be dealt with using simple Monte Carlo simulation. The evaluation of geometric tolerance is a very complicated problem, as it is a 3D problem. We will first look at the evaluation of dimensional tolerance.

2.4.1.1 Evaluation of Dimensional Tolerance Stack-up

With respect to the objective of minimizing tolerance stack-up, one can say that a setup plan is better than another one if the variation of the resultant part dimension is smaller. Therefore, to evaluate the effectiveness of a setup plan, the variances of the resultant part dimensions should be examined. This can be achieved through simple Monte Carlo simulation. Monte Carlo simulation is a popular device for studying an artificial stochastic model of a physical or mathematical process (Rubinstein 1981). Under the Monte Carlo simulation, a system is examined through repeated evaluations of the mathematical model as its parameters are randomly varied according to their true behavior.

Figure 2.42 shows a block diagram of a simulation model for a setup plan evaluation. Given a part design, a setup plan can be generated. Based on the generated setup plan, an NC program can be created. Machining of the part (or parts) based on the NC program is then simulated. The simulation derives part dimensions. Also, based on the simulation, sample variances for resultant part dimensions can be calculated. The smaller the sample variances, the better the setup plan in terms of minimizing tolerance stack-up.

Figure 2.42. Block diagram of a simulation model for setup plan evaluation.

The part shown in Figure 2.43 is a typical rotational part. Features A, B, C, D, and E are to be machined. Because the part consists of five surfaces that need to be machined, four dimensions (and their tolerance requirements) are needed.

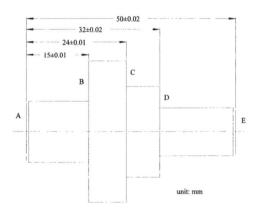

Figure 2.43. An example of a typical rotational part (unit: inch).

Usually the part is machined in two setups with features A and B in one setup and features C, D, and E in the other setup. Either setup can be machined first. When machining features C, D, and E, either feature A or feature B can be used as a setup datum. Similarly, any one of features C, D, and E can be used as a setup datum when machining features A and B. Therefore, there are a total of $2\binom{2}{1}\binom{3}{1} = 12$ possible setup plans, as shown in Table 2.7. Based on the setup plans, NC programs were made and the machining processes simulated. It is assumed that the setup error follows a normal distribution, with a mean of 0 mm and a standard deviation of 0.006 mm (μ = 0 mm and σ = 0.006 mm); whereas the machine motion error follows a normal distribution, with a mean of 0 mm and a standard deviation of 0.003 mm (μ = 0 mm and σ = 0.003 mm). The four resultant dimensions are normally distributed random variables denoted X_1, X_2, X_3, and X_4, respectively. Let σ_Y^2 be an indicator of the quality of the part, in which $Y = X_1 + X_2 + X_3 + X_4$. Therefore, $\sigma_Y^2 = \sum_{i=1}^{4} \sigma_i^2$, in which σ_i^2 is the variance of X_i, (i = 1, 2, 3, 4). The smaller σ_Y^2 is, the better the quality of the part.

Let $\sigma_{Y_i}^2$ denote the σ_Y^2 resulting from the ith setup plan, in which i = 1, 2, ..., 12. Let e be the setup plan selected by the setup planning algorithm. To show the effectiveness of the setup planning algorithm, one needs to test the equality of $\sigma_{Y_i}^2$ (i = 1,2, ...,12) and show that $\sigma_{Y_i}^2$ is small. To perform the test, 5,000 samples were taken. The sample variance ($S_{Y_i}^2$, a point estimator of the population variance σ_Y^2) resulting from each setup plan was calculated (Table 2.8). Let

$$\sigma_{Y_a}^2 = \min\left[S_{Y_i}^2\right],$$
$$\sigma_{Y_b}^2 = \operatorname{mid}\left[S_{Y_i}^2\right] \text{ (the middle, or the sixth smallest among the } S_{Y_i}^2\text{'s}$$
$$\text{except } S_{Y_e}^2 \text{), and}$$
$$\sigma_{Y_c}^2 = \max\left[S_{Y_i}^2\right],$$

in which i = 1, 2, ..., 12, $i \neq e$. In other words, the ath, bth, and cth setup plans are the best, the middle, and the worst setup plans, respectively, among those setup plans that were not selected by the setup planning algorithm. The σ_Y^2 resulting from the eth setup plan is compared, respectively, with those resulting from the ath, the bth, and the cth setup plans. The hypotheses are as follows:

- Comparing $\sigma_{Y_e}^2 = \sigma_{Y_a}^2$

  H₀: $\sigma_{Y_e}^2 = \sigma_{Y_a}^2$

  H₁: $\sigma_{Y_e}^2 > \sigma_{Y_a}^2$

- Comparing $\sigma_{Y_b}^2 = \sigma_{Y_e}^2$

  H₀: $\sigma_{Y_b}^2 = \sigma_{Y_e}^2$

H_1: $\sigma_{Y_b}^2 > \sigma_{Y_e}^2$

- Comparing $\sigma_{Y_c}^2 = \sigma_{Y_e}^2$

H_0: $\sigma_{Y_c}^2 = \sigma_{Y_e}^2$

H_1: $\sigma_{Y_c}^2 > \sigma_{Y_e}^2$

Table 2.7. Setup methods for the rotational part shown in Figure 2.43.

Setup Plan	Method
1	Machine features C, D, and E using feature A as a setup datum and then machine features A and B using feature C as a setup datum.
2	Machine features A and B using feature C as a setup datum and then machine features C, D, and E using feature A as a setup datum.
3	Machine features C, D, and E using feature A as a setup datum and then machine features A and B using feature D as a setup datum.
4	Machine features A and B using feature D as a setup datum and then machine features C, D, and E using feature A as a setup datum.
5	Machine features C, D, and E using feature A as a setup datum and then machine features A and B using feature E as a setup datum.
6	Machine features A and B using feature F as a setup datum and then machine features C, D, and E using feature A as a setup datum.
7	Machine features C, D, and E using feature B as a setup datum and then machine features A and B using feature C as a setup datum.
8	Machine features A and B using feature C as a setup datum and then machine features C, D, and E using feature B as a setup datum.
9	Machine features C, D, and E using feature B as a setup datum and then machine features A and B using feature D as a setup datum.
10	Machine features A and B using feature D as a setup datum and then machine features C, D, and F using feature B as a setup datum.
11	Machine features C, D, and E using feature B as a setup datum and then machine features A B using feature E as a setup datum.
12.	Machine features A and B using feature E as a setup datum and then machine features C, D, and E using feature B as a setup datum.

For the example of rotational part, setup plan 2 is selected by our setup planning algorithm. Therefore, $e = 2$. From Table 2.8, one can see that $a = 4$, $b = 7$, and $c = 8$. The comparison of the setup plans is shown in Table 2.9. From Table 2.9, one can conclude that $\sigma_{Y_2}^2$ is not significantly different from $\sigma_{Y_4}^2$, $\alpha = 0.01$, BOE (based on evidence). However, $\sigma_{Y_2}^2$ is significantly different from both $\sigma_{Y_7}^2$ and $\sigma_{Y_8}^2$, $\alpha = 0.01$, BOE. In other words, the setup plan selected by our setup planning algorithm is not significantly different from the best setup plan. However, it is significantly better than both the middle setup plan and the worst setup plan, $\alpha = 0.01$, BOE. Therefore, our setup planning algorithm is effective.

Table 2.8. Results of the simulation for the setup plans shown in Table 2.7.

Setup Method	S_Y^2 (10^{-6} mm²)
1	215.41
2 (e)	180.12
3	219.67
4 (a)	179.01
5	217.53
6	179.57
7 (b)	216.72
8 (c)	236.30
9	210.55
10	233.11
11	217.07
12	227.74

Table 2.9. Comparison of the setup plans for the part shown in Figure 2.42.

	$\sigma^2_{Y_e} = \sigma^2_{Y_a}$	$\sigma^2_{Y_b} = \sigma^2_{Y_e}$	$\sigma^2_{Y_c} = \sigma^2_{Y_e}$
F- value	$S_{Y_2}^2 / S_{Y_4}^2 = 1.006$	$S_{Y_7}^2 / S_{Y_2}^2 = 1.205$	$S_{Y_8}^2 / S_{Y_2}^2 = 1.312$
P-value	.414	2.2×10^{-11}	0

2.4.1.2 Evaluation of Geometric Tolerance Stack-up

When evaluating dimensional tolerance, we simply sum the variances of individual dimensions under consideration. This strategy will not work for geometric tolerances, however, because there are different categories of geometric tolerance, namely, (1) form tolerances that include straightness, flatness, roundness, and cylindricity, (2) orientation tolerances that include parallelism, angularity, and perpendicularity, (3) location tolerances that include concentricity, symmetry, and position, (4) runout tolerances that include circular runout and total runout, and (5) profile tolerances that include profile of a line and profile of a surface. In addition, evaluating geometric tolerance is a 3D problem that renders the method of adding a random component to a dimension useless. To solve this problem, an evaluation approach has been proposed based on feature discretization, manufacturing error analysis, Monte Carlo simulation, and virtual inspection (Musa 2004). Specifically, the approach involves the following steps:

- A set of discrete points is used to represent the surface whose tolerances are involved in the analysis.
- Monte Carlo simulation is used to study the effect of various manufacturing errors on the spatial locations of these points.
- Virtual inspection is then conducted based on the coordinates of these points, which allows the analysis of any type of tolerance (geometric as well as dimensional).

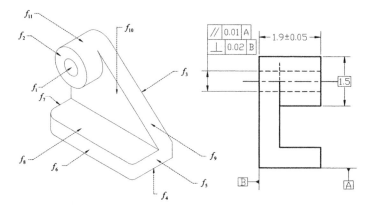

Figure 2.44. An example of a shaft support (unit: inch).

Consider the shaft support shown in Figure 2.44. Four surfaces require machining; namely, f_1, f_2, f_3, and f_4. It is obvious that a total of three setups are needed when using universal fixtures. At the setup formation stage, there are only two grouping alternatives; namely, $\{(f_1, f_2), (f_3), (f_4)\}$ and $\{(f_2), (f_1, f_3), (f_4)\}$. Note that the accuracy of machining is mainly affected by the primary locating surface since the shaft support is machined on a vertical milling center. Therefore, only variation of primary locating surface is taken into consideration in datum selection. For grouping alternative $\{(f_1, f_2), (f_3), (f_4)\}$, only setup (f_3) has two primary locating surface options; namely, feature f_2 or feature f_{10} (feature f_6 is not suitable as a primary locating surface because of its narrow shape). Likewise, only setup (f_1, f_3) in grouping alternative $\{(f_2), (f_1, f_3), (f_4)\}$ has two options, feature f_2 or feature f_{10}. At the setup sequencing stage, two grouping alternatives (each having three setups) will result in 12 different sequences. Therefore, there are 24 possible setup plans for the shaft support, as outlined in Table 2.10. Note that setup plan 9 is the one obtained using our setup planning algorithm.

Table 2.10. Possible setup plans for the shaft support shown in Figure 2.44.

Setup Plan	Setup 1				Setup 2				Setup 3			
	Machined Feature	1st Datum	2nd Datum	3rd Datum	Machined Feature	1st Datum	2nd Datum	3rd Datum	Machined Feature	1st Datum	2nd Datum	3rd Datum
1	f_1, f_2	f_3	f_4	f_5	f_3	f_2	f_4	f_5	f_4	f_8	f_3	f_5
2	f_3	f_2	f_4	f_5	f_1, f_2	f_3	f_4	f_5	f_4	f_8	f_3	f_5
3	f_4	f_8	f_3	f_5	f_1, f_2	f_3	f_4	f_5	f_3	f_2	f_4	f_5
4	f_1, f_2	f_3	f_4	f_5	f_4	f_8	f_3	f_5	f_3	f_2	f_4	f_5
5	f_3	f_2	f_4	f_5	f_4	f_8	f_3	f_5	f_1, f_2	f_3	f_4	f_5
6	f_4	f_8	f_3	f_5	f_3	f_2	f_4	f_5	f_1, f_2	f_3	f_4	f_5
7	f_2	f_3	f_4	f_5	f_1, f_3	f_2	f_4	f_5	f_4	f_8	f_3	f_5
8	f_1, f_3	f_2	f_4	f_5	f_2	f_3	f_4	f_5	f_4	f_8	f_3	f_5
*9	f_4	f_8	f_3	f_5	f_2	f_3	f_4	f_5	f_1, f_3	f_2	f_4	f_5
10	f_2	f_3	f_4	f_5	f_4	f_8	f_3	f_5	f_1, f_3	f_2	f_4	f_5
11	f_1, f_3	f_2	f_4	f_5	f_4	f_8	f_3	f_5	f_2	f_3	f_4	f_5
12	f_4	f_8	f_3	f_5	f_1, f_3	f_2	f_4	f_5	f_2	f_3	f_4	f_5
13	f_1, f_2	f_3	f_4	f_5	f_3	f_{10}	f_4	f_5	f_4	f_8	f_3	f_5
14	f_3	f_{10}	f_4	f_5	f_1, f_2	f_3	f_4	f_5	f_4	f_8	f_3	f_5
15	f_4	f_8	f_3	f_5	f_1, f_2	f_3	f_4	f_5	f_3	f_{10}	f_4	f_5
16	f_1, f_2	f_3	f_4	f_5	f_4	f_8	f_3	f_5	f_3	f_{10}	f_4	f_5
17	f_3	f_{10}	f_4	f_5	f_4	f_8	f_3	f_5	f_1, f_2	f_3	f_4	f_5
18	f_4	f_8	f_3	f_5	f_3	f_{10}	f_4	f_5	f_1, f_2	f_3	f_4	f_5
19	f_2	f_3	f_4	f_5	f_1, f_3	f_{10}	f_4	f_5	f_4	f_8	f_3	f_5
20	f_1, f_3	f_{10}	f_4	f_5	f_2	f_3	f_4	f_5	f_4	f_8	f_3	f_5
21	f_4	f_8	f_3	f_5	f_2	f_3	f_4	f_5	f_1, f_3	f_{10}	f_4	f_5
22	f_2	f_3	f_4	f_5	f_4	f_8	f_3	f_5	f_1, f_3	f_{10}	f_4	f_5
23	f_1, f_3	f_{10}	f_4	f_5	f_4	f_8	f_3	f_5	f_2	f_3	f_4	f_5
24	f_4	f_8	f_3	f_5	f_1, f_3	f_{10}	f_4	f_5	f_2	f_3	f_4	f_5

A simulation study based on the approach described in Musa (2004) is used to evaluate the setup plans in terms of parallelism and perpendicularity tolerances of the machined component. The simulation, implemented with MATLAB, takes to-be-machined part specifications and process capability as inputs, generates sample points representing features formed by machining processes, and conducts virtual inspection of the toleranced feature based on sample points generated. The errors involved in a machining process include machine tool error, workpiece location error, workpiece geometry error, thermal error, and the like. The effects of these errors can be generally classified into two categories: cutting tool deviation from ideal position and workpiece deviation. A process capability database, established from historical data and experiments, that captures and lumps error sources provides the simulation with probability distributions that represent the underlying error forming mechanism in manufacturing processes.

Figure 2.45 shows sample points generated for the shaft support. Only features f_1, f_2, f_3 and f_4 are shown, in that other features are irrelevant as far as tolerance specifications are concerned. The results of all possible setup plans are simulated under the same machining conditions. The input parameters for the simulation are as follows: the flatness of all flat surfaces in raw cast part is 0.05; the z-axis error for the milling operation follows $N(0, 0.0005^2)$; translation errors for the x, y and z axes all follow $N(0, 0.001^2)$; and rotational errors for the x, y and z axes all follow $N(0, 0.001^2)$. The simulation result (Table 2.11) shows that the parallelism between f_1 and f_4 does not vary much among setup plan alternatives. However, the perpendicularity between f_1 and f_3 does display systematical variation among setup plan alternatives. Among all alternatives, setup plan 9, which is the one generated by our algorithm, has the overall best precision.

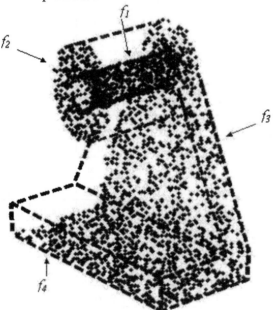

Figure 2.45. Generated sample points of the shaft support.

Table 2.11. Simulation results*.

Setup Plan	Parallelism				Perpendicularity			
	Sample Mean	STD	95%CI of μ		Sample Mean	STD	95%CI of μ	
			Lower Limit	Upper Limit			Lower Limit	Upper Limit
1	0.0072	0.0014	0.0070	0.0075	0.0101	0.0025	0.0096	0.0106
2	0.0067	0.0020	0.0063	0.0071	0.0099	0.0021	0.0094	0.0103
3	0.0064	0.0013	0.0061	0.0066	0.0104	0.0028	0.0099	0.0110
4	0.0072	0.0020	0.0068	0.0076	0.0100	0.0024	0.0095	0.0105
5	0.0068	0.0012	0.0065	0.0070	0.0099	0.0020	0.0095	0.0103
6	0.0066	0.0012	0.0063	0.0068	0.0100	0.0022	0.0095	0.0104
7	0.0073	0.0011	0.0071	0.0075	0.0067	0.0011	0.0065	0.0069
8	0.0073	0.0011	0.0071	0.0075	0.0078	0.0018	0.0075	0.0082
9	0.0066	0.0010	0.0064	0.0068	0.0060	0.0011	0.0058	0.0062
10	0.0071	0.0015	0.0068	0.0074	0.0065	0.0013	0.0063	0.0068
11	0.0080	0.0012	0.0078	0.0082	0.0076	0.0012	0.0073	0.0078
12	0.0065	0.0019	0.0061	0.0069	0.0082	0.0019	0.0078	0.0086
13	0.0062	0.0020	0.0058	0.0066	0.0136	0.0039	0.0128	0.0144
14	0.0061	0.0014	0.0058	0.0064	0.0098	0.0021	0.0094	0.0103
15	0.0065	0.0018	0.0062	0.0069	0.0130	0.0025	0.0125	0.0135
16	0.0068	0.0012	0.0066	0.0071	0.0140	0.0029	0.0134	0.0145
17	0.0072	0.0018	0.0068	0.0075	0.0092	0.0024	0.0087	0.0097
18	0.0074	0.0012	0.0071	0.0076	0.0091	0.0018	0.0087	0.0094
19	0.0071	0.0015	0.0068	0.0074	0.0081	0.0012	0.0078	0.0083
20	0.0066	0.0017	0.0063	0.0070	0.0088	0.0014	0.0085	0.0091
21	0.0073	0.0014	0.0070	0.0076	0.0092	0.0013	0.0090	0.0095
22	0.0075	0.0016	0.0071	0.0078	0.0090	0.0013	0.0088	0.0093
23	0.0078	0.0012	0.0075	0.0080	0.0092	0.0014	0.0089	0.0095
24	0.0071	0.0010	0.0069	0.0073	0.0084	0.0011	0.0082	0.0086

* Simulation run number, $n=100$; $t_{0.975}=1.98$.

2.4.1.3 Multi Attribute Utility Analysis

A part usually has multiple tolerance specifications. It is very likely that a setup plan is better than others with respect to its ability to minimize stack-up on certain tolerances. However, it might be inferior when considering other tolerances. In other words, the optimality of a setup plan cannot be readily determined, as it depends on what is preferred by a decision maker. Under this circumstance, a setup plan evaluation approach must take into account different preference settings and generate a quantitative measure. This can be achieved using multiple attributes utility analysis (MAUA).

The notion of utility originated in the domain of economics (von Neumann 1947) as a measure of individuals' subjective preferences (satisfaction level) among alternatives. Therefore, in a way similar to the quantification of heat and light in physics, one should be able to develop a numerical measurement of utility. Quantification of utility is implemented by the combination of two "natural" concepts; namely, (1) one alternative is preferable to another, which implies a "greater" operator, and (2) one alternative is equally desirable to either of two other alternatives associated with probability of occurrence, (α and $1-\alpha$, respectively), which implies an "equal" operator. A set of axioms were developed (von Neumann 1947) to formulate these concepts into a mathematically rigorous theory entailing lengthy proofs. Utility theory was further developed (Savage 1954; Luce 1957; and Fishburn 1970) in the two decades that followed.

The tolerances a setup plan can maintain might be viewed as its attributes. The setup plan can be evaluated by mapping its attributes to a numerical overall utility value with a multiple attribute function, which is realized by capturing a decision maker's preference structure through interviews. The common procedure in developing a multiple-attribute utility function is illustrated in Figure 2.46. The decisionmaker should first be identified, and then an analyst with good knowledge in decision analysis is supposed to guide the decision maker through the procedure. To identify an appropriate multiple-attribute function form, two concepts are of great importance. One is *preferential independence*, which suggests that the direction of preference of an attribute remains unchanged regardless of the levels of other attributes. The other is *utility independence*, which suggests that a single attribute function form remains unchanged regardless of the levels of other attributes. If both types of independence measure hold, a simple additive multiple-attribute function is valid. Otherwise, other function form — such as a more complicated multiplicative multiple-attribute function — could be applied. Most setup planning cases satisfy both measures of independence.

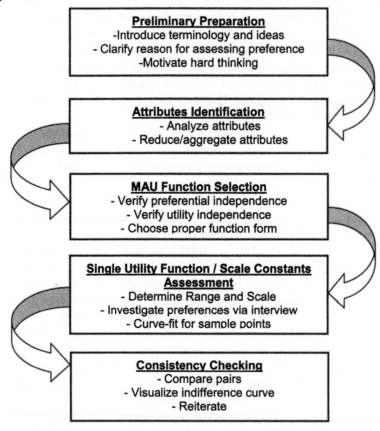

Figure 2.46. Common procedure for developing a multi-attribute utility function.

The assessment of single-utility functions and constant scales (as well as subsequent consistency checking) follows a relatively fixed procedure regardless of different application areas (Keeney 1976). The implementation of a setup plan evaluation might be illustrated using the simplified bearing bracket shown in Figure 2.47. Three features need to be machined; namely, (1) the round flat surface A, (2) the cylindrical surface B, and (3) the through hole C. The functionality of the bearing bracket is as follows. A bearing is to be positioned in feature C, that allows the part to rotate around a shaft going through it. Feature A is to be used as a locating plane for another part to sit on. Two tolerance specifications result: a perpendicularity of 0.02 between features A and C and a runout of 0.03 between features B and C.

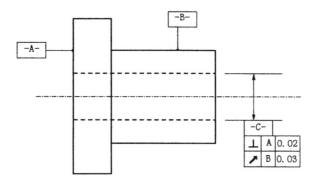

Figure 2.47. A simplified bearing bracket (unit: inch).

The part geometry dictates that two separate setups are required for machining features A and B due to their differing tool-approach directions. Feature C shares tool-approach directions with both feature A and feature B. A decision has to be made about which setup feature C should be grouped to. The setup sequence is not an issue in that there is no direct interaction between feature A and feature B, and a chuck can easily assume the fixturing task for both setups. Therefore, there are two setup plan alternatives, with the only difference between them being that of grouping feature C. Setup plan 1 machines features A and C in one setup, and then machines feature B in another. Setup plan 2 machines features B and C in one setup, and then machines feature A in another. Based on tolerance analysis principles previously described, setup plan 1 will achieve tighter control regarding perpendicularity, whereas setup plan 2 will result in better runout precision. The attributes applicable here are the perpendicularity and the runout tolerances, which are random variables that follow certain distributions. The distributions are obtained using the simulation methodology previously described, as shown in Figure 2.48. The parametric forms of these distributions are denoted $f_{1p}(x)$, $f_{1r}(y)$, $f_{2p}(x)$, and $f_{2r}(y)$. The subscript of the function symbol is to be interpreted as follows. The first letter is the index of the setup plan number, and the second is the attribute index, either p for perpendicularity or r for runout. Letter x is assigned as variable in perpendicularity attribute space, and letter y in runout attribute space.

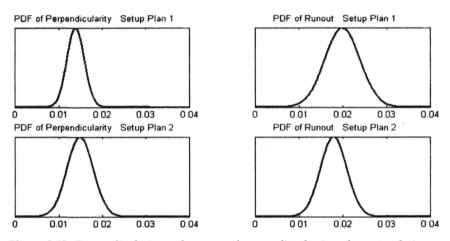

Figure 2.48. Perpendicularity and runout tolerance distributions from simulation.

We will use the additive multi-attribute utility model in the form of $U(X) = \sum_{i=1}^{n} k_i U_i(X_i)$ to incorporate attributes into a single function. In this model, X represents an alternative, which is to be evaluated against n criteria simultaneously. With respect to these n criteria, X is broken down to n attributes. X_i is the level of the *ith* attribute against the *ith* criterion. $U_i(X_i)$ is the subjective value (utility) of how a particular level of the *ith* attribute X_i satisfies the decision maker with respect to the *ith* criterion. Its range is generally scaled to (0,1). The summation of k_i is equal to 1. The overall utility of alternative X is the weighted (by k_i) sum of each attribute's utility. The level of the *ith* attribute X_i has a twofold meaning. First, if the level X_i is a fixed value, then $U_i(X_i)$ computation is straightforward (often referred to as a *value function* in that there is no uncertainty involved). Second, if the level X_i represents a distribution describing the probability of different values the level X_i might assume, then $U_i(X_i)$ is called a *utility function* and its computation is as follows:

- For discrete cases, $U_i(X_i) = \sum p_j x_{ij}$, where x_{ij} is the *jth* value X_i might assume, with a probability of p_j.
- For continue cases, $U_i(X_i) = \int f(x_i) U_i(x_i) dx_i$, where x_i is the value X_i might assume, with a probability density function of $f(x_i)$.

We can now proceed to utility function construction and constant scale probing. This step is of the most importance if the decision maker's preferences are to be captured and represented adequately and accurately by utility functions and constant scales. A successful investigation of preference via interview requires skill and devotion from both analyst and decision maker, and in a sense the investigating process is more an art than a science. Various methods of investigation have been developed (Keeby 1976). Here, we apply a *mid-value splitting technique* for utility function construction. The basic idea is to guide the decision maker to choose utility values for certain tolerance levels, thus obtaining a set of sample points. These sample points are then used to construct a utility function. Assume for the perpendicularity tolerance the sample points obtained are {(0.015, 0.990), (0.018, 0.750), (0.020, 0.500), (0.022, 0.250), (0.025, 0.010)}. The first number in a sample point is the perpendicularity tolerance level, whereas the second number is the utility of that tolerance level. The first sample point, for example, means that a perpendicularity of 0.015 has a utility of 0.990 for the decision maker. Similarly, assum the sample points obtained for the runout tolerance are {(0.020, 0.990), (0.024, 0.750), (0.025, 0.500), (0.028, 0.250), (0.038, 0.010)}. Sketches made based on these two sets of sample points both resembled sigmoid functions, except that the curve for runout tolerance utility exhibited unbalanced tails. We decide to fit the sample points to a sigmoid function family with the form $y = 1 - \dfrac{1}{1 - e^{-a(x+b)}}$.

Because the runout utility function has unbalanced tails, we decide to fit two different sigmoid functions connected at the midpoint. The two utility functions, shown in Figure 2.49, have the following forms:

$$U_p(x) = 1 - \frac{1}{1 - e^{-868(x-0.02)}}, \text{ and}$$

$$U_r(y) = \begin{cases} 1 - \dfrac{1}{1 - e^{353(x-0.025)}} & y < 0.025 \\[2mm] 1 - \dfrac{1}{1 - e^{906(x-0.025)}} & y \geq 0.025. \end{cases}$$

The next step is to determine the scale constants: k_p (for perpendicularity) and k_r (for runout). Assuming the decision maker regards two setup plans equally desirable, one achieves a perpendicularity of 0.015 and a runout of 0.038 and the other a perpendicularity of 0.020 and a runout of 0.020. Then, we have the following:

$$k_p U_p(x_p = 0.015) + k_r U_r(y_r = 0.038) = k_p U_p(x_p = 0.02) + k_r U_r(y_r = 0.02)$$
$$k_p + k_r = 1$$

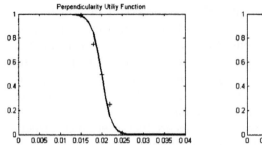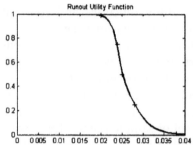

Figure 2.49. Utility functions.

Solving these two simultaneous equations yields k_p=0.6667 and k_r=0.3333 scale constants for perpendicularity and runout, respectively. A 3D diagram was built with MATLAB for better visualization of the tentative two-attribute utility function, as shown in Figure 2.50. This utility function is accepted after consistency checking. Finally, we reach the stage of computing overall utility for the two setup plan alternatives:

For setup plan 1, $U_1 = k_p \int f_{1p}(x)U_p(x)dx + k_r \int f_{1r}(y)U_r(y)dy$ = 0.6875

For setup plan 2, $U_2 = k_p \int f_{2p}(x)U_p(x)dx + k_r \int f_{2r}(y)U_r(y)dy$ = 0.6354

This result shows that setup plan 1 is better than setup plan 2, given the decision maker's preference. Note that with a different decision maker and different preferences (reflected in different utility functions and/or different scale constants) setup plan 2 might be more desirable.

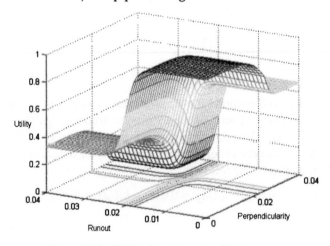

Figure 2.50. Utility function visualization.

2.4.2 Tolerance Assignment and Inspection Planning

Once a setup plan is selected and released to production, the working dimensions and production tolerances to be maintained in each setup need to be determined. This problem is commonly referred as *tolerance assignment* and is commonly solved using tolerance chart analysis (Wade 1967). The idea behind a tolerance chart is to identify the dimensional chain for the final blueprint dimensions and then allocate individual tolerances so that the blueprint tolerance is not violated even in the worst-case scenario. Tolerance chart analysis does not ensure that the workpiece has enough machining stock at each stage for it to be machined per assigned tolerances. Moreover, there is no guarantee that the available production equipment will be able to achieve the allocated tolerances. Nonetheless, most researchers regard tolerance assignment as an optimization problem based on the principle of tolerance chart analysis. Methods proposed in the literature differ from one another in the choice of objective function and constraints. Most methods consider cost minimization or tolerance maximization as the objective. Design tolerance requirements and process capability are normally considered as constraints. Table 2.12 summarizes various tolerance assignment methods. It can be seen that machining stocks and production tolerances are determined based on process knowledge and experience rather than rigorous mathematical analysis. These methods typically evolve from the experience of individual machinists and involve trial and error. This is a hindrance in achieving computer automation. To solve this problem, a systematic approach, based on process capability analysis, has been proposed (as shown in Figure 2.51).

The proposed approach uses setup plan information as input. It takes two routes, depending on the availability of process capability data for the selected machining processes. If process capability data is available, a capability-based method is used. Note that process capability is part dependent. Therefore, for a new part the corresponding process capability needs to be extrapolated from existing data on similar parts (Jain 2003). In the absence of process capability data (e.g., a brand new part that requires new production equipment), a simulation-based method is used. The probability distributions for locating and machining errors to be included in the simulation are then determined based on the production equipment to be used. All sets of error distributions that meet part design specifications are feasible solutions. Among these feasible solutions a desirable one is selected based on cost and cycle time considerations. The selected solution is then used to estimate the process capability based on the data obtained in multiple simulation runs.

Table 2.12. Summary of tolerance assignment methods.

Method	Stack-up Analysis	Machining Stock Calculation	Tolerance Assignment
Gadzala 1959, Wade 1967	Manual identification of dimensional chains in a tolerance chart	Experience and process knowledge	Process knowledge
Xiaoqing and Davies 1988	Matrix tree chain method for tracing dimensional chains	Experience and process knowledge	Industrial standard and subsequent iterations
Mittal 1990	Graphical method of representing dimensional chains	Experience and process knowledge	Process capability used as a constraint
Ngoi and Ong 1993	Path tracing technique to trace process links	Experience and process knowledge	Process capability used as a constraint
Ji 1993, 1994	Based on tolerance chart	Experience and process knowledge	Determined from tolerance grade table
Ji 1995	Tree representation of tolerance chart to identify dimension chains	Experience and process knowledge	Industrial standard and subsequent iterations

With process capability data (readily available, extrapolated from existing data, or estimated through simulation), working dimensions for each machining cut are determined based on the stock removal required for each operation. The main reason for leaving machining stocks in pre-finishing operations is to have enough material for the finishing operation, which results in closer dimensional and geometrical control of the feature. Therefore, machining stock required for an operation can be calculated based on required precision and the process dispersion of the pre-finishing operation. An algorithm for determining working dimensions based on this idea can be found in Jain (2003). Once working dimensions are determined, production tolerances can be assigned based on desired process capability level (C_p). For example, if a C_p of 1 is sufficient, then the production tolerance would be set at $\pm 3\sigma$ process dispersion. This tolerance assignment method is pragmatic, as there is no point in assigning production tolerances that cannot be achieved with available production equipment.

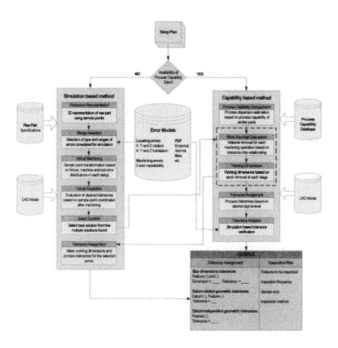

Figure 2.51. A systematic approach to tolerance assignment.

At this point, an inspection plan should be developed to check the parts being produced and to make process adjustments when necessary to ensure that no nonconforming parts are being produced. Two key components of an inspection plan are sample size and sampling frequency. Assuming the process tolerance follows a normal distribution, from the confidence interval computation of a single mean the following equation holds:

$$P\left\{|\bar{x} - \mu| \le z_{\alpha/2}\frac{\sigma}{\sqrt{n}}\right\} = 1 - \alpha.$$

Here, $\bar{x}$ is the average of inspected dimensions, μ is the nominal dimension, σ is the standard deviation of inspected dimensions, n is the sample size, and $(1-\alpha)$ is the confidence level.

Therefore, if it is desirable to detect a process shift of δ with a confidence level of $1-\alpha$ (i.e., $P\left\{|\bar{x} - \mu| \le \delta\right\} = 1 - \alpha$), the sample size can be determined as

$$n = \left(\frac{z_{\alpha/2}}{\delta}\right)^2 \sigma^2.$$

When a process shift of δ occurs, the probability of detecting the shift with $1-\alpha$ confidence is $p = \Phi\left(-\dfrac{\delta + Z_{\alpha/2} \times \sigma/\sqrt{n}}{\sigma/\sqrt{n}}\right) + \left(1 - \Phi\left(\dfrac{Z_{\alpha/2} \times \sigma/\sqrt{n} - \delta}{\sigma/\sqrt{n}}\right)\right)$. Therefore, the average number of sampling inspections needed to detect the shift, called average run length (ARL), is ARL = $1/p$. If samples are taken every h hours, the time needed to detect the shift — called average time to signal (ATS) — is ATS = ARL$\times h$ = h/p. Therefore, if it is desirable to detect a process shift of δ in ATS hours, the sampling frequency should be every $\text{ATS} \times \left[\Phi\left(-\dfrac{\delta + Z_{\alpha/2} \times \sigma/\sqrt{n}}{\sigma/\sqrt{n}}\right) + \left(1 - \Phi\left(\dfrac{Z_{\alpha/2} \times \sigma/\sqrt{n} - \delta}{\sigma/\sqrt{n}}\right)\right)\right]$ hours.

2.4.3 Fixture Design Integration

Setup planning and fixture design are two closely related tasks. To set up a workpiece is to locate the workpiece in a desired position on the machine table. A fixture is then used to provide some type of clamping mechanism to maintain the workpiece in the position and to resist the effects of gravity and/or operational forces. Although setup planning is constrained by fixtures to be applied, it provides guidelines for fixture design. As a result, we are facing the chicken or egg dilemma. Most researchers circumvent this problem by focusing on either setup planning or fixture design. Those working on setup planning address the issues of feature orientation analysis (cutting-tool-approach direction) and precedence constraint analysis, whereas the issues of kinematics analysis and force analysis are commonly addressed in fixture design research. Because both setup planning and fixture design aim to ensure the precision of the machining process, we propose an integrated solution methodology driven by tolerance analysis and augmented by product, process, and production modeling.

This integrated methodology is depicted in Figure 2.52. First, product design and manufacturing requirement information is automatically extracted from the CAD model and represented by FTG. Once the design datum frames — which consist of geometric datums in three perpendicular directions — are identified, an initial DMG (DMG1) can be constructed, which forms an initial design of setups. Having taken into consideration production scheme (integrated, distributed, or combined operations), machine tool capability, and tolerance decomposition, DMG1 can be converted into a detailed DMG (DMG2) in which the machining features are grouped into setups and operations are sequenced in terms of machine configuration. When a DMG2 has been formed, tolerance stack-up analysis and setup plan verification follow, to provide information for fixture design and machine tool requirement specifications. Using a DMG2, production cycle time and costs can be optimized based on various objectives.

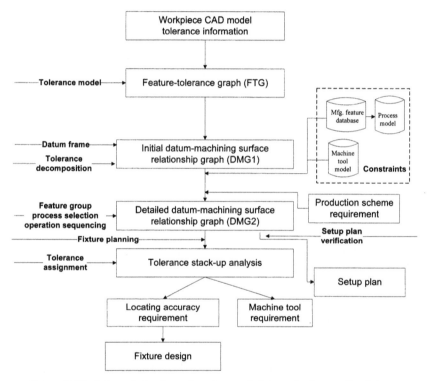

Figure 2.52. Integrated setup planning and fixture design methodology.

Fixture configuration techniques have been developed whereby structural design can be generated automatically when fixturing surfaces and points are specified on the workpiece fixture (Rong 1997; An 1999; Brost 1996). However, the fixture design has to be started over on a trial-and-error basis if one of the following happens: (1) the workpiece geometry is changed, (2) the setup plan is changed, or (3) the fixture design does not satisfy process requirements. To solve this problem, we adopt a flexible fixture design methodology that can be truly integrated into setup planning. First, the workpiece model and workpiece variation model are established with fixturing information (including setup information). Then, a fixture design variation model is derived from the fixture design generated before the condition change. Key techniques used are geometric reasoning and fixture design verification. The former is the extension of the fixture design technique developed by Rong (1997). The latter is based on the results from a project on fixture design verification (Kang 2004), including the verification of fixturing accuracy, accessibility, interference, stability, and stiffness effects. Fixture design involves fixture planning, which determines locating datums and clamping positions, during part-level setup planning; conceptual fixture design, which determines the selection of fixture base and workpiece layout on the fixture during stage of the station-level setup planning; and the fixture detail design, which creates fixture structural designs (with verifications) at each setup stage. Computerized fixture design techniques and systems are explored in later chapters.

References

Alting, L., and H.-C. Zhang. "Computer-aided Process Planning: The State-of-the-Art Survey," *International Journal of Production Research* 27:4, pp. 553–585, 1989.

An, Z., S. Huang, J. Li, Y. Rong, and S. Jayaram. "Development of Automated Fixture Design Systems with Predefined Fixture Component Types: Part 1, Basic Design," *International Journal of Flexible Automation and Integrated Manufacturing* 7:3/4, pp. 321–341, 1999.

ANSI (American National Standards Institute). *Dimensioning and Tolerancing*. ANSI Y14.5M-1994, New York: American Society of Mechanical Engineers, 1995.

Bai, Y., and Y. Rong. "Machining Accuracy Analysis for Computer-aided Fixture Design," in *Manufacturing Science and Engineering*, PED-64, pp. 507–512, New York: American Society of Mechanical Engineers, 1993.

Boerma, J. R., and H.J.J. Kals. "FIXES, A System for Automatic Selection of Set-ups and Design of Fixtures," *Annals of the CIRP* 37:1, pp. 443–446, 1988.

Buckingham, E. *Dimensions and Tolerances for Mass Production*. New York: Industrial Press, 1954.

Chang, T.-C. *Expert Process Planning for Manufacturing*. Reading, MA: Addison-Wesley, 1990.

Chang, T.-C., and R. A. Wysk. *An Introduction to Automated Process Planning Systems*. Englewood Cliffs, NJ: Prentice-Hall, 1985.

Chen, C.L.P. "Setup Generation and Feature Sequencing Using Unsupervised Learning Algorithms," in *Proceedings of the 1993 NSF Design and Manufacturing Systems Conference* 1, pp. 981–986, Dearborn, MI: Society of Mechanical Engineers, 1993.

Chen, J., Y. F. Zhang, and A.Y.C. Nee. "Setup Planning Using Hopfield Net and Simulated Annealing," *International Journal of Production Research* 36:4, pp. 981–1000, 1998.

DeGarmo, E. P., J. T. Black, and R. Al Kohser. *Materials and Processes in Manufacturing* (8th ed.). Upper Saddle River, NJ: Prentice-Hall, 1997.

Delbressine, F.L.M., R. de Groot, and A.C.H. van der Wolf. "On the Automatic Generation of Set-ups Given a Feature-based Design Representation," *Annals of the CIRP* 42:1, pp. 527–530, 1993.

Demey, S., H. Van Brussel, and H. Derache. "Determining Set-ups for Mechanical Workpieces," *Robotics and Computer-Integrated Manufacturing* 12:2, pp. 195–205, 1996.

Doyle, L. E. *Metal Machining*. New York: Prentice-Hall, 1953.

Fishburn, P. C. *Utility Theory for Decision Making*. New York: Wiley, 1970.

Goldberg, K. Y., and R. Brost. "A Complete Algorithm for Synthesizing Modular Fixtures for Polygonal Parts," in *IEEE Transactions on Robotics and Automation* RA-12:1, pp. 31–46, 1996.

Halevi, G., and R. D. Weil. *Process Planning Principles: A Logical Approach*. New York: Elsevier, 1991.

Ham, I., and S. C.-Y. Lu. "Computer-aided Process Planning: The Present and the Future," *Annals of the CIRP* 37:2, pp. 1–11, 1988.

Hayes, C., and P. Wright. "Automating Process Planning: Using Feature Interactions to Guide Search," *Journal of Manufacturing Systems* 8:1, pp. 1–14, 1989.

Huang, S. H. "Automated Setup Planning for Lathe Machining," *Journal of Manufacturing Systems* 17:3, pp. 196–208, 1998.

Huang, S. H., and Q. Liu. "Rigorous Application of Tolerance Analysis in Setup Planning," *International Journal of Advanced Manufacturing Technology* 21:3, pp. 196–207, 2003.

Huang, S. H., and N. Xu. "Automatic Setup Planning for Metal Cutting: An Integrated Approach," *International Journal of Production Research* 41:18, pp. 4339–4356, 2003.

Huang, S. H., and H.-C. Zhang. "Tolerance Analysis for Setup Planning in CAPP," in *Advanced Tolerancing Techniques*, pp. 427–460, New York: Wiley, 1997.

Huang, S. H., and H.-C. Zhang. "Tolerance Analysis in Setup Planning for Rotational Parts," *Journal of Manufacturing Systems* 15:5, pp. 340–350, 1996.

Huang, S. H., H.-C. Zhang, and W.J.B. Oldham. "Tolerance Analysis for Setup Planning: A Graph Theoretical Approach," *International Journal of Production Research* 35:4, pp. 1107–1124, 1997.

Huang, X., and P. Gu. "Tolerance Analysis in Setup and Fixture Planning for Precision Machining," *Proceedings of the Fourth International Conference on Computer Integrated Manufacturing and Automation Technology*, pp. 298–305, Tory, NY: Rensselaer Polytechnic Institute, 1994.

Jain, A. "Process Capability Analysis for Tolerance Assignment in Discrete Part Manufacturing," M.S. thesis, University of Cincinnati, OH, 2003.

Jain, A., N. Xu, S. H. Huang, and Y. K. Rong. "Process Capability Analysis for Production Tolerance Assignment," *Transactions of the North American Manufacturing Research Institution of SME* 31, pp. 531–538, 2003.

Ji, P. "A Linear Programming Model for Tolerance Assignment in a Tolerance Chart," *International Journal of Production Research* 31:3, pp. 739–751, 1993.

Ji, P. "An Automatic Tolerance Assignment Approach for Tolerance Charting," *International Journal of Advanced Manufacturing Technology* 9, pp. 362–368, 1994.

Ji, P., M. Ke, and R. S. Ahluwalia. "Computer-aided Operational Dimensioning for Process Planning," *International Journal of Machine Tools and Manufacture* 35:10, pp. 1353–1362, 1995.

Joneja, A., and T.-C. Chang. "Setup and Fixture Planning in Automated Process Planning Systems," *IIE Transactions* 31, pp. 653–665, 1999.

Kim, I.-H., J.-S. Oh, and K.-K. Cho. "Computer Aided Setup Planning for Machining Processes," *Computers and Industrial Engineering* 31:3/4, pp. 613–617, 1996.

Kiritsis, D. "A Review of Knowledge-based Expert Systems for Process Planning: Methods and Problems," *International Journal of Advanced Manufacturing Technology* 10:4, pp. 240–262, 1995.

Krulikowski, A. *Fundamentals of Geometric Dimensioning and Tolerance* (2d ed.). Albany, NY: Delmar, 1998.

Lee, C. L. *Digital Circuits and Logic Design.* Englewood Cliffs, NJ: Prentice-Hall, 1987.

Lin, A. C., S.-Y. Lin, D. Diganta, and W. F. Lu. "An Integrated Approach to Determining the Sequence of Machining Operations for Prismatic Parts with Interacting Features," *Journal of Materials Processing Technology* 73, pp. 234–250, 1998.

Lin, L., Y. F. Zhang, and A.Y.C. Nee. "An Integrated Setup Planning and Fixture Design System for Prismatic Parts," *International Journal of Computer Applications in Technology* 10:3/4, pp. 198–212, 1997.

Luce, R. D., and H. Raiffa. *Games and Decisions.* New York: Wiley, 1957.

Ming, X. G., and K. L. Mak. "Intelligent Setup Planning in Manufacturing by Neural Networks Based Approach," *Journal of Intelligent Manufacturing* 11:3, pp. 311–331, 2000.

Mittal, Ro. O., S. A. Irani, and E. A Lehtihet. "Tolerance Control in the Machining of Discrete Components," *Journal of Manufacturing Systems* 9:3, pp. 233–246, 1990.

Model, M. L. *Data Structures, Data Abstraction: A Contemporary Introduction Using C++.* Englewood Cliffs, NJ: Prentice-Hall, 1994.

Musa, R., S. H. Huang, and Y. K. Rong. "Simulation-based Tolerance Stack-up Analysis in Machining," *Transactions of the North American Manufacturing Research Institution/SME* 32, pp. 533-540, 2004.

Ong, S. K., and A.Y.C. Nee. "Application of Fuzzy Set Theory to Setup Planning," *Annals of the CIRP* 43:1, pp. 137–144, 1994.

Öztürk, F., N. Kaya, O. B. Alankuş, and S. Sevinş. "Machining Features and Algorithms for Set-up Planning and Fixture Design," *Computer Integrated Manufacturing Systems* 9:4, pp. 207–216, 1996.

Parker, A. *Algorithms and Data Structures in C++.* New York, CRC Press, 1993.

Rong, Y., and Y. Bai. "Machining Accuracy Analysis for Computer-aided Fixture Design," *Journal of Manufacturing Science and Engineering* 118, pp. 289–300, 1996.

Rong, Y., and Y. Bai. "Automated Generation of Modular Fixture Configuration Design," *Journal of Manufacturing Science and Engineering* 119, pp. 208–219, 1997.

Rong, Y., and Y. Zhu. *Computer-Aided Fixture Design.* New York: Marcel Dekker, 1999.

Rong, Y., X. Liu, J. Zhou, and A. Wen. "Computer-Aided Setup Planning and Fixture Design," *International Journal of Intelligent Automation and Soft Computing* 3:3, pp. 191–206, 1997.

Rubinstein, R. Y. *Simulation and the Monte Carlo Method.* New York: Wiley, 1981.

Sakurai, H. "Automatic Setup Planning and Fixture Design for Machining," *Journal of Manufacturing Systems* 11:1, pp. 30–37, 1992.

Sarma, S. E., and P. K. Wright. "Algorithms for the Minimization of Setups and Tool Changes in 'Simply Fixturable' Components in Milling," *Journal of Manufacturing Systems* 15:2, pp. 95–112, 1996.

Savage, L. J. *The Foundations of Statistics.* New York: Wiley, 1954.

SME (Society of Mechanical Engineers). "Manufacturing Technology Drives U.S. Productivity," *Manufacturing Engineering* 125:5, pp. 20–28, 2000.

Trappey, J. C., and C. R. Liu. "An Automatic Workholding Verification System," *Robotics and Computer-integrated Manufacturing* 9:4, pp. 321–326, 1992.

van Houten, F.J.M. "Strategy in Generative Planning of Turning Processes," *Annals of the CIRP* 35:1, pp. 331–335, 1986.

von Neumann, J., and O. Morgenstern. *Theory of Games and Economic Behavior* (2d ed.). Princeton, NJ: Princeton University Press, 1947.

Wade, O. R. *Tolerance Control in Design and Manufacturing.* New York: Industrial Press, 1967.

Wang, H.-P., and J.-K. Li. *Computer-Aided Process Planning.* Amsterdam: Elsevier, 1991.

Xiaoqing, T., and B. J. Davies. "Computer Aided Dimensional Planning," International *Journal of Production Research* 26:2, pp. 283–297, 1988.

Wilson, F. W., and P. D. Harvey. *Manufacturing Planning and Estimation Handbook.* New York: McGraw-Hill, 1963.

Wu, H.-C., and T.-C. Chang. "Automated Setup Selection in Feature-based Process Planning," *International Journal of Production Research* 36:3, pp. 695–712, 1998.

Young, R.I.M., and R. Bell. "Fixturing Strategies and Geometric Queries in Set-up Planning," *International Journal of Production Research* 29:3, pp. 537–550, 1991.

Zhang, H.-C., and E. Lin. "A Hybrid-Graph Approach for Automated Setup Planning In CAPP," *Robotics and Computer Integrated Manufacturing* 15:1, pp. 89–100, 1999a.

Zhang, H.-C., and E Lin. "A 'GATO' Algorithm for the Setup Planning of Prismatic Parts," *Transactions of the North American Manufacturing Research Institution of SME* 27, pp. 245–250, 1999b.

Zhang, H.-C., S. H. Huang, and J Mei. "Operational Dimensioning and Tolerancing in Process Planning: Setup Planning," *International Journal of Production Research* 34:7, pp. 1841–1858, 1996.

Zhang, Y. F., W. Hu, Y. Rong, and D. W. Yen. "Graph-based Set-up Planning and Tolerance Decomposition for Computer-aided Fixture Design," *International Journal of Production Research* 39:14, pp. 3109–3126, 2001.

Zhang, Y. F., G. H. Ma, and A.Y.C. Nee. "Modeling Process Planning Problems in an Optimization Perspective," *Proceedings of the 1999 IEEE International Conference on Robotics and Automation*, Detroit, MI, May 1999.

Zhang, Y. F., A.Y.C. Nee, and S. K. Ong. "A Hybrid Approach for Set-up Planning," *International Journal of Advanced Manufacturing Technology* 10, pp. 183–190, 1995.

Zhou, F., T.-C.Kuo, S. H. Huang, and H.-C. Zhang. "Form Feature and Tolerance Transfer from a 3D Model to a Setup Planning System," *International Journal of Advanced Manufacturing Technology* 19:2, pp. 88–96, 2002.

CHAPTER 3

Computer-aided Fixture Design

3.1 AN OVERVIEW OF COMPUTER-AIDED FIXTURE DESIGN

A fixture is a device that locates and holds workpieces in position during manufacturing processes. Fixtures play an important role in shortening production cycle time and ensuring production quality, and thus reducing production cost. Fixture design, fabrication, and testing consume a major portion of production development time.

In a manufacturing system, it is desirable that fixtures be flexible so that the turn-around time can be reduced. Flexible fixturing involves a design that allows for rapid conversion of existing fixture designs into those that meet new production requirements with little change of hardware and without extensive testing. Therefore, flexible fixturing may include flexible fixture hardware and fixture design and analysis software. The hardware may include modularized fixture systems for part families in customized mass production and modular (and other flexible) fixture systems for small-volume production. The software may include the fixture planning, design, and verification functions for a quick and quality fixture design with verification, which can be applied in the concurrent engineering environment.

3.1.1 Fixturing Technology

3.1.1.1 A Historical Review

The history of fixturing technology is as long as that of manufacturing technology. Since the first production line was developed, dedicated fixtures have been developed for performing many functions, such as locating, supporting, guiding, graduating, and so on, so that repeatable production quality and high production rates based on short workpiece loading time can be achieved in mass production. In the long-time practice of fixturing technology, fixture components have been highly standardized to reduce the cost of fixture production. At the same time, standard and general-purpose fixtures have been developed, such as vises, chucks, clamping straps, and so on. However, fixture structural design is far from standardized because of the nature of multiple solutions for the same fixturing requirement.

Dedicated fixtures are designed for the production of specific workpieces. The design, fabrication, and testing of fixtures require significant time and cost in the production development cycle. To reduce fixture development time and cost, modular fixtures were developed during World War II. A modular fixture system is a set of standard reusable fixture components that can be used to build different fixture configurations for different fixturing requirements. The use of modular fixtures enhances fixture flexibility and reduces the time and cost of fixture development, which is especially beneficial to small-volume production and new product prototyping. However, modular fixtures are not very popular in daily production because of the requirements of specific fixturing performance in mass production, such as fixturing stiffness, compact spacing, long life of fixturing components, and ease of fixturing operations, and because of the insignificant benefit from the reuse of fixtures.

When group technology (GT) was introduced, adjustable fixtures were developed for part families based on the concept of similarity. Adjustable fixtures are single fixtures with certain functional components that can be adjusted in position or quickly changed to meet different shape and size requirements. This technology takes advantage of both dedicated and modular fixtures, and is especially beneficial to batch production and mass customization. The further development of adjustable fixtures into modularized and standard designs may even increase the advantages of flexibility, as well as specified fixturing performance, although it might involve the complications of computerized planning, design, and verification.

As CNC techniques and machines have been developed and widely applied in production, fixturing requirements have changed. With advances in the performance (motion accuracy, stiffness, and controllability) of machine tools and cutting tools, it is very common now to combine rough and finish machining in a single setup and to machine multiple features from different directions in one setup. This requires the fixture to be more rigid and accurate, as well as spatially efficient. On the other hand, the requirements of some other functions (such as guiding and graduating) are no longer necessary in fixture design because of the high performance of machine tools.

It should be mentioned that although there are special and highly innovated flexible fixtures developed for specific applications, the mechanical fixtures, particularly modular and modularized adjustable fixtures, are still the main forms of flexible fixtures used in real production today and this will remain the case in the near future. The forgoing discussion is not only to the fixtures in machining operations, but also applies to the fixtures of other manufacturing processes such as assembly, welding, inspection, straightening, and the like.

3.1.1.2   Fixture Design

Fixture design is a highly experience-based process, which usually requires 10 and more years of manufacturing practice to design quality fixtures. It is also a tedious and time-consuming task. However it is crucial to issues of quality, cycle time, and cost of production.

Computer-aided fixture design (CAFD) with verification has become a means of providing solutions in fixture design. Although fixtures can be designed using CAD functions, a lack of scientific tools and a systematic approach to the evaluation of design performance leads to trial-and-error strategies that result in several problems: (1) the overdesign of functions, which is very common and often depredates performance (e.g., unnecessary heavy design), (2) the inability to ensure quality of design before production, (3) the long cycle time of fixture design, fabrication, and testing, which may take weeks if not months, and (4) a lack of technical evaluation of fixture design in the business quoting processes in the business cycle. Motivation of CAFD has been derived from the demand of rapid generation of conceptual and detail fixture designs even in product and production design stages, providing tools for fixture design and process verification, and CAD/CAM integration.

In the past 15 years, CAFD and analysis has been recognized as an important area that includes fixture planning, fixture design, and fixturing verification. Fixture planning seeks to determine locating datum surfaces and locating/clamping positions on workpiece surfaces for totally constrained locating and reliable clamping. Fixture design seeks to generate a blueprint of fixture assembly according to different production requirements, such as production volume and machining conditions. Fixture design verification seeks to evaluate fixture design performances for satisfying production requirements, such as completeness of locating, tolerance stack-up, accessibility, fixturing stability, and ease of operation (Rong 1999).

3.1.2   Computer-aided Fixture Design

CAFD involves setup planning, fixture planning, fixture design, and computer-aided fixture design verification, as diagramed in Figure 3.1.

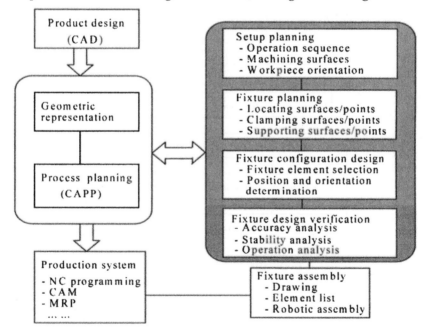

Figure 3.1. Computer-aided fixture design and verification.

For many years, fixture planning has been the focus of academic research with significant progress in both theoretical (Chou 1989; Xiong 1998; Wu 1998; Brost 1996; Asada 1985; Martin 2002; DeMeter 1998; Wang 1999) and practical (Ma 1999; Fuh 1994) studies. Most analyses are based on strong assumptions (e.g., frictionless smooth surfaces in contact with rigid-fixture objective and single-objective functions for opti-mization.

Fixture design is a complex problem that involves consideration of many operational requirements. Four generations of CAFD techniques and systems have been developed. GT-based part classification for fixture design and on-screen editing exists (Grippo 1987; Rong 1992).

Based on geometric reasoning, an automated modular fixture design technique and system has been developed (Rong 1997; Kow 1998). Once the locating and clamping positions are determined in the fixture planning stage, modular fixture design becomes a process of selecting proper fixture components from a database and assembling them into a desired configuration in which the assembly relationships of fixture components are maintained. Figure 3.2 shows a block diagram of an automated modular fixture design system. Figure 3.3 shows an example of a modular fixture designed via CAFD.

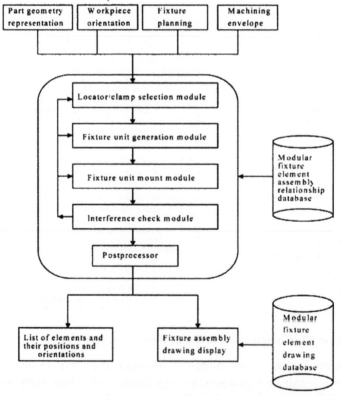

Figure 3.2. A block diagram of an automated modular fixture design system.

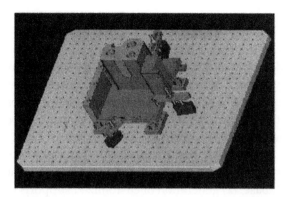

Figure 3.3. A modular fixture designed with CAFD.

In many cases, permanent fixtures with simple geometry are used in mass production. However, there is often no requirement that standard fixture components be used. This is particularly true of the supporting components, which can be customer made to achieve certain performance requirements. Geometric constraint-based CAFD tech-niques and systems have been developed and applied in industry. In these system fixture designs can be generated automatically using CAD functions.

For more complicated fixture design, new design requirements are typically met by adapting an existing design, especially in regard to the concept of part families in mass customization. First, the workpiece is modeled with fixturing information and the existing fixture is modeled with internal constraints. When the design variation of the workpiece is identified, a redesign of the existing fixture is achieved automatically for the new workpiece. In fixture design, utilizing best-practice knowledge is a consideration in quality design. CAFD allows redesigns to inherit fixture-design knowledge from an existing design. Together with fixture verification, this makes it possible to achieve fixture structural design standardization.

This chapter explores computer-aided dedicated fixture design with predefined fixture component types, adaptive fixture design for part families, case-based reasoning (CBR), and sensor-based fixture design. The development of CAFD tools enhances both the flexibility and performance of workholding systems by providing a more systematic and analytic approach to fixture design. CAFD increasingly becomes a necessary system component of concurrent engineering, flexible manufacturing, computer-integrated manufacturing, CAD/CAM inte-gration, and reconfigurable manufacturing systems.

3.2   AUTOMATED DEDICATED FIXTURE DESIGN: BASIC DESIGN

An automated modular fixture design system was developed in our previous study (Rong 1997). Because dedicated fixtures are commonly used in mass production, this section explores a technique of automated permanent fixture design involving predefined fixture component types. The design methodology is divided into two stages: basic design and detail design. Basic design activities include (1) selection of functional fixture components such as locators and clamps from a standard fixture component database, (2) generation of customized supports with variable dimensions for different fixture design requirements, and (3) assembly of fixture components into a final configuration on a fixture base. To implement the fixture design procedure, models are developed to represent the standard fixture components and customized supports. Assembly relationships among fixture components are established based on compatibility analysis. Detail design includes fixture unit combination, connection design, interference avoidance modification, and technological rule-based modification. Principles and implementation of basic design are presented first, with issues of detail design following.

3.2.1   Introduction

Fixtures are used to locate and hold workpieces during manufacturing processes to ensure production quality, productivity, and low cost. Fixtures can generally be divided into two categories: modular fixtures and dedicated fixtures (Hoffman 1991). Modular fixtures are standard fixture components such as standard locators, clamps, supports, and baseplates that can be assembled into a variety of configurations for different workpieces and used in low-volume production applications (Rong 1999). Dedicated fixtures are specially designed and fabricated for a given workpiece and are used in mass production due to the advantages of specially designed performance, such as convenient operation, stiff support in desired directions, and efficient structural space utilization. In that fixture design and fabrication significantly influence manufacturing quality and lead time, it is desirable to put in place automated design and verification of dedicated fixtures at the product design and manufacturing planning stages so that alternative designs can be compared for optimal solutions. In addition, an automated fixture design process accommodates flexible manufacturing systems (FMSs) and computer-integrated manufacturing systems (CIMSs) (Thompson 1986).

Fixture design can be divided into three phases: setup planning, fixture planning, and fixture configuration design (Rong 1999). The objective of setup planning is to determine the number of setups, the position and orientation of workpieces in each setup, and the machining surfaces in each setup. Fixture planning seeks to determine the locating, supporting, and clamping points on workpiece surfaces. The task of fixture configuration design is to select or generate fixture components and place them into a final configuration to fulfill the functions of locating and clamping the workpiece. The research presented in this section concentrated on fixture configuration design. Input to the automated dedicated fixture configuration system includes the fixture planning for a certain setup, the CAD model of the workpiece, and CAD models of the fixture components. The output of the system is a dedicated fixture configuration design.

Previous research on automated fixture design concentrated on modular fixtures. Such studies include the following

- Automated modular fixture configuration design with the assistance of fixture component assembly relationships (Rong 1997, 1998)

- Fixture design retrieval based on group technology (Grippo 1987; Rong 1992) and case-based reasoning (Kumar 1995; Sun 1995) techniques

- Fixture design based on kinematics analysis (Chou 1989; Asada 1985; Mani 1988), expert systems (Nee 1991; Markus 1984; Pham 1990), and geometric analysis (Brost 1996; Wu 1998)

- Fixturability (Ong 1995) and fixturing surface accessibility analysis (Li 1998)

- A preliminary work on automated generation of dedicated fixture designs (Wu 1997; An 1999)

There are certain key differences between dedicated and modular fixtures. Modular fixture design includes a component library with pre-designed and dimensioned standard fixture components. Thus, modular fixture configuration design works toward an assemblage of fixture components. In designing dedicated fixtures, fixture components can be designed from geometric shapes with variable dimensions and connections. Numerous uncertainties are involved in regards to dedicated fixture design tasks. Thus, to streamline the design process automated dedicated fixture configuration design is divided into two stages: basic design and detail design (Figure 3.4).

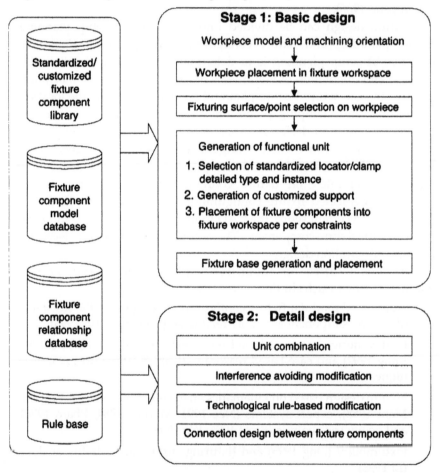

Figure 3.4. Automated dedicated fixture configuration design process.

Basic design mainly includes the generation of an initial result of the dedicated fixture configuration, including standard fixture component selection, support-type selection and dimensioning, and position and orientation determination for fixture components. Detail design includes fixture unit combination, interference avoidance modification, connection design, and technological rule-based modification. Fixture unit combination seeks to optimize the fixture structure by combining two or more functional units into a multi-purpose unit based on functional and spatial conditions. Interference avoidance modification seeks to access any interference that may exist in the design result and then make proper modifications. Connection design seeks to finalize connection features between the fixture supports and the fixture base and between the standardized locator/clamp and supports.

3.2.2   Structural Analysis of Dedicated Fixtures

Based on the study on dedicated fixture structures, fixtures can be classified into three types of components

- Locators and clamps that directly interact with the workpiece to perform certain fixturing functions.
- Fixture base, which holds all functional components and supports together to make one integrated structure. It is also connected to the machine table.
- Supports for locators and clamps connected to the fixture base.

Another way to represent a dedicated fixture is to decompose it into several units that on a fixture base fulfill fixturing functions such as locating and clamping. Typical fixturing functions are illustrated in Figure 3.5. A functional unit usually consists of locators/clamps and a support. In principle, all fixture components of dedicated fixtures can be customized. However, locators/clamps are typically standard components, as they come in direct contact with the workpiece and are easily worn out. Standard components also permit exchangeability, ensure high resistance to wear, and shorten manufacturing leadtime. On the other hand, supports are used to support locators/clamps on the fixture base. They are usually customized in shape and dimensions to satisfy the different requirements of the workpiece and fixture configurations. Therefore, supports need to be designed to adapt to various applications.

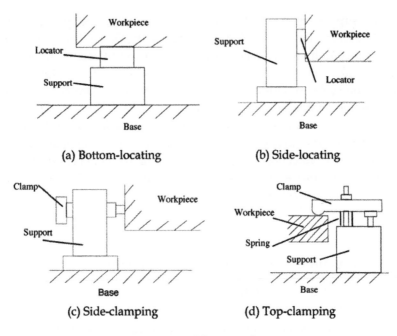

Figure 3.5. Typical fixturing functions.

In many cases, the types of supports are predefined based on existing practice and experience. To design and fabricate fixture supports quickly, basic shape templates for supports can be predefined and stored in a computer database, with detailed shape dimensions and assembly positions remaining adaptable to various fixture designs. Therefore, the major tasks of an automated dedicated fixture design are to select the appropriate fixture component types, determine dimensions, and place these types and dimensions into a configuration on the fixture base.

3.2.3   Basic Design of Dedicated Fixtures

Based on structural analysis of dedicated fixtures, the following five representations and models should be established toward an automated dedicated fixture design.

- A fixturing requirement representation to identify fixture design input information

- A standard fixture component model for the type and dimension selection of locators/clamps
- A customized support template model for the generation of supports
- A fixture component relationship database for determining the compatibility between locators/clamps and basic support templates
- A constraint-based fixture component assembly model for positioning fixture components on fixture base

In the following subsections these representations and models are introduced as the foundation of an automated dedicated fixture configuration design system.

3.2.3.1 Fixturing Requirement Representation

Fixturing requirements are identified in the fixture planning stage, at which fixturing accuracy, accessibility, and stability are considered in determining fixturing surfaces and points (Ma 1999). The results of fixturing requirement identification include fixturing positions, the number and the types of fixture units required, and the properties of fixturing surfaces and points. A functional fixture unit is originated from the specified fixturing surface/point on the workpiece. Thus, the representation of the fixturing surface/point should provide information for the generation of fixture units. Fixturing requirement representation can be defined by a set of geometric and machining data as follows:

$$M_{FS\_WP} = \left\{ F_{id}, F_{geo\_type}, F_{func\_type}, F_{finish}, F_{DOF}, \bar{n}, \ p_{1,} \ p_{2,} S_{accs}, P_{accs}, \text{Stiff} \right\}, \qquad (3.1)$$

where F_{id} is an integer number representing the fixturing surface ID; F_{geo_type} is the geometric type of the fixturing surface, which could be a plane or inner or outer cylindrical surface; F_{func_type} is the fixturing-function type of the fixturing surface, which could be side-locating, side-clamping, bottom-locating, or top-clamping; F_{finish} is the surface finish of the fixturing surface; F_{DOF} is the number of degrees of freedom to be restricted; $\bar{n}$ is a normal vector of the fixturing surface if the F_{geo_type} is a plane, or axis vector if F_{geo_type} is a cylindrical surface; p_1 is the primary fixturing point; p_2 is the optional assistant point, which may be needed to determine the orientation of the locator/clamp; S_{accs} is the surface accessibility value; P_{accs} is the value of local fixturing point accessibility; and Stiff is the stiffness requirement for the fixture unit.

Among the variables at Eq. 3.1, F_{func_type}, F_{finish}, F_{DOF} and Stiff are user-specified, S_{accs} and P_{accs} can be obtained by using the discretization algorithm (Li 1998), and the other variables can be directly extracted from the workpiece CAD model when the fixturing surface/point is specified.

3.2.3.2   Standard Fixture Component Model

In dedicated fixture design, one important issue is to use standard fixture components as much as possible for the purpose of reducing fixture fabrication time and cost. Because functional fixture components (locators/clamps) are in direct contact with workpieces and are subject to wear, such components should be made of hard materials and should be replaceable. Therefore, locators/clamps are usually standard and commercially available in certain dimension series. A standard fixture component model is established for selecting and dimensioning standard fixture components in dedicated fixture design, which leads to the establishment of standard fixture component databases.

A standard fixture component can be described by its component type, functional surfaces, and dimensions. The information is retrieved during standard fixture component selection. Figure 3.6 shows some examples of standard locators and clamps. The fixture component type information is used to determine how the component is used in fixture design.

Functional surfaces used to locate or clamp the workpiece are defined as contact faces and associated functional points are defined as contact points. Surfaces in contact with fixturing supports are defined as supported faces, and associated points are defined as the supported points. For example, in Figure 3.7a, the highlighted top surface is the contact face and its center point PNT1 is the contact point, whereas the highlighted surface in Figure 3.7b is the supported face and its center point PNT2 is the supported point. A functional surface can be represented as follows:

Func_Surf = {Surface_Id, Surface_Type, $\vec{n}$, Point_Id, Point_Type,

$\qquad$ If_ct_above_spted, Surface_Prop},   (3.2)

where Surface_Id is an integer number representing the functional surface ID; Surface_Type is the functional type of the functional surface, which could be contact or supported; $\overset{\text{v}}{\mathbf{n}}$ is the normal direction of the functional surface; Point_Id is an integer number representing the functional point ID; Point_Type is the type of the functional point, which could be on plane, on cylinder, on sphere, on cylinder axis, on hole axis, or on slot; If_ct_above_spted describes the relative positional relationship between the contact surface and the supported surface, which could be +1 or -1; and Surface_Prop is the surface finish of the locating surface.

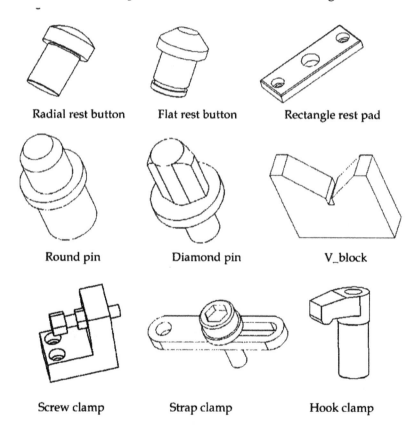

Figure 3.6. Examples of standardized locators/clamps.

Dimensional information is an important part of the standard fixture component model. For a particular component, primary design dimensions play a major role in determining how a component fits the workpiece. The primary design dimensions may determine the primary size, functional height, and/or primary connection size of a standard fixture component. For example, the diameter dimension of the rest button (ϕdia_a) in Figure 3.8 and the linear dimension of the rest pad (oal_b) in Figure 3.9 are such primary design dimensions. The dimensions of thk_b in Figure 3.8 and pad_thk in Figure 3.9 are functional height dimensions. And ϕdia_d in Figure 3.8 and ϕdia_c in Figure 3.9 are primary connection dimensions. The locators/clamps are selected according to their primary dimensions. Other dimensions may be defined with certain relationships with the primary design dimensions.

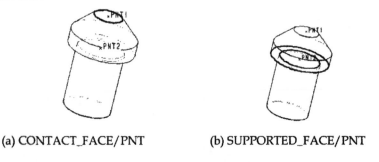

(a) CONTACT_FACE/PNT   (b) SUPPORTED_FACE/PNT

Figure 3.7. Functional surfaces/points of a flat rest button.

Typically, the locators/clamps of the standard component library are composed of families of components (also called table-driven components or instances). A family of components is a collection of similar components varying in different sizes or slightly different detailed features. Every family has a generic basic model that all instances of the family resemble. Concepts such as class, inheritance, and group technology can be applied here for an effective way of retrieving large numbers of standard components. Figure 3.8 is an example of a generic model with dimension names displayed. The instances of the model are listed in the form of a family table, as shown in Table 3.1. Thus, each dimension can be described as follows:

DIMENSION = {dim_name, dim_type, func_type, famtab_attribute,

default_value, min_value, max_value},   (3.3)

where dim_type is the dimension type, which could be a diameter, radius, length, or angle; func_type could be a primary design dimension, functional height, supported connection dimension, or contact height; and famtab_attribute indicates if the dimension is obtained from a family table or not, which could be 1 or 0.

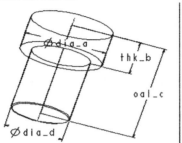

Figure 3.8. Generic model of a rest button.

Table 3.1. Example of a family table.

Instance	dia_a	thk_b	oal_c	dia_d
4_41475	9.7	4.78	12.7	6.375
4_45065	10	6	12	6.025
4_45066	10	8	14.	6.025
4_45067	13	6	14	8.025
4_45068	13	8	16	8.025
4_45455	13	7	20	10
4_45060	19	10	25	12.025
4_45061	19	12	28	12.025

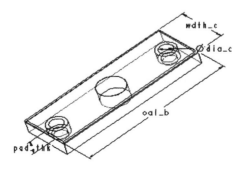

Figure 3.9. Generic model of a rest pad.

3.2.3.3   Customized Support Template Model

In fixture design, the function of supports is to connect locators/clamps to the fixture base and to make the height of the fixture unit. Although the support may vary greatly in shape and size due to the diversity of the workpiece and different fixturing requirements, in many cases the basic shapes of the supports are predefined in shop practice and stored in a computer database. Typical support shapes are shown in Figure 3.10, in which the dimensional relationships are different for different types of supports.

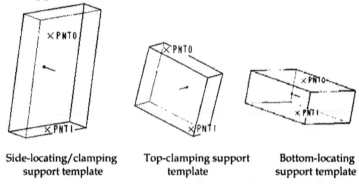

| Side-locating/clamping support template | Top-clamping support template | Bottom-locating support template |

Figure 3.10. Typical support templates.

The detailed shapes, dimensions, and assembly positions of the supports may be changed to obtain different fixture designs. Therefore, several basic support templates are employed as the basis for obtaining a suitable support component through the modification of shape and dimensions. Similar to the standard fixture component, each support template can be modeled with its type, functional surfaces, and dimensions. Functional surfaces used to support the locator/clamp are defined as supporting faces, and associated functional points are defined as supporting points. Surfaces used to connect the support to the fixture base are defined as supported faces and associated functional points are defined as supported points. Figure 3.11 shows a bottom-locating support in which the top surface is the supporting face and PNT0 is the supporting point, whereas the bottom surface is the supported face and the center PNT1 is the supported point. A functional surface-of-support template can be modeled as follows:

$$\text{Func\_Surf=\{Surface\_Id, Surface\_Type, } \vec{n} \text{, Point\_Id, Point\_Type\}} \qquad (3.4)$$

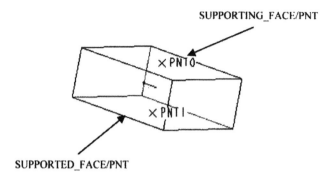

Figure 3.11. Supporting and supported face/point of a bottom-locating support.

Dimensional information is also important in the support template model. Unlike the standard fixture component, most of the dimensions are relation-driven. The primary design dimension of a support template is the dimension that represents its functional height. The functional height satisfaction is the top priority among the design rules. When the value of a functional height dimension changes to fit the clearance between the locator/clamp and the fixture base, other dimensions need to be changed accordingly. The changes of other dimensions can be realized either by recommended relations with the primary design dimensions or by user specification. Some dimensions are constant when the primary dimension is in a certain range. For example, among all dimensions of the support template shown in Figure 3.12, d4 represents the functional height and hence is the primary design dimension. Other dimensions may vary according to the recommended relations. Each dimension of the support template can be modeled as follows:

$$\text{DIMENSION} = \{\text{dim\_name, dim\_type, func\_type, attribute,}$$
$$\text{default\_value, min\_value, max\_value}\}, \qquad (3.5)$$

where func_type specifies the function of the dimension, which could be the functional height, supporting connection dimension, or supported connection dimension; and attribute indicates whether the dimension is generated by relations or not.

3.2.3.4 Fixture Component Relationship Database

Fixture component relationships provide the information for determining if a particular support can be used with a selected locator/clamp. The fixture component relationship database can be described in a table, as shown in Table 3.2. The rows of the table index the detailed types of the standardized locators/clamps, while the columns index the detailed types of the customized support templates. The table element at position (i, j) represents whether the jth locator/clamp can be supported by the ith support.

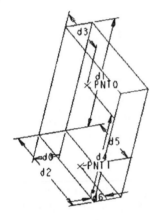

Primary design dimension: d4

Dimension relationships:

1) d0 = ceil (4/100)*15

2) d1 = ceil (1.3*d4)

3) d2 = ceil (d4/100)*60

4) d3 = 0.5* d2

5) d5 = d3

6) d6 = 0.5*d0

Note: ceil is a function to obtain the smallest integer not less than the real value.

Figure 3.12. Example of a support template with dimension relationships.

Table 3.2. Fixture component relationship database.

	jth locator/ clamp type ith Support type	1 Flat rest button	2 Radial rest button	3 Rectangle pad	4 Screw clamp	⋯	m Strap clamp
1	Horizontal support for locating button and screw clamp	1	1	0	1	⋯	0
2	Vertical support for locating button	1	1	0	0	⋯	0
3	Vertical support for locating pad	0	0	1	0	⋯	0
⋮	⋮	⋮	⋮	⋮	⋮	⋱	⋮
n	Vertical support for strap clamp	0	0	0	0	⋯	1

The fixture component relationship database can be constructed either interactively or automatically. In the interactive mode, an operator decides what types of locators/clamps can be used with what types of support templates, as well as the corresponding values in the table. With the automated method, feature recognition and relationship inference techniques are used to determine the database, as illustrated in Figure 3.13. To construct the fixture component relationship database through feature recognition and relationship inference, the functional type (such as bottom-locating or side-clamping) and geometric match conditions (such as component size and contact area) between locators/clamps and supports are considered.

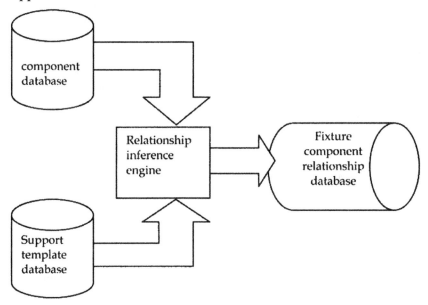

Figure 3.13. Auto-construction of a fixture component relationship database.

3.2.3.5 Constraint-based Fixture Component Assembly Model

The constraint-based fixture component assembly has been applied to modular fixture design (Kang 1998). It can be also applied to the dedicated fixture design. To assemble components properly, three rules are applied: (1) the constraints must be complete, (2) the component cannot be overconstrained, and (3) the dependence on other components should be reduced as much as possible.

To specify the spatial constraints between the locators/clamps and the workpiece, a fixture workspace is defined that provides a global datum for fixture assembling. Based on the fixture workspace, four relationships are defined for the constraint-based fixture component assembly: (1) the relationship between workpiece and fixture workspace, (2) the relationship between locator/clamp and fixture workspace, (3) the relationship between support and fixture workspace, and (4) the relationship between fixture base and fixture workspace.

First, to assemble the workpiece into the fixture workspace a coordinate system is defined on the workpiece model to position and orient the workpiece with respect to the z axis of the machining tool axis. To position a locator/clamp in the fixture workspace, the constraint relationships are specified between the locator/clamp and the fixture workspace. In Figure 3.14, NEW_DATUM_PNT represents the new datum point feature created in the fixture workspace, at the same position as the fixturing point on the workpiece. DATUM_PLANE represents the datum plane feature in the fixture workspace whose normal direction is parallel to the normal/axis direction of the fixturing surface on the workpiece. Therefore, the locator/clamp can be positioned into the fixture workspace where the constraint relationship with the workpiece is maintained, as shown in Figure 3.15.

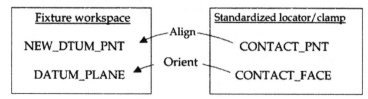

Figure 3.14. Constraint model of assembling standardized locator/clamp.

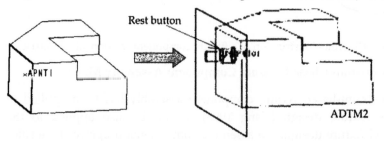

Fixturing surface/point on workpiece   Result of assembling a rest button

Figure 3.15. Example of assembling a standardized rest button.

A complete functional fixture unit in the fixture workspace is composed of a locator/clamp and a support. After the locator/clamp has been positioned according to the constraint-based relationship between the locator/clamp and fixture workspace, the corresponding support needs to be assembled by the constraint-based relationship between the support and fixture workspace, as shown in Figures 3.16 and 3.17. Finally, the fixture base is placed into the fixture workspace by following the same principles as assembling the workpiece. That is, the alignment constraint of the datum coordinate systems is kept the same between the fixture base and the fixture workspace. The size of the fixture base is determined in terms of the overall dimensions of the workpiece and fixture units after they are assembled. It is noted that all assembly functions are implemented automatically by the fixture design program.

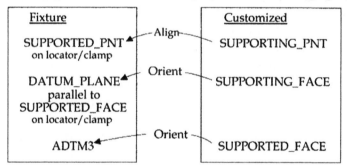

Figure 3.16. Constraint model of assembling customized support.

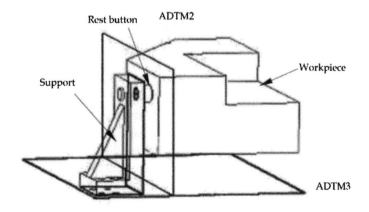

Figure 3.17. Example of assembling a rest button.

3.2.4   Implementation of the Basic Design of Dedicated Fixtures

Based on the fixture workspace relationships defined previously, automatic generation of dedicated fixture configuration design can be implemented. The procedure of generating a fixture unit is as follows: (1) generate a fixture workspace, (2) position the workpiece in the fixture workspace according to the constraint-based assembly relationship, (3) select the type and dimensions of a locator/clamp and assemble it into the fixture workspace according to the constraint-based assembly relationship, (4) select the support template according to the fixture component relationship, determine dimensions of the support by matching the primary design dimension with functional requirements such as the supporting height, and then mount it into the fixture workspace according to the constraint-based assembly relationship, and (5) select or generate a fixture base and assemble it into the fixture workspace according to the constraint-based assembly relationship.

In selecting the type of locator/clamp, a rule base needs to be established in advance. Figure 3.18 diagrams typical rules for selecting the detail type of a locator. Next, an instance with proper dimensions needs to be determined based on the dimensional information of the workpiece. The type of support template is determined according to the fixture component relationship pre-stored in a database. To dimension the support, the primary design dimension is determined such that the functional height of the selected locator/clamp instance is reached with respect to the fixture base. Other dimensions of the support template can be adjusted accordingly per recommended relationships. After the dimension adjustment, the required support model can be obtained by regenerating the corresponding support template based on parametric CAD modeling technique.

A prototype of a basic design module of an automated dedicated fixture configuration design system has been developed on a commercial CAD platform by incorporating C++ with an application programming interface (API). The system consists of four basic modules: (1) component info input, (2) automated design, (3) design modification, and (4) design output. Figure 3.19 shows the system functions included in each module and Figure 3.20 shows a fixture design example designed using this prototype system.

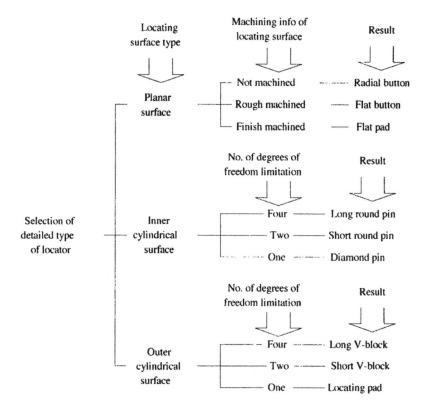

Figure 3.18. Rules for the detailed type selection of locators.

Component Info Input	Automated Design	Design Modification	Design Output
• Info input of standardized locator/clamp • Info input of customized support template • Construction of component relationship database • L/C type selection rule base management	• Loading workpiece • Input of fixturing surface/point • Functional unit generation and placement • Base generation and placement	• Change unit • Copy unit • Delete unit • Change position of unit • Modify support dimension in unit	• Bill of material • Layout of fixture configuration

Figure 3.19. Basic modules in the prototype system.

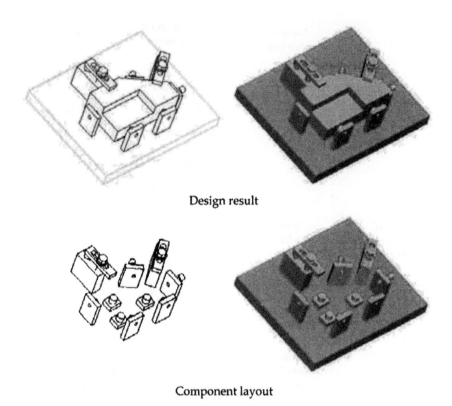

Design result

Component layout

Figure 3.20. Example of dedicated fixture configuration design.

3.3 AUTOMATED DEDICATED FIXTURE DESIGN: DETAIL DESIGN

Dedicated fixtures, which employ predefined fixture components, are widely used in mass production due to the advantages of specially designed performance, such as convenient operation, stiff support in desired directions, and efficient structural space utilization. Because fixture design and fabrication significantly affect manufacturing quality, cost, and lead time, automatic design and verification of dedicated fixtures is desirable at the product design stage. In addition, an automated fixture design process is required to accommodate flexible manufacturing systems (FMSs) and computer-integrated manufacturing systems (Thompson 1986). To bridge this gap, research has been conducted to study the techniques of automated dedicated fixture configuration design with predefined fixture component types. Based on the research, a prototype system has been developed on a commercial CAD system with using an API to automate the dedicated fixture configuration design process.

The design methodology is divided into two stages: basic design and detail design. The objective of basic design is to obtain a preliminary configuration design result. Basic design activities include (1) selection of standard fixture components such as locators and clamps, (2) generation of customized supports, and (3) assembly of fixture components into a final configuration. Detail design includes interference avoidance modification, fixture unit combination, connection design, and technological rule-based modification. Interference avoidance modification seeks to access any interference that may exist as a result of basic design and then make proper modifications. Fixture unit combination seeks to optimize the fixture structure by combining two or more functional units into a multi-purpose unit based on functional, spatial, and other practical conditions. Connection design seeks to finalize the connection features between locators/clamps and fixture supports and between fixture supports and the fixture base. Technological rule-based modification seeks to make modifications to the result of basic design, especially the customized support, based on technological knowledge that improves fixture performance (as regards such stiffness, stability, and ease of workpiece loading/unloading).

Once a basic design is complete, a preliminary configuration of the dedicated fixture can be created. However, to a large extent this design is not the best one or even necessarily usable in practical applications. In addition, some detailed features need to be finished. The objective of detail design is to refurbish the preliminary configuration design result for a final complete and optimized configuration design of the dedicated fixture. The tasks in detail design include:

- Interference avoidance modification
- Fixture unit combination
- Connection design
- Technological rule-based modification

3.3.1   Interference Avoidance Modification

Geometric interference is frequently encountered in the design of fixture assemblies. When this problem occurs, the components in the assembly have to be modified or redesigned, increasing the cost of production. Interference avoidance modification seeks to access any interference that may exist as a result of basic design and then make proper modifications to remove the interference at the design stage. Thus, interference avoidance modification can be divided into two phases: (1) interference checking to discover existing interference, and then (2) making modifications to remove the interference if interference does exist.

To check interference, geometric and topological data are necessary. Any interference between two objects is checked by computer analysis of the solid models that represent the objects and thus the data structure of the model has a strong influence on the efficiency of the interference checking (Yang 1994). Many commercial CAD systems supply solid interference checking functions while encapsulating and hiding the internal representation of the solid models. Thus, for the convenience of program development, these solid interference checking functions provided by CAD software are used to check whether one solid interferes with another without regard to the internal data structure of the solid model.

In automated fixture design, the fixture components should be properly located on the workpiece surface or around the workpiece perimeter, without interfering with each other or with the machining tool-path envelope. Generally, there are three types of interference in fixture design: (1) interference among fixture components, including locators, clamps, and supports; (2) interference between fixture components and a workpiece; and (3) interference between fixture components and a machining envelope (Hu 2000; Ngoi 1997).

Once the interference has been detected with interference checking functions, proper modifications should be applied to the basic design to avoid or eliminate interference. Possible solutions include changing the size of fixture components, changing the shape of fixture components, changing the orientation of fixture components, changing the distance between interfering parts in the assembly, and combining fixture components. Figure 3.21 depicts three methods of making modifications to eliminate interference.

(a) Distance changed between interfering parts to avoid interference

(b) Fixture components combined to remove interference

(c) Orientation of fixture unit changed to avoid interference

Figure 3.21. Several practical methods of avoiding or eliminating interference.

Other possible modifications are depicted in Figure 3.22. When the interference occurs on the side of the support where the support collides with the protrusion of the workpiece, the solution may be decreasing the side dimension of the support (Figure 3.22a). If the interference occurs on the bottom of the support, there are several possible solutions. When the interference area is not large, the low cross portion of the support can be cut to eliminate the interference (Figure 3.22b). If the interference is relatively large, the support may be moved backward and separated into two pieces, as shown in Figure 3.22c. Similar analysis can be performed on the supports that connect two locators to the fixture base, as shown in Figures 3.22d, e, and f.

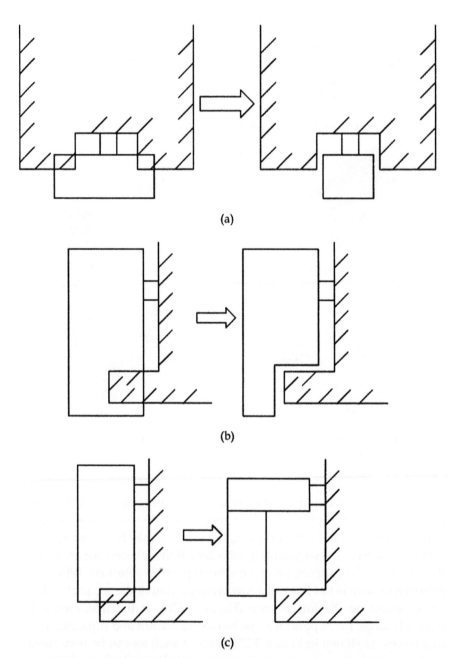

Figure 3.22. Fixture design modification for interference avoidance.

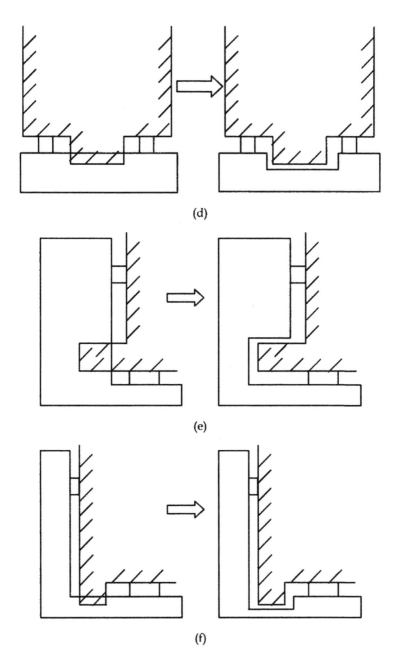

(d)

(e)

(f)

Figure 3.22. Fixture design modification for interference avoidance (Continued).

3.3.2   Fixture Unit Combination

In basic design, a fixture unit is generated for each fixturing function. Fixture units may need to be combined toward lowering the risk of interference, erasing interference, improving manufacturability and fixturing performance, and reducing cost. The objective of fixture unit combination is to optimize the fixture structure by combining two or more functional units into a multipurpose unit based on functional, spatial, and other practical considerations while keeping the fixturing functions unchanged. In our system, because locators/clamps are standard fixture components and supports are customized components, fixture unit combination can be confined to the combination of supports.

To support the fixture unit combination process, a fixture unit combination rule base is needed. There are two fundamental rules for unit combination: the distance rule and the functional rule (Wu 1997). The distance rule states that fixture units that are close in distance can be combined into one. The functional rule states that the combination types depend on the nature of fixturing functions, especially the orientation (side, bottom, or top) of functional components.

However, these two rules may be ambiguous and not operable. In fact, in establishing a fixture unit combination rule base two factors should be considered: interference and manufacturability.

Manufacturability is a function of tolerance requirements, equipment of a certain manufacturing enterprise, materials, number of parts, shape of product, volume of product, and other variables. Between these two factors, interference is ranked as the higher priority. Thus, three operable rules have been put forward and employed in our system:

- If fixture unit combination is the only way to eliminate or avoid interference, then combine without conditions.

- If fixture unit combination is not the only way to eliminate or avoid interference and if fixture unit combination can improve manufacturability, then combine.

- If fixture unit combination is not the only way to erase or avoid interference and if fixture unit combination cannot improve manufactruability, then do not combine.

In implementing the last two rules, manufacturability evaluation is the key. Manufacturability evaluation is a complex problem in two respects:

(1) manufacturability is a function of multiple variables and (2) the weighted importance of these variables is different between different manufacturing enterprises. Thus, it is difficult to establish a unified model to evaluate manufacturability for different manufacturing enterprises. However, for a specific manufacturing enterprise the model for manufacturability evaluation can be established by assigning weights to all these variables. Figure 3.21b shows an example of fixture unit combination.

3.3.3   Connection Design

A product is only as strong as its weakest link. For many products, the weakest area is at the joints, where the components are assembled or connected. Well-connected assemblies result from good design, quality parts, and properly executed engineering and connection technology.

In basic design, preliminary support templates are selected according to the fixture component relationship database and are used as a basis for generating detailed supports (including connection features for practical applications). The objective of connection design is to finalize the connection features between locators/clamps and fixture supports and between fixture supports and a fixture base.

3.3.3.1   Connection Design Between Locators/Clamps and Supports

Because locators/clamps are standard fixture components, their connection feature types and connection dimensions are decided when they are selected and loaded into the system. The connection feature information of locators/clamps is included in the standard fixture component model and thus it can be retrieved from the standard fixture component database. Given the known connection features of a locator/clamp, the corresponding connection features of the matching fixture support can be generated correspondingly.

3.3.3.2   Connection Design Between Fixture Supports and Fixture Base

There are many types of connection approaches. However, in real industrial dedicated-fixture-design practice screw connections and welding are widely employed. Thus, in our research and system only these two connection methods are considered in the connection design between fixture supports and a fixture base.

There are advantages to both screw connections and welding. The advantages of screw connections are as follows.

- Commercially available in a wide range of styles, materials, and sizes
- Capable of joining same or dissimilar materials in uniform as well as in unusual joint configurations
- Can be easily installed in factories or in the field with both standard manual or power installation tools, with a maximum of safety
- Easily removed and replaced

The advantages to welding connections are as follows.

- High production efficiency
- Shorter production cycle
- Better joint quality
- Lower cost
- Easily automated

The first step of connection design is to choose the connection method (i.e., screw connection or welding). Here, reusability should be the focus. If our dedicated fixture system never needs to be disassembled or seldom disassembled, a permanent connection such as welding may represent our best option, provided strength and appearance criteria are satisfied. With a greater need for disassembly, screw connections might be a better option.

Because both screw connections (layout design and dimension determination) and welding have been standardized, two standard databases (a screw connection standard database and a welding standard database) can be established to support the connection design process. The retrieving indexes include the dimensional information of preliminary support template, machining force information, and so on.

Figure 3.23 shows examples of the connection design of a support. Figure 3.24 shows examples of welding joint designs for a user to select.

Before connection design   After connection design

Figure 3.23. An example of connection design.

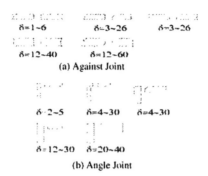

δ=1~6   δ=3~26   δ=3~26

δ=12~40   δ=12~60

(a) Against Joint

δ=2~5   δ=4~30   δ=4~30

δ=12~30   δ=20~40

(b) Angle Joint

Figure 3.24. Welding joints for fixture connection design.

3.3.4   Technological Rule-based Modification

All generated fixture supports should be checked using process technology knowledge for verification. The purpose of technological rule-based modification is to make modifications to the result of basic design, especially the generated customized support, based on process technology knowledge. As a result, dedicated fixture performance shows improvement in stiffness, stability, reduced weight, ease of workpiece loading/unloading, and so on. For example, if the stiffness requirement of a fixture unit is high, one or more ribs might need to be added to fixture supports. To improve the stability, the base of fixture supports may be enlarged. Figure 3.25 compares two supports: one without any rib because of a low stiffness requirement and one with a rib because of a high stiffness requirement. Figure 3.26 shows several other examples of fixture design modifications based on predefined rules. To support such modifications, a process technology knowledge base is established.

Support without rib   Support with a rib due to stiffness requirement

Figure 3.25. Design modification of enhancing fixturing stiffness.

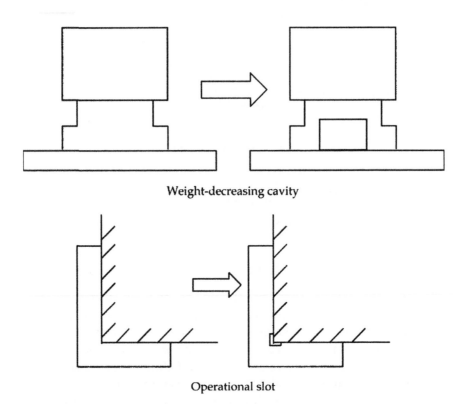

Weight-decreasing cavity

Operational slot

Figure 3.26. Examples of technological modification of fixture design.

3.3.5   System Implementation and Example

A prototype of an automated dedicated fixture configuration design system has been developed on a CAD platform. Detail design modules have been integrated with basic design modules, leading to a complete prototype system of automated dedicated fixture configuration design. The detail design consists of four basic modules: (1) interference-avoiding modification, (2) fixture unit combination, (3) connection design, and (4) technological rule-based modification. Figure 3.27 shows the system functions in each module. Figure 3.28 shows the final dedicated fixture design after detail design, corresponding to the basic fixture design shown in Figure 3.20 in the previous section. Figure 3.29 shows another example of a dedicated fixture design.

Unit Combination	Interference-avoiding Modification	Connection Design	Technological Rule-Based Modification
• Unit combination rule base management • Interactive unit combination • Automatic unit combination	• Partial interference checking • Global interference checking • Interference avoidance modification	• Connection design generation	• Process knowledge base management • Automatic analysis and modification • Interactive analysis and modification

Figure 3.27. Basic modules in the detail design system.

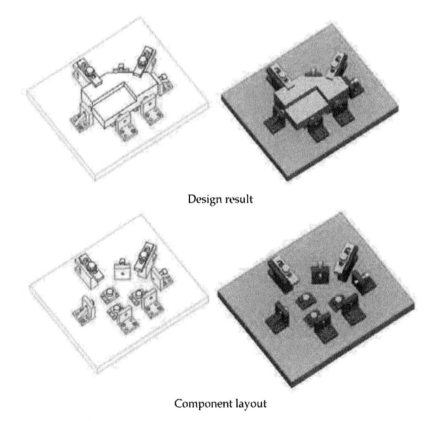

Design result

Component layout

Figure 3.28. Fixture configuration design result after detail design.

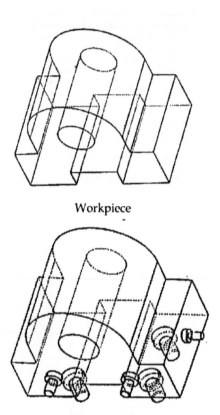

Workpiece

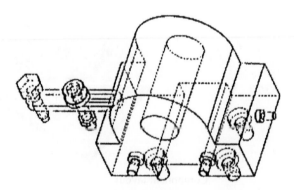

Locator design and placement

Clamp design and placement

Figure 3.29. Procedure of dedicated fixture design.

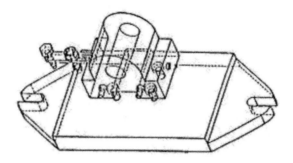

Fixture base design

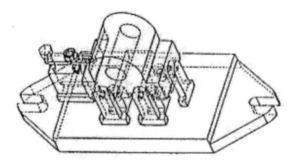

Individual connector design

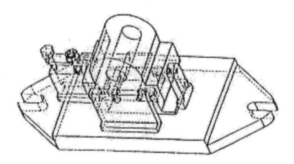

Connector combination

Figure 3.29. Procedure of dedicated fixture design (Continued).

3.4   ADAPTIVE FIXTURE DESIGN FOR PART FAMILY

3.4.1   Introduction

In the area of computer-aided fixture design (CAFD), the "auto-design" of modular fixtures is discussed in the literature. All components of modular fixtures, including fixture base plate, locators, and supports, are standardized. Fixture generation in this case is relatively easy to carry out by simply specifying a number of components and by specifying definitions of assembly relationships.

In mass production, however, dedicated fixtures play a unique role and thus cannot be replaced by modular fixtures. Previous work in the dedicated fixture design is largely based on the predefined support template, wherein entire components and the fixture base have relatively simple geometry. Such templates therefore do not deal well with complicated fixture components. An adaptive fixture design system is presented in this section in an attempt to accommodate complicated fixture structures, including fixture bases.

The basic problem to be addressed can be described as follows. As for a class of parts (family of parts), their typical fixture configuration and fixture assembly have been well defined and implemented based on current best practice. The typical fixture design is so called a common fixture. The fixture configuration and assembly of a given part (family member) is determined from a predefined fixture configuration and assembly. More specifically, part analysis is based on certain assumptions. First, the common fixture exists. Second, the shape and size of new parts are changed somehow, yet it still belongs to the same class of part as the existing part. Parts that belong to the same family usually have a variation range of dimensions and similar structures, as well as similar manufacturing processes to produce these parts. A new fixture can thus be generated according to differences identified from com-parison with existing fixtures.

The purpose of adaptive fixture design is to design fixtures rapidly to reduce new product development time and to reuse existing fixture data as much as possible toward facilitating reconfigurable manu-facturing systems. Adaptive fixture design is realized on the basis of part families.

Fixture design is a highly experience-based process involving solutions that are often not unique. A good design with impressive performance can often be generated by modifying a design with proven performance. The expert knowledge in this case is associated with the existing design. When a fixture needs to be designed, often the best way is to identify an existing fixture that is designed to work with similar parts in the same family and then minimally modify the existing fixture design to accommodate the requirements of the new part. If there are no changes required to the fixture base, the new fixture development cost and time is reduced significantly because the production line and existing fixtures, with minor modification, can be still used to manufacture the new part.

In summary, an adaptive fixture design can be generated via the following procedure. For a given new part, search for a suitable fixture in the existing fixture library, provide multilevel modification suggestions and corresponding cost estimates, and finalize the design with minimum modification cost.

Based on concurrent engineering principles, fixture design is part of manufacturing systems design. Therefore, fixture design requirement information comes from setup planning, which specifies design constraints. A conceptual fixture design is usually conducted first, in which part layout design is the first step.

3.4.2   Conceptual Fixture Design

In manufacturing systems design, the way parts are laid out on the fixture base affects production cycle time and needs to be determined before fixture design. A conceptual fixture design is usually carried out with a comparison of several alternative solutions in order to achieve an optimal one. Conceptual fixture design is performed once the manufacturing sequence is determined at the setup planning stage and the locating/clamping method is determined at the fixture planning stage. Figure 3.30 shows a block diagram of a conceptual fixture design.

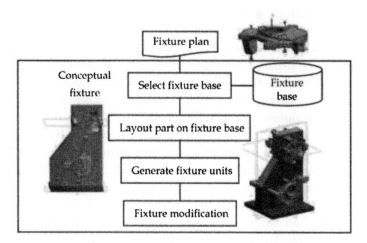

Figure 3.30. Conceptual fixture design.

Once the setup is determined, including part orientations, process sequences, and the machine tools required at each setup, a fixture base is selected based on process requirements and manufacturing knowledge. The parts may be placed in different ways on the fixture base. Figure 3.31 shows CAD interface for placing parts on a fixture base and modifying the layout position and orientation.

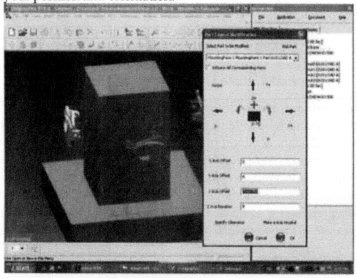

Figure 3.31. CAD interface for part layout on fixture base.

Figure 3.32 shows several other examples of part layouts on various fixture bases.

Figure 3.32. Examples of part layouts on fixture bases.

The fixture plan includes a set of locating and clamping points on the part surfaces of The workpiece in each setup. Conceptual fixture design can thus be generated following the principles of fixture unit generation presented in Section 3.2. Figure 3.33 shows the conceptual fixture design for the part shown in Figure 3.31. All conceptual fixture designs can be implemented within a CAD environment.

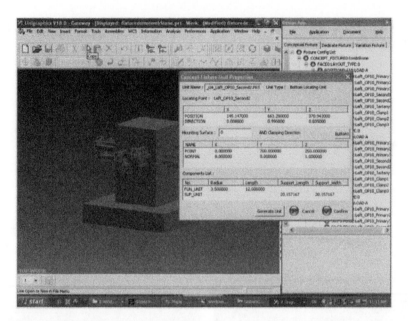

Figure 3.33. Example of a conceptual fixture design with CAD interface.

3.4.3   Fixture Modeling for Adaptive Fixture Design

The basic idea of adaptive fixture design is to first model a typical part in the family, including fixturing information and a common fixture for that family of parts. Then variations between the new part and the typical part in the family are identified. Finally a new fixture design is generated by modifying the common fixture design.

3.4.3.1   Family Part Modeling

A part can be modeled according to its 3D data, manufacturing features, and fixturing fixtures, as indicated in Figure 3.34. Each feature of the part is specified by position and orientation as well as the feature's shape parameters. A work part model can be expressed as

$$WP=\{CAD\ 3\text{-}D\ data,\ MF\_SET,\ FIX\_SET\},\qquad(3.6)$$

where MF_SET is a set of manufacturing features and FIX_SET is a set of fixturing features in the workpiece. Geometric data for manufacturing features and the cutting tools used to produce them are useful in fixture design. The geometrical information is extracted from CAD models and the tooling information is acquired from the results of setup planning. Fixturing features are regarded as a set of locating features and clamping features described as

$$FIX\_SET = \{L, C\}, \tag{3.7}$$

where {L} is a locator set and {C} a clamp set. These sets are represented, respectively, as the positional and orientation vectors $L = \{r_i, n_i\}$ and $C = \{r_j, n_j\}$. Here, i and j are the indexes of the number of locators and clamps.

Figure 3.35 diagrams a workpiece and a location associated with three coordinate systems — the global coordinate system OXYZ, part—local coordinate system O'X'Y'Z', and fixture-local coordinate system QUVW. A transformation matrix, T, can be used to describe the relationship between r_k and r'_k

$$\begin{cases} r_k = T \cdot r'_k \\ n_k = T \cdot n'_k \end{cases}. \tag{3.8}$$

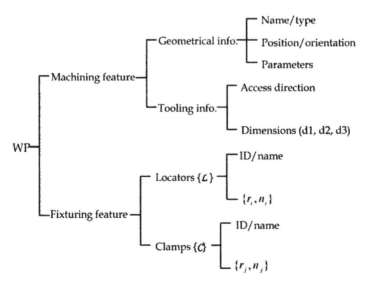

Figure 3.34. Representation of a manufacturing feature.

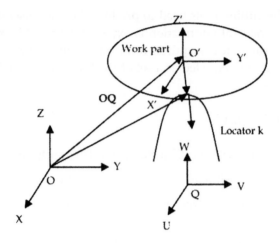

Figure 3.35. Fixture coordinate systems.

3.4.3.2   Part Variation Model

Variations in workpieces are mainly identified as differences in feature dimensions and orientations. When a feature is specified, such differences can be identified as

$$\begin{cases} \bar{r}' = \bar{r} + \Delta\bar{r} \\ \bar{n}' = \bar{n} + \Delta\bar{n} \end{cases} \tag{3.9}$$

and $$\begin{cases} \Delta r' = \Delta Tr + T\Delta r \\ \Delta n' = \Delta Tn + T\Delta n, \end{cases}$$

where $\Delta r' = (\Delta x', \Delta y', \Delta z', 1)^T$ and $\Delta n' = (\Delta x', \Delta y', \Delta z', 1)^T$ are in the part co-ordinate system while $\Delta r = (\Delta x, \Delta y, \Delta z, 1)^T$ and $\Delta n = (\Delta x, \Delta y, \Delta z, 1)^T$ are in the global coordinate system. Here, r' and n' can be acquired directly from a part's CAD model data, and Δr and n can be solved when the part's layout on the fixture base is determined.

3.4.3.3   Fixture Modeling

A typical fixture consists of several subassemblies, or units. Each fixture unit performs one or more fixturing functions. The components of a fixture unit are directly connected with one another and typically only one element is mounted directly on the fixture base. One or more components are contacted directly with the workpiece serving as the locator, clamp, or support. Let F denote the fixture. We then have

$$F = \{U_i \mid i \in n_u\}$$
$$U_i = \{e_{ij} \mid j \in n_{ei}\} \tag{3.10}$$
$$F = \{\{e_{ij} \mid j \in n_{ei}\} \mid i \in n_u\},$$

where U_i denotes a fixture unit in a fixture, n_u is the number of units in fixture F, e_{ij} is the jth component in unit i, and n_{ei} is the number of components in unit U_i.

Figure 3.36 shows a fixture unit in which support 0 is directly mounted on the fixture base. Support n is connected with a clamp that is in contact with the part. The clamping point remains at the same location for both part surface and clamp surface. The clamping point is transformed to the global coordinate system. Any difference in clamping point will thus be passed to the fixture model if a new part has a clamping point that varies from the typical part. This difference can be used in modification of the fixture design for the new part.

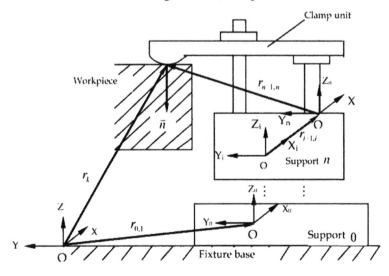

Figure 3.36. A clamping unit in fixture design.

3.4.4 Similarity Identification

When a new part belongs to an existing family, a common fixture can be selected from a fixture database. Data associated with the fixture might include fixture base, part layout, functional points positions and orientation, and fixture units (with variable ranges specified), as shown in Figure 3.37. According to how similar the common fixture is to the new part, the fixturing requirement for the new part needs to be analyzed to determine how to modify the design and how much modification needs to be performed. It is desirable in this process to upgrade existing fixtures to expand fixture utility database toward being able to accommodate multiple parts with minimum modification and toward developing quick-change unit structures.

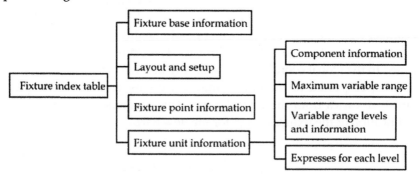

Figure 3.37. Fixture information for similarity analysis.

Similarity analysis in this case seeks to compare two sets of locating/clamping points on the surfaces of the family part and the new part, as sketched in Figure 3.38. These points exist as pairs corresponding to fixturing functions and are determined at the fixture planning stage. Conventional similarity analysis (based on continuous distance func-tions) may not be valid here because in this case fixture modification time and cost are determined by the change/replacement of fixture units and the fixture base. The problem is that such changes may not be a continuous function of the variable distances.

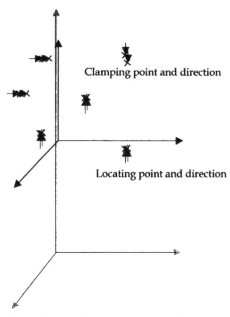

Figure 3.38. Fixturing points in the respective coordinate systems of the family part and the new part.

Similarity analysis is performed between the fixture units in two fixture designs. The fixture unit similarities are then considered together for a fixture similarity. The fixture unit similarity can be classified into four levels. At the first level, modification is considered minimal if, for example, a pair of a locating/clamping points are so close in proximity difference that there is no need to change the support of the unit or the fixture base (the locator/clamp may be replaced by a different one). At the second level, the difference in locating/clamping points requires that the shape and size of the fixture support be changed/modified but the position of the unit and the connection to the fixture base do not need to be changed. Modification cost may not be significant. At the third level, the difference in locating/clamping points requires that the position of the support be changed. Although the same fixture base can be used, the connection location for the support needs to be modified. The modification effort in this case is greater than in the first two cases. Then if the difference of the requirements of the fixture units in new fixture design is too much that the whole fixture base needs to be rebuilt, the modification effort would be very significant. This is the fourth level of similarity, the least similar case. When the fixture model is established for the family part, these four levels of modification are represented as four variable ranges (measured by boxes in Figure 3.39), with respective associated cost values of modification.

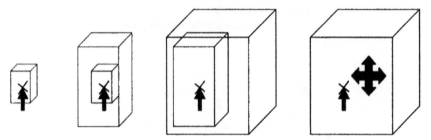

Figure 3.39. Variable ranges of a fixture unit at different levels.

3.4.4.1 Fixture Plan Matching Algorithm

For two sets of points $\{P_i\}$ and $\{Q_j\}$, Eq. 3.11 represents a transformation matrix , $\{m_{ij}\}$, that best maps the points of P onto the points of Q.

$$F(t,R,m) = \sum_{i-1}^{I}\sum_{j-1}^{J} m_{ij}(\|P_i - t - RQ_j\|^2 - \partial_j),\qquad(3.11)$$

where $\forall i \ \sum_{j-1}^{J} m_{ij} \leq 1 \quad \forall j \ \sum_{i-1}^{I} m_{ij} \leq 1 \quad \forall ij \ m_{ij} \in \{0,1\}$, t is a translation vector, R is rotation matrix, and ∂_j works as a threshold distance, indicating how far the two points can be before they must be treated as outliers beyond the specified area (i.e., before the fixture unit must be redesigned). For example, if the number of outliers is greater than two, one can say that the fixture does not match the new part.

Since this is a discrete problem, the deterministic annealing method can be used to transform the discrete problem into a continuous one in order to reduce the local minima trap (Bridle 1991), where m_j can be defined as

$$m_j = \frac{e^{\beta X_j}}{\sum_{k-1}^{J} e^{\beta X_k}}.\qquad(3.12)$$

The problem is then equivalent to the problem of finding M to maximum S via

$$S(M) = \sum_{i=1}^{I}\sum_{j=1}^{J} m_{ij} D_{ij} \tag{3.13}$$

Such that $D_{ij} = -(\left\| P_i - t - RQ_j \right\|^2 - \partial_j)$. The following standard procedure can be applied to solve the problem.

1.   Assume R = E, t = zero, and $\beta = \beta_0$.
2.   Calculate all members of D via $D_{ij} = -(\left\| P_i - t - RQ_j \right\|^2 - \hat{\partial}_j)$.
3.   Calculate all members of M via $m_{ij} = e^{\beta D_{ij}}$.
4.   Make M work with $\forall j \ \sum_{i=1}^{I} m_{ij} = 1 \ \ \forall i \ \sum_{j=1}^{J} m_{ij} = 1$ by iteratively normalizing in rows and columns.
5.   Get new R and t according to M.
6.   Increase β.
7.   If $\beta < \beta_{max}$ repeat step 1 through 6.

It should be noted that step 5 is the only unsolved problem in the algorithm. To match set P_1 to P_2, let t equal the center difference of the two point sets and R equal UV^T. UV is then the singular decomposition of matrix $P_1^T P_2$ (Kong 2003). If the modification cost is considered as the relationship between modification cost and the distance C_j, Eq. 3.11 may be written as,

$$F(t, R, m) = \sum_{i=1}^{I}\sum_{j=1}^{J} C_j (m_{ij} \left\| P_i - t - RQ_j \right\|^2 - \partial). \tag{3.14}$$

Thus we see that by applying the same algorithm the modification cost can be optimized.

3.4.5   Implementation Examples

Figure 3.40 shows two parts in a family, and Figure 3.41 shows a part in the family with fixturing and machining features specified. Figure 3.42 shows a common fixture design for the family of parts.

Figure 3.40. Two parts in a family.

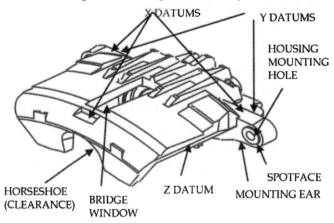

Figure 3.41. Fixturing (locating datum) and machining features.

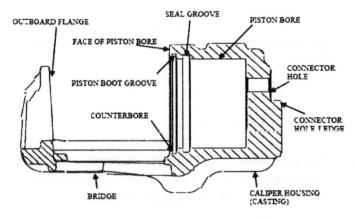

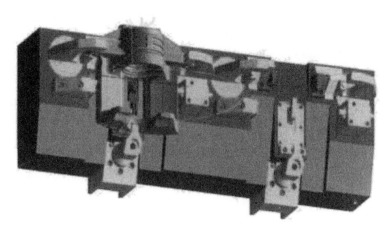

Figure 3.42. Common fixture for caliper.

When the need for a new part in the family is recognized with fixturing information, differences between the new part and any existing part are identified by analysis of data as listed in Tables 3.3 and 3.4.

Table 3.3. Locating feature positions and variations.

		LOC_X1	LOC_X2	LOC_X3	LOC_Y1	LOC_Y2	LOC_Z1
Calp-1	X	86.5	86.5	-19.5	74.5	74.5	42.5
	Y	87	-87	0	43	-43	76
	Z	0	0	-24		-17	
Calp-2	X	86.5	86.5	-11.5	79.5	79.5	48
	Y	77	-77	0	43	-43	71.3
	Z	0	0	-10	-6	-6	-4

Table 3.4. Clamping feature positions and variations.

		CL_EAR1	CL_EAR2	CL_WIN	CL_TOP	CL_Z	CL_SIDE	CL_SIDE
	X	86.5	86.5	60.5	-79.5	42.5	NONE	NONE
	Y	87	-87	0	0	-76	NONE	NONE
	Z	27(31)	27(31)	-19	-13.6	-2	NONE	NONE
Calp-2	X	86.5	86.5	57	NONE	48	-22	-22
	Y	77	-77	0	NONE	-71.3	59	-59
	Z	24	24	-17.5	NONE	-4	24	24

According to variation levels identified in the overall fixture structure and each unit, the same fixture base can be used for the new part (Calp-2) and no modification to the fixture base is required. However, the units may need to be redesigned to incorporate changes in locating positions. Figure 3.43 shows the modification of one unit of the fixture. The connection of the unit to the fixture base is maintained. Therefore, modification cost is minimized.

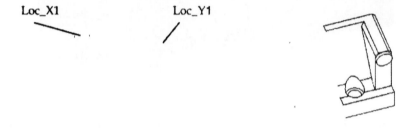

Fixture unit in Calp-1 fixture   Fixture unit in Calp-2 fixture

Figure 3.43. Modification of fixture unit.

In summary, the adaptive fixture design starts with comparative analysis among new parts in a part family, and common fixture for the family. Differences in fixturing requirements are identified for the overall fixture structure and for each unit. An algorithm is then developed that matches fixture design requirements to the common fixture design in order to minimize modification effort. The new fixture is designed and fixture design modification is conducted at three levels. If the variation is not significant and the fixture base can be reused, modification cost is reduced significantly and automated modification is possible. Figure 3.44 depicts the procedure for the adaptive fixture design.

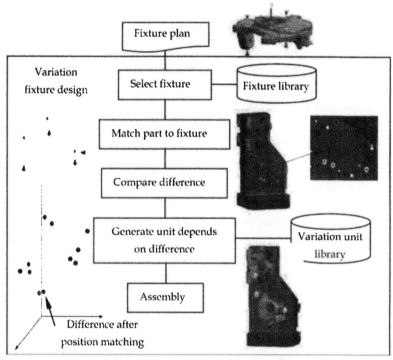

Figure 3.44. Procedure for variation fixture design.

3.5   CASE-BASED REASONING FIXTURE DESIGN

Fixtures accurately locate and secure a part during machining operations such that the part can be manufactured to design specifications. To reduce design costs associated with fixturing, various CAFD methods have been developed through the years to assist the fixture designer. One approach is to use a case-based reasoning (CBR) method whereby relevant design experience is retrieved from a design library and adapted to provide a new fixture design solution. Indexing design cases is a critical issue in any CBR approach, and CBR systems can suffer from an inability to distinguish between cases if indexing is inadequate. This section presents a CAFD methodology, CAFixD, that adopts a rigorous approach to defining indexing attributes in which axiomatic design functional requirement decomposition is adopted. Thus, a design requirement is decomposed in terms of functional requirements, physical solutions are retrieved and adapted for each individual requirement, and the design is then reconstituted to form a complete fixture design. Furthermore, adaptability is used as the basis upon which designs are retrieved in place of the normal attribute similarity approach, which can sometimes return a case that is difficult or impossible to fix. This section explains the CAFixD framework and operation and discusses in detail indexing mechanisms used in CAFixD.

3.5.1   Introduction

A key concern of any manufacturing company is its ability to manufacture high-quality products in as short a time as possible. The rapid release of a product into the marketplace, ahead of competitors, is crucial to securing a higher percentage of the marketplace. Fixtures play an important role in this regard within many manufacturing processes. They accurately locate and secure a workpiece during machining such that the part can be manufactured to design specifications. Thus, fixtures have a direct effect upon machining quality, productivity, and the cost of products. The workpiece rests on locators that accurately locate the workpiece, and clamps are used to hold the workpiece securely in this position during machining. The typical structure of a fixture consists of a baseplate to which the clamping and locating units are attached. Locating and clamping units consist of a supporting unit and either a locator or clamp. Fixtures may contain different numbers and types of clamping and locating units, but units generally always follow this same basic format.

Costs associated with the design and manufacture of fixtures are significant, accounting for some 10 to 20% of the total cost of a manufacturing system (Bi 2001). Various approaches have therefore been pursued with the aim of reducing fixturing costs. One approach has been to develop flexible modular fixture systems that can be used in a variety of situations. An alternative approach has been to examine options of simplifying and shortening the fixture design process. Various CAFD methods employing artificial intelligence in design techniques have been studied through the years to assist the fixture designer. This section proposes the CAFixD methodology, a new fixture design method based upon a CBR approach. Initially, various CAFD approaches are critiqued to clarify the need for the development of the CAFixD approach. The

CAFixD methodology is then discussed. The overall CAFixD framework is presented, with significant emphasis given to describing indexing mechanisms to be employed. A worked example of its operation is also developed.

Much effort in the CAFD community has concentrated on developing systems that generate optimal fixture configuration layouts. These layouts specify the optimum positions at which the fixture should contact the workpiece being machined. Rule-based (Nee 1991; Roy 1997; Trappey 1992) and genetic algorithm (Krishnakumar 2000; Wu 1996) methods are typical approaches that have been used to develop CAFD systems. However, such systems do not specify the actual physical form of the individual units of a fixture. Their output takes the form of a list of coordinates stating where the fixture should contact the workpiece. Conceptual work has focused on designing individual fixture units using a combined genetic-algorithm/neural-network approach (Kumar 1999). However, the output of this approach is essentially a high-level conceptual design of a fixture unit that specifies its basic type and the nature of its components. Attempts at designing complete fixture units have been largely based on geometric approaches (Wu 1998; An 1999) in which the basic concept is to identify the critical dimension of a particular fixture unit (normally its height) and via existing mathematical relationships to then relate all other dimensions to this critical dimension.

CAFD approaches to date typically do not consider the complete operational requirements of a fixture, but instead focus on some specific aspect of a fixture (most often the fixture configuration layout). There remains the need to develop CAFD techniques that incorporate all aspects related to fixture design, as has been recognized (Bi 2001). Other requirements may include loading/unloading time, fixture weight, cost, assembly time, and so on. What is still missing from the CAFD field therefore is a system that can fully design a complete fixture using all operational requirements to guide the design process.

CBR has also been used to develop CAFD systems (Bi 2001; Kumar 1995). CBR is an example of analogical reasoning — a technique whereby knowledge from similar design experiences is retrieved and adapted to provide a solution for a new design problem. CBR appears well suited to fixture design because a good fixture design is largely dependent on the designer's experience (Kumar 1995). Essentially, there are two main stages in a CBR approach: case recall and case adaptation (Maher 1997). Design case recall, divisible into the following major tasks, is concerned with finding a relevant case within the case library.

- Indexing: Features required by the current problem are identified.

- Retrieval: Cases that have all or some of the required features are identified.

- Selection: Retrieved cases are evaluated and then ranked by order of similarity.

The adaptation process recognizes the differences between the selected design and the new problem for which a design solution is sought. Once the necessary changes are identified, they are then made. With regard to indexing cases, inseparability is an important issue (McSherry 2002). Inseparability occurs when a CBR system is unable to distinguish between two cases (i.e., two cases have the same values for all attributes). However, it is unlikely that both cases will be equally suited to the current design requirements. Inseparability is caused by either having too few indexing attributes or by selecting a poor choice of attributes (i.e., the attributes and/or their values are common to all or many designs and do not distinguish among designs). The indexing approach adopted by many has been to define attributes associated with the design problem. For example, fixture design cases have been indexed using attributes that described the workpiece for which a fixture is to be designed. Such attributes include machining features, inter-feature relationships, surface information, machining direction, and so on (Kumar 1995). However, there are few guidelines on choosing appropriate indexes, and the norm is for the designer to determine appropriate indexes via experience. Thus, there is a need to develop a formal methodology for determining case indexes that clarify design requirements.

Many CBR systems base case recall on attribute similarity. That is, they employ a nearest-neighbor approach using standard weighting techniques (Chang 2000; Varma 1999; Liao 2000). However, high similarity between cases does not necessarily result in a case that can be easily adapted. Indeed, a less similar design case may in fact be more readily adapted in certain design situations (Leake 1996). Attempts have therefore been made to tie other components of the CBR process more closely to adaptation (Smyth 1995, 1996; Rosenman 2000). For example, Smyth and Keane (1995) have based retrieval directly on the likelihood of adaptability. They determined what adaptation strategies were required to fix a particular case, and retrieved the design with the most favorable adaptation strategy. However, adaptability-based retrieval is a computationally expensive approach for two reasons. One, the approach requires the CBR system to determine what changes need to be made to fix a case, to then decide how this change can be achieved (as there may be several means of affecting a change), and to check how making a change will affect the rest of the design. This is a far more complicated process than merely checking attribute similarity, as it requires the prophetic ability to determine the effect of design decisions. Second, the approach can require the navigation of a very large search space, leading to control problems. Thus, some method of initially constraining the search space needs to be defined. Keeping the library small is very important in adaptability-based retrieval, which is in contrast to similarity-based retrieval in which a large number of similar design cases are desired to improve the likelihood of finding a suitable design solution.

During case retrieval, the attributes of several designs are assessed and ranked to determine which design is best suited to the current design requirements. A standard method employed by many CBR systems (Chang 2000; Liao 2000) is to use a weighting approach. This involves determining the difference between a required attribute, i, represented as $(P_{i,\max})$ and the recalled attribute (P_i), attaching a weighting factor stating the importance of this attribute (w_i) and then calculating an overall figure of merit (FOM) for the recalled attribute. This process is repeated for n attributes until an FOM for the complete recalled case can be computed using the equation

$$\text{FOM} = \left[\sum_{i=1}^{n} W_i (P_i / P_{i,\max}) \right] \bigg/ \sum_{i=1}^{n} W_i \qquad (3.15)$$

This is a fairly simple technique widely used in CBR systems and in many decision making operations as a whole. However, a significant limitation is that a linear relationship is assumed between the performance level of a feature and the worth of that level to the designer. However, a recalled attribute may only become worthwhile to a designer when it is very close to the required level. Consider the example shown in Figure 3.45. A clamping unit is required to exert a force of 114 lbs. A recalled case may be able to exert a force of 60 lbs, giving it a worth value of 0.55 using the linear weighting approach. However, in reality such a low force capability may be of little value to the designer. A parabolic-style curve, represented by the dotted line in Figure 3.45, may be a more accurate reflection of the worth of various force values. The worth of the recalled attribute is in fact much lower than 0.55 at 0.18 (weighting attribute runs from 0 to 1). Thus, the linear weighting approach offers a limited means of expressing a designer's preferences or the importance of a design attribute.

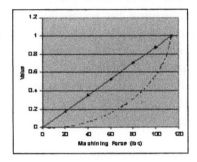

Figure 3.45. Linear-based weighting.

On a similar note, an inherent feature of any CBR system is its learning capability. The case base is constantly growing, with new cases being added to it. Obviously, this can exacerbate the control problem highlighting the need for an effective vetting method and a carefully controlled learning mechanism that restricts case library growth. Keeping the library small is very important in adaptability-based retrieval, which is in contrast to similarity-based retrieval (for which a large number of similar design cases is generally desired to improve the likelihood of finding a suitable design solution). Thus, the criteria by which cases are considered for addition to the case base need to be developed carefully and need to focus on whether a case contributes essential or valuable knowledge to the library.

A further approach to evaluating design cases can be found within axiomatic design (Suh 2001). There are two fundamental aspects to axiomatic design. The first is the systematic decomposition of design requirements, known as functional requirements (FRs) the design must perform, and corresponding solutions, known as design parameters (DPs). The second aspect relates to the evaluation criteria used to judge design alternatives. Selection among design options is the following two axioms:

■ *Axiom 1, The Independence Axiom*: Maintain the independence of FRs so that each FR can be controlled independently with a DP such that no other FR is affected.

- *Axiom 2, The Information Axiom*: Minimize the information content of the design where the information content is a measure of the probability of success in achieving specified FRs.

One important point to note is that Axiom 2 is only applied when Axiom 1 has been satisfied. The systematic decomposition of design requirements/solutions advocated by axiomatic design is a powerful technique and is incorporated into CAFixD as a method of expressing FRs and DPs. However, certain concerns arise from the axioms used to evaluate designs. First, the second axiom requires that a probability density function (pdf) for each DP be known. These pdfs illustrate the likelihood of a particular DP achieving its FR. In reality, however, these pdfs are unknown or are difficult to obtain. As a result, Axiom 2 is often not used during the design process and possible design solutions are judged against one another purely on the basis of Axiom 1. Where Axiom 2 is used, the pdf is often simplified. Jang et al. (2002) used Axiom 2 to evaluate potential solutions when designing a foil-strut system. Their FRs were to minimize the bending moment experienced by the system and to minimize the weight. Various designs were evaluated on the basis of simplified pdfs that for convenience were assumed linear. However, questions remain regarding how accurate these simplifications can be and how accurate they need to be without having a negative impact on the validity of the evaluation of designs using the second axiom. In essence therefore, two issues arise from Axiom 2. The first is how the necessary pdfs can be generated, and the second is how assumed pdfs can be verified as accurate.

A second concern arises from the limited role constraints play in the design process. Constraints tend to be aspects of design considerations such as cost, weight, and so on. No formal method is proposed in axiomatic design for handling constraints. Essentially, as long as constraints are not violated then decisions are made purely on the basis of the two axioms. However, involving constraints more actively in the design process can alter the design solution, as found by Pena-Mora and Li (2001). They applied axiomatic design as a method of designing workplans for fast-track construction projects. Designs were evaluated on the basis of their constraints using Axiom 2. The constraints considered were the financial cost associated with a certain course of action and the time it would take to perform a task by following this course of action. Using Axiom 2, Pen-Mora and Li evaluated two design alternatives in terms of their envisaged time and cost performance levels. The net result was that after applying the second axiom they were unable to make a choice between the two design alternatives. They concluded that the choice of design solution depended on which constraint criterion was more important, although they did not offer any specific technique for making the final decision. Thus, one feature of axiomatic design is that all constraints and indeed FRs are considered equally important.

In reality, however, this is rarely the case. Some functions or aspects of a design will always be more important than others, and compromise and trade-offs are intrinsic to the design process. Standard weighting techniques are often used to assign relative importance to specific aspects of a design requirement, but axiomatic design does not offer this possibility because the incorporation of weights violates the integrity of Suh's equation for calculating information content (Suh 2001). Suh recommends that to highlight the importance of a particular functional requirement the designer should simply assign a tighter design range (i.e., a tighter acceptable range of performance values for an FR). However, the actual performance value of an FR and the importance of that FR are two fundamentally different aspects of a design requirement that cannot be arbitrarily lumped together by tightening the acceptable performance range. Tightening the range will indeed highlight and increase the impact of that FR's effect on the information content of a design, but this impact is not an accurate reflection of how important this particular FR is to the designer. This method essentially involves artificially altering a design requirement to suit a particular design method. However, a designer should not have to alter a design requirement just to suit a particular design approach.

3.5.2   Design Methodology

The previous section presented a critique of various CAFD approaches. This section summarizes the objectives of the CAFixD methodology, details how these objectives are met, and discusses the CAFixD approach. This section also provides an overview of the CAFixD framework and an in depth description of the indexing mechanisms used therein.

3.5.2.1   CAFixD Objectives

The objectives of the CAFixD approach are as follows.

- To develop a CAFD technique that incorporates a full understanding of the complete operational requirements of a fixture into the design process.
- To use a full understanding of design requirements to generate complete fixture designs that fully detail the physical structure of locating/clamping units.
- To address the inseparability issue within CBR by developing a formal method for determining the indexing attributes of a design case.
- To develop a retrieval method that is computationally feasible, has a well-defined control mechanism to restrict control of the search space, and has the greatest probability of returning a satisfactory design solution.
- To develop a method that can effectively measure adaptability and gauge the effect of potential design decisions.

To achieve these objectives, the following needs to be studied.

- The concept of design requirement decomposition is used to produce a complete list of functional requirements of the fixture design problem. This accomplishes two goals: (1) functional requirements can be used as a thorough indexing mechanism for design cases thus alleviating the problem of inseparability and (2) the thoroughness of this requirement decomposition technique allows the designer to fully define the total operational requirements of a fixture and sub-sequently use this comprehensive specification to drive and guide the design process.
- Emphasis is given to adaptability-based retrieval to help ensure that a satisfactory design solution is achieved. However, similarity-based retrieval is still used to vet possible design solutions and to help constrain the search to prevent control problems arising during navigation of the solution space.
- A data structure called the "second layer of the design matrix" is proposed as a means of identifying possible adaptations required to fix a design and of gauging the effect of potential design decisions when evaluating designs in terms of their adaptability.

Overall, the CAFixD methodology (Figure 3.46) decomposes the design problem into a series of smaller problems, searches the case base for a solution to each individual problem, and then reconstitutes the individual solutions to form one complete solution.

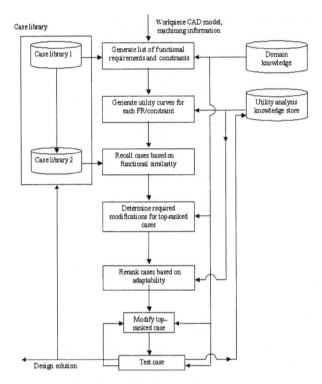

Figure 3.46. The CAFixD methodology.

The approach is similar to that adopted by a human designer, who would initially generate a conceptual design solution and subsequently fill in the details of that solution at a detailed design stage. Thus, CAFixD has two design case libraries. One contains conceptual design solutions and is used during the conceptual design stage, and the second contains detailed designs of individual fixture units and supports the detailed design stage. During retrieval, emphasis is given to evaluating the adaptability of design cases.

Initially, a series of design rules selects the appropriate conceptual design from case library 1. Workpiece and machining information are then processed to generate a list of FRs and constraints the design must satisfy. Utility analysis (Thurston 1991) is then used to guide the decision-making process during retrieval from case library 2. Utility analysis is similar to standard linear weighting but is considered a more expressive and accurate method of capturing a designer's preferences because it allows the designer to state nonlinear preferences.

Similar to weighting approaches, the output of utility analysis is a numerical measure of merit or utility (U) representing the relative desirability of a design alternative in regard to several attributes. Thus in CAFixD for each FR and constraint the designer must record preferences in the form of a utility curve, an example of which is shown in Figure 3.47. These graphs illustrate the utility of a given attribute based on its value. In Figure 3.47 for example, a weight of 6 lbs is of considerably greater utility than that of 16 lbs. The method by which these curves are generated is outside the scope of this section, but interested readers are directed to Boyle et al. (2003) for a description of the process. Using utility curves and the FR/constraint list, candidate fixture units are retrieved on the basis of functional similarity. Case library 2 contains design cases indexed by their FRs.

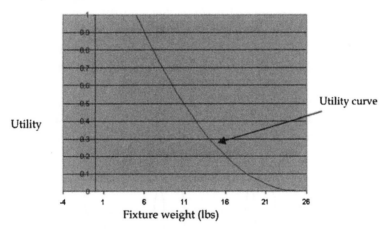

Figure 3.47. A utility curve.

The top-ranked cases are then reevaluated in terms of their adaptability. Specifically, cases are reevaluated in terms of the design decisions that will have to be made in order to meet the new design situation. The effect of these changes on the overall worth of the design is then used as the basis for retrieving the most suitable case. The case most favorable in terms of required modifications (i.e., modifications that result in the design of highest utility) is then proposed as the most suitable. Modifications are subsequently executed and the design tested to ensure that FRs are met. The design is also evaluated for possible addition to the case library.

3.5.2.2   Indexing Design Cases

Axiomatic design decomposition principles are used to determine the indexing of design cases and their solutions, as illustrated in Figure 3.48. Axiomatic design decomposition (Suh 2001) involves the processing of information across four domains. Mapping occurs among the customer domain, the functional domain, the physical domain, and the process domain. The needs of the customer are listed as customer attributes (CAs) in the customer domain and are subsequently formulated into a set of FRs. A design solution is then created through mapping between FRs and DPs, which exist in the physical domain. These DPs are mapped onto process variables (PVs). A fundamental aspect of the mapping process is the idea of decomposition. The design process progresses from the highest level of abstraction down to a more detailed level. This results in the formation of design hierarchies in the FR, DP, and PV domains.

In fixture design, CAs represent workpiece and machining information. This maps onto a list of FRs that explicitly state the functions the fixture design must perform. These FRs relate to all desired fixture operational requirements. In addition, constraints (e.g., fixture cost) may also be included here. The FRs map onto DPs, which are the individual fixture units used to achieve the FRs. The PVs in turn are base design parameters (for example, the thickness of a locator) used to achieve the DPs. The significant difference between this and standard CBR approaches is that the functions of the sought design solution are explicitly stated. Normal CBR approaches map directly from CAs directly onto DPs.

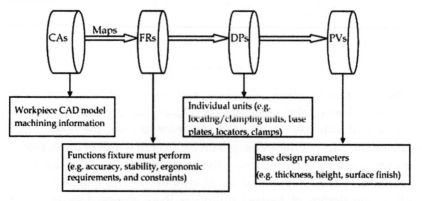

Figure 3.48. Axiomatic domains applied to fixture design.

0.0.0.1 Design Case Libraries

The high-level design of the case library is diagramed in Figure 3.49. The case base consists of two libraries. Case library 1 is related to fixture planning. It stores conceptual fixture designs largely in terms of their locating principles. The second case base holds the individual units that constitute the fixture design. Examples include locator support units and clamp types. The approach adopted is to navigate through case library 1 to retrieve a conceptual design, before proceeding to the second case library to retrieve appropriate fixture units. Thus, the output from case library 1 constrains the search through case library 2 in that only units that can be used in the retrieved locating principle are considered potential solutions.

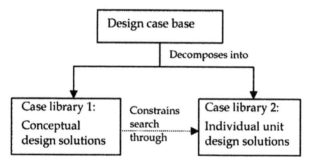

Figure 3.49. The design case base.

Case Library 1. The structure of case library 1 is depicted in Figure 3.50a. It contains cases that are conceptual in nature (i.e., they contain information related to locating principles in terms of locating methods and locating point distributions). There are three basic locating methods: plane, pin-hole, and external profile locating. For each method there are subsequent decompositions and refinements of the root locating method. For example, plane locating (3-2-1) has seven variations (Figure 3.50b).

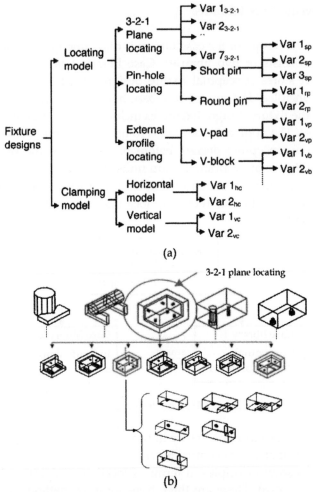

(a)

(b)

Figure 3.50. Case library 1 decomposition (a) and
the seven variations of plane location (b).

The third variation is locating using six points of location. Three
locators provide primary location, two provide secondary location, and
one provides tertiary location. The primary points all act in the same
direction but can act on different planar surfaces, as can be the case with
the secondary locating points. Once the conceptual design has been
found in case library 1, the search for a design solution can proceed to the
second case library, where appropriate individual fixture units can be

retrieved for modification. To each leaf of case library 1 are attached skeleton FR sets.

FR Decomposition. Skeleton FR sets can be generated for each of the design cases in case library 1. A partial decomposition of the format of a design requirement for a simple 3-2-1 locating scheme is shown in Figure 3.51.

FR_1 – Locate workpiece to required accuracy
    $FR_{1.1}$ – Locate the workpiece
        $FR_{1.1.1}$ – Provide location directions – *6 FRs*
        $FR_{1.1.2}$ – Provide contact between locator and workpiece – *6 FRs*
    $FR_{1.2}$ – Control accuracy of locations
        $FR_{1.2.1}$ – Locate workpiece to required drawing tolerances – *6 FRs*

FR_2 – Support workpiece against machining forces experienced during machining
    $FR_{2.1}$ – Hold workpiece in situ during machining – *6 FRs*
    $FR_{2.2}$ – Support workpiece during machining – *12 FRs*

FR_3 – Ergonomic requirements
    $FR_{3.1}$ – Prevent damage at the fixture workpiece interface
    $FR_{3.2}$ – Channel coolant flow during machining
    $FR_{3.3}$ – Ease the loading/unloading of the workpiece into/from the fixture
    $FR_{3.4}$ – Assist tool positioning during machining
    $FR_{3.5}$ – Error proof the fixture

Constraints:
C_1 – fixture cost
C_2 – fixture weight
C_3 – workpiece loading time
C_4 – workpiece unloading time
C_5 – fixture assembly time

Figure 3.51. A partial FR decomposition.

FRs are grouped into three main categories. One group deals with locating accuracy requirements, the second with stability requirements of the fixture, and the third with ergonomic issues related to fixturing. The first two groups are the simplest to handle in terms of automating their generation. The locating principle determines the number of units in the fixture design. As there are six units associated with variation 3 of 3-2-1 locating, there are therefore six FRs related to the accuracy of the locating units ($FR_{1.2.1}$), six FRs related to the clamping forces required to hold the workpiece against the locators ($FR_{2.1}$), and 12 FRs related to the stiffness of the six locating and six clamping units ($FR_{2.2}$). A tolerance analysis of the workpiece is performed to determine the performance values of the locating accuracy FRs. Similarly, a simple force analysis of the machining forces allows the performance values for the stability FRs to be defined. The most significant problem is related to the third group of FRs, which involves ergonomic considerations. These FRs include design requirements such as chip shedding, error proofing, workpiece surface protection at the locator/workpiece interface, and assisting tool positioning during machining. These need to be user specified and are created interactively with the designer.

Case Library 2. The second case base contains information related to individual fixture units (i.e., information related to an individual clamping or locating unit, including where it can be used). Figure 3.52 presents a partial breakdown of the case library, which contains locating units, clamping units, locator types, and fixture base types that can be combined to create a complete fixture for a workpiece.

Each step down the hierarchy represents a refinement of the unit design. For example, locators can be organized as two types (horizontal locators and vertical locators) depending on the direction of support they provide. Horizontal locators can be subsequently decomposed into two types, designated HL01 (a requirement for step-over locating units) and HL02 (a requirement for simple locating units). HL01 represents the situation in which a face exists below and extends beyond the locating face. HL02 represents the situation in which there is a face above or no face above the locating plane. Each of these locating situations requires specific types of locating units. HL01 requires the use of L-shaped locating units, whereas HL02 allows tower side-locating units and possibly L-shaped locator units to be used, an example of which is shown in Figure 3.52.

To each leaf of the case base are attached previous instances of locating and clamping units together with the FRs the unit is intended to fulfill. Also linked are the relevant PVs.

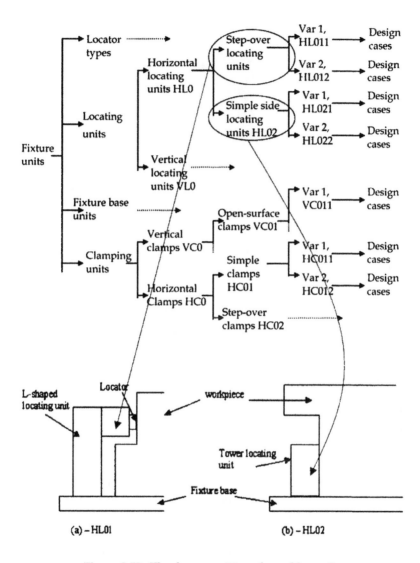

Figure 3.52. The decomposition of case library 2.

Indexing Design Cases: An Example. In this section, a fixture design for a caliper is used to illustrate how cases are stored in the case library. Figure 3.53 illustrates the caliper part and its corresponding fixture design.

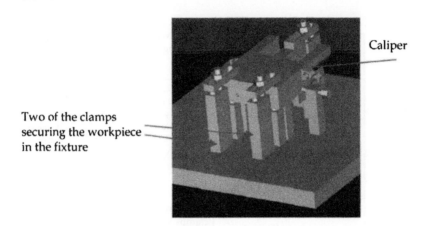

Figure 3.53. The fixture for a caliper.

The fixture has a 3-2-1 locating principle, by which two of the primary locators (P1 and P2) act on the same planar surface on the bottom surface of the caliper, whereas the remaining primary locator (P3) acts on a different surface. Three strap clamps are used to clamp the caliper against the primary locators, whereas simple screw clamps are used to provide secondary and tertiary clamping. The secondary clamps sit entirely beneath the caliper.

Storing the Design in Case Library 1. The design example is a 3-2-1 fixture design. The two secondary locators act on the same planar surfaces, but only two of the primary locators act on the same planar surface. The third acts on a different surface. Thus, navigation through case library 1 is a 3-2-1 plane-locating design of type 3. The primary locators are split, and thus the designation of the primary location scheme as $XSplit1_{Var3,3\text{-}2\text{-}1}$ in the decomposition of case library 1 shown in Figure 3.54a. The two secondary locators (S1 and S2) act on the same surface, and thus their designation as $YSplit0_{Var3,3\text{-}2\text{-}1}$. The tertiary location (T1) is indexed as $ZSplit0_{Var3,3\text{-}2\text{-}1}$. Figure 3.54b illustrates the locating points on the part.

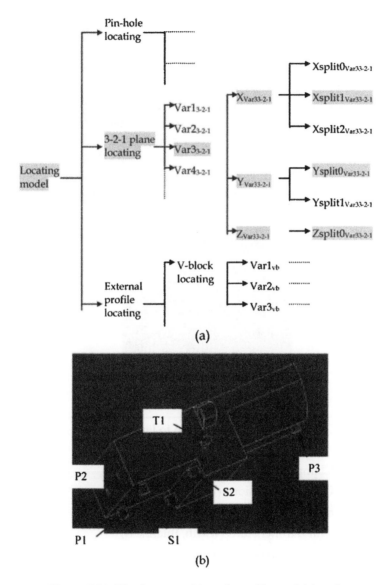

(a)

(b)

Figure 3.54. The decomposition of case library 1 (a) and
the workpiece locating points (b).

Storing the Design in Case Library 2. Case library 2 contains the designs
of the individual units that constitute the complete fixture design. This
section explores the manner in which these individual units are stored in
the case library. One of the primary clamping units (Figure 3.55a) is used

as an example. The clamp consists of the clamping support unit, the fixture base, the clamp arm, the nut/screw assembly, and the mounting pin for the arm. Within case library 2, it is denoted as a vertical clamping unit, type 01, variation 1 (i.e., VC01Var1).

The unit has two FRs associated with it, one of which is related to stability and assumes the form shown in Figure 3.55b. The FR requires that the clamping unit be capable of withstanding a force of 100 lbs acting against it such that displacement of the clamp surface is restricted to less than 0.002 inch. The corresponding DP is the clamping stiffness, which can be achieved through multiple PVs related to the physical and geometric properties of the individual entities that comprise the clamping unit.

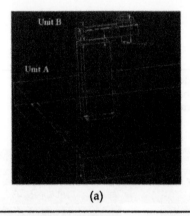

(a)

FR$_{2.1.2.1}$ – Control w/piece translation to less than 0.002 in.
        (+ve Z direction) at loc P1 under 100 lbs
DP$_{2.1.2.1}$ – Clamp C1 stiffness of 16.67E6 lb/in.^{2}
PVs are: Unit A dimensions (width, thickness)
         Base plate dimensions (thickness)
         Unit B dimensions (width, thickness)
         Material properties of base plate
         Material properties of main units

(b)

Figure 3.55. A primary clamping unit (a) and its FR (b).

Storing Relationships Among the FRs, DPs, and PVs. During retrieval and adaptation, relationships among FRs, DPs, and PVs must be known. Axiomatic design uses one matrix to store the relationship between the FR and DP domains and another matrix for the DP and PV domains. The CAFixD approach, however, advocates a format called "the second layer" that explicitly shows the relationship among the FRs, DPs, and PVs in a single representation while simultaneously highlighting the effect various types of constraints have on the design solution.

Figure 3.56 presents this second layer. The second layer illustrates how the PVs, DPs, and FRs are related. Thus, $PV_{2.17}$ has an effect on $DP_{2.7}$ determined by the relationship $RelDP_{2.7.1}$. $DP_{2.7}$ then acts as the input to $RelFR_{2.7.1}$, which controls the level of $FR_{2.7}$. However, $DP_{2.7}$ can also be controlled by $PV_{2.18}$ and $PV_{2.19}$ through their respective relationships $RelDP_{2.7.2}$ and $RelDP_{2.7.3}$. The second layer also illustrates clearly the roles of the constraints. Two types of constraints exist. Local constraints exist on PVs. For example, C_{10} and C_{11} are local constraints that limit the value a PV can have. Thus, if $PV_{2.19}$ is a thickness, C_{10} will be the minimum allowed thickness and C_{11} the maximum. These local constraints act upon the PVs.

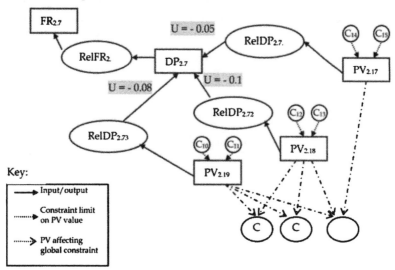

Figure 3.56. The second layer of the design matrix.

Global constraints pertain to the fixture design as a whole and include items such as total cost, weight, and so on. The PVs may or may not have an effect on these constraints, but these are limits on the fixture design as a whole. For example, altering $PV_{2.19}$ will increase cost (C_1), but there is no limit on the cost of the thickness change itself. It may be possible to offset the increased cost of the thickness change elsewhere in the design. In this way, the second layer allows the designer to gauge the global effect of local decisions. The effect of design changes on global constraints is measured in terms of the change in utility (U) of a design caused by making such a change. This change in utility is determined during the adaptability-based retrieval stage.

3.5.3   Retrieving Design Cases

The decision-making process within CAFixD's retrieval stage is guided by utility analysis. It attempts to provide a more quantitatively accurate numerical technique for making decisions than is possible with standard weighting techniques. The significant advantage utility analysis has over using weighting factors as the basis for evaluation is that it creates a utility curve that is nonlinear, which provides greater freedom with which to express preferences. Thus it allows for the fact that a designer's preferences may change depending on the particular value of an attribute. Figure 3.57 illustrates the differences among the standard weighting techniques, axiomatic design, and utility curves in terms of how accurately each approach can represent a designer's preferences.

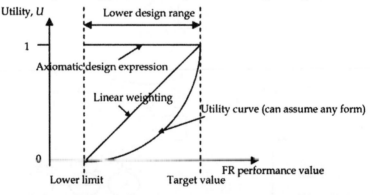

Figure 3.57. Expressing a designer's preferences.

A design range has been specified around a target FR performance value. According to axiomatic design principles, as long as a design performs within the specified design range the designer has no varying preference. Essentially, it makes no difference if the design performs at the target value or at either end of the design range. Thus, a black-and-white view of the world is assumed. Linear weighting allows the designer to say that the closer to the target value a design option gets (i.e., the higher its perceived utility) the more highly the designer rates that design. However, only a linear relationship between utility and performance is possible. Utility analysis, however, provides the designer with an opportunity to fully express a preference. The utility curve can assume any shape, including linear, quadratic, logarithmic, exponential, or indeed any curved or linear form. Thus, it is a powerful tool with which to express a designer's preferences.

The process adopted for utility analysis is to determine the utility curve *for each FR* and then combine all utility functions using the following multiplicative (Thurston 1991):

$$U(X) = \frac{1}{K}\left[\left[\prod_{i=1}^{n}(Kk_iU_i(X_i)+1)\right]-1\right]$$

(3.16)

$U(X) =$	overall utility of the complete set of FRs
$X_i =$	performance level of each FR (e.g. tolerance requirement performance level is 1.5mm with respect to X, Y, and Z)
$X =$	set of FRs at levels (x_1, x_2, ... x_n)
$K_i =$	assessed single FR scaling constant (i.e. the importance of each FR relative to the others)
$U_i(X_i) =$	assessed single FR utility curve
$i =$	1, 2, ... n FRs
$K =$	scaling constant

K is obtained by normalizing $U(X)$ in the standard way:

$$1 + K = \prod_{i-1}^{n}(1 + Kk_i)$$ (3.17)

The key tasks are to generate the single FR utility curves and the assessed single FR scaling constants. The utility curve $U_i(x_i)$ expresses the worth of varying levels of each FR in isolation. The scaling factor k_i relates to the trade-off between attributes the decision maker is prepared to make. Both are created through the decision maker answering a series of lottery questions.

Case retrieval occurs in two stages. Initially, cases are retrieved on the basis of functional similarity. Cases are retrieved that have FRs similar to the current design requirement. The utility curves generated for each FR are used to determine the worth of each design case in terms of its FR performance value. To recap, a design case is an individual unit, such as a locating or clamping unit. A complete fixture design consists of a number of these units. The design cases with the highest utility are functionally the most similar. This is determined at what is essentially a vetting stage that removes cases unlikely to provide a suitable solution. The second stage of the retrieval process then reevaluates the top-ranked cases in terms of the necessary changes that must be made to a case and how this will affect the final design solution in terms of its ability to meet the constraints specified by the designer at the start of the process. Adaptation knowledge in conjunction with utility analysis is used to evaluate possible design solutions in this way.

During retrieval and adaptation, the relationships among FRs, DPs, and PVs must be known. Axiomatic design uses a matrix to store the relationship between the FR and DP domains and a second matrix for the DP and PV domains. The CAFixD approach, however, advocates a format called "the second layer" that explicitly shows the relationship between the FRs, DPs, and PVs in a single representation, while simultaneously highlighting the effect various types of constraints have on the design solution.

Figure 3.58 shows this second layer. The second layer illustrates how PVs, DPs, and FRs are related. Thus, $PV_{2.17}$ has an effect on $DP_{2.7}$ determined by the relationship $RelDP_{2.7.1}$. $DP_{2.7}$ then acts as the input to $RelFR_{2.7.1}$, which controls the level of $FR_{2.7}$. However, $DP_{2.7}$ can also be controlled by $PV_{2.18}$ and $PV_{2.19}$ through their respective relationships $RelDP_{2.7.2}$ and $RelDP_{2.7.3}$. The second layer also illustrates roles of the constraints. Two types of constraints exist. Local constraints exist on the PVs. For example, C_{10} and C_{11} are local constraints that limit the value a PV can have. Thus, if $PV_{2.19}$ is a thickness, C_{10} will be the minimum allowed thickness and C_{11} the maximum. These local constraints act upon the PVs.

Global constraints pertain to the fixture design as a whole and include items such as total cost, weight, and so on. The PVs may or may not have an effect on these constraints, but these are limits on the fixture design as a whole. For example, altering $PV_{2.19}$ will increase cost (C_1), but there is no limit on the cost of the thickness change itself. It may be possible to offset the increased cost of the thickness change elsewhere in the design. In this way, the second layer allows the designer to gauge the global effect of local decisions. The effect of design changes on global constraints is measured in terms of the change in utility (U) of a design caused by making such a change. This change in utility is determined during the adaptability-based retrieval stage.

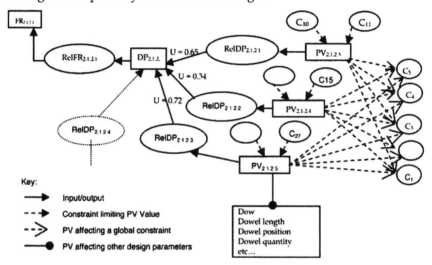

Figure 3.58. Updated second layer of the design matrix.

During this stage, the change that must be made to a PV is determined using the second layer of the design matrix. Consider Figure 3.58. It has been determined that to meet $FR_{2.1.2.1}$, a case in the case base must alter the corresponding design parameter $DP_{2.1.2.1}$. There are various PVs by which this DP can be altered by the required amount. These include $PV_{2.1.2.3}$, $PV_{2.1.2.4}$, and $PV_{2.1.2.5}$. The objective is to determine which of these PVs is the most desirable with which to control the DP.

To achieve the required DP level, the change in $PV_{2.1.2.3}$ (the thickness of a locator for example) is 1 inch. From the design matrix, the relationship between this PV and the constraints on the fixture design are known. Thus, for example, the new weight of the fixture as a result of increasing the locating unit thickness is known. A utility curve is associated with each constraint. This utility curve states the worth of this attribute to the design based on the value of the attribute. In this case, the attribute is the weight of the fixture.

A utility value can therefore be obtained for a design change based on its effect on the constraints existing in the design. The process is repeated for all constraints or attributes affected by a single design change, and the resulting single-attribute utility functions are then combined to determine the overall utility $U(X)$. This combined figure then represents the worth of that design change to the designer in terms of the overall worth of the adapted design.

This can be repeated for all possible design changes ($PV_{2.1.2.3}$, $PV_{2.1.2.4}$, and $PV_{2.1.2.5}$) and associated utilities recorded. The second layer of the design matrix can then be updated (as illustrated in Figure 3.58) to show the worth of the fixture design for each of the possible design changes. Thus, the calculated utilities for each design change are 0.65 for $PV_{2.1.2.3}$, 0.34 for $PV_{2.1.2.4}$, and 0.72 for $PV_{2.1.2.5}$. The most desirable design change is therefore to select $PV_{2.1.2.5}$ as the PV by which to control $DP_{2.1.2.1}$.

The second layer can also show the design decisions affected by making this change (e.g., the dowel diameter must change), in addition to the length, position, and quantity of dowels. For example, altering $PV_{2.1.2.5}$ results in other design decisions, such as dowel diameter modifications (as illustrated in Figure 3.58). By knowing the future design decisions that will have to be made, a knowledge base can be constructed of design decisions favorable in terms of preferences regarding the design process itself and not just the designed artifact. For example, although one modification to a retrieved case may be favorable in terms of the designed product's attributes it may be unfavorable in terms of the amount of time it takes to design it, or it may result in a large number of propagated design decisions that will have to be made. Thus, from a product performance point of view a decision may be desirable, but from a design process point of view it may be of little worth. Thus, one possible extension of the approach outlined previously would be to formulate utility curves that express preferences regarding the design process, and not just preferences regarding the designed artifact and its performance. Currently, such factors are only considered under the umbrella of the design cost of the fixture.

Through listing the possible ways of adapting a design to meet a specific FR in this way, the second layer presented in Figure 3.58 acts as a record of design rationale. The various possible courses of action have been outlined, and the reasons for choosing one over another have been recorded in terms of the utility value associated with each major design decision. It is therefore possible to explain why a specific option was chosen from a number of possible alternatives.

3.5.4 Worked Example

A worked example has been conducted in which a fixture was generated for the part shown in Figure 3.59a. Features to be machined are the two channels and the hole. Initially, the appropriate conceptual design is retrieved from case library 1 based on workpiece geometry. Standard fixture design rules support this process, and for this workpiece a standard 3-2-1 locating model is retrieved along with its skeleton FR set defining tolerance and force stability FRs.

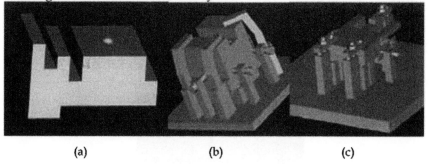

(a) (b) (c)

Figure 3.59. The workpiece (a) and two complete fixture designs (b and c).

Upon generation of the list of functional requirements, the designer defines design preferences in the form of utility curves for each FR and constraint. Case library 2 contains all individual fixture units of the two complete fixture designs presented in Figures 3.59b and 3.59c. Fixture design 1 is shown in Figure 3.59b. It is a simple 3-2-1 locating principle fixture with three identical primary locators acting on the bottom surface of the workpiece. Three hydraulic clamps perform primary clamping. Two identical screw clamps perform the secondary clamping requirements and act against two secondary locators, both of which act upon the same surface and are identical. A step-over side locator provides tertiary location, and a screw clamp provides tertiary clamping. Fixture design 2 (Figure 3.59c) is a fixture that holds a caliper in place during machining. The fixture has a 3-2-1 locating principle whereby two of the primary locators act on the same planar surface beneath the caliper, whereas the remaining primary locator acts on a different surface. Three strap clamps are used to clamp the caliper against the primary locators, and simple screw clamps are used to provide secondary and tertiary clamping. The secondary locators and clamps sit entirely beneath the caliper.

Retrieval occurs in two stages. Initially, retrieval is performed on the basis of functional similarity (i.e., how similar the FRs of the stored design cases are to the current design requirement). The utility curves are used to determine the worth of each design case on the basis of its FR value. The worth of a complete fixture design can be obtained by combining the utility values for each FR of that design. This stage is essentially a vetting process. The most functionally similar cases are then reranked on the basis of their adaptability, as described previously. The results from the study are presented in Figure 3.60a.

	Complete		Synthesized	
	Design 1	Design 2	Design 3	Design 4
Overall utility (similarity)	0.898	0.925	0.939	0.898
Overall utility (adaptability)	0.62	0.524	0.602	0.64

(a)

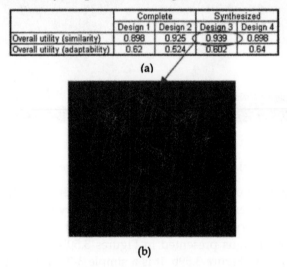

(b)

Figure 3.60. Retrieval results (a) and design solution (b).

The methodology would normally generate designs by searching for the most appropriate design solution (individual fixture units) for each FR before combining them to form one complete fixture design. Designs 3 and 4 represent the results of the standard approach of the method. For purposes of comparison, however, the two complete fixture designs presented in Figure 3.59 were each considered as a whole during retrieval (i.e., they were not decomposed such that the best parts of each design could be assembled to form a new design). Fixture design 3 is a synthesized design found by combining the most functionally similar units of fixture designs 1 and 2. Fixture design 4 (Figure 3.60b) represents a complete design compiled from individual units that each had the highest utility in terms of adaptability. This new synthesized case was the best case in terms of retrieval by adaptability.

To illustrate the retrieval process, following explores an example that focuses on one particular FR for which a design solution is sought, with both functional similarity and adaptability-based retrieval mechanisms detailed. CAFixD has generated a conceptual design and a list of FRs and constraints the design solution must satisfy for the workpiece presented in Figure 3.61.

The conceptual design requires a fixture that contacts the workpiece at six positions. Therefore, the design solution requires details of six individual locating units. The designer's preferences have also been recorded in the form of utility curves. The number of FRs associated with a fixture design can vary, but it is not unusual for there to be approximately 30 FRs for each design. For each of those FRs, CAFixD now determines those cases in the library that are most similar to the current design requirement. The FR that will be used to illustrate the retrieval process is related to the stiffness of one of the locating units.

$FR_{2.1.2.4}$ – Provide a supporting stiffness of 6.67Mlb/in at locating point 5 (one of the six locating points)

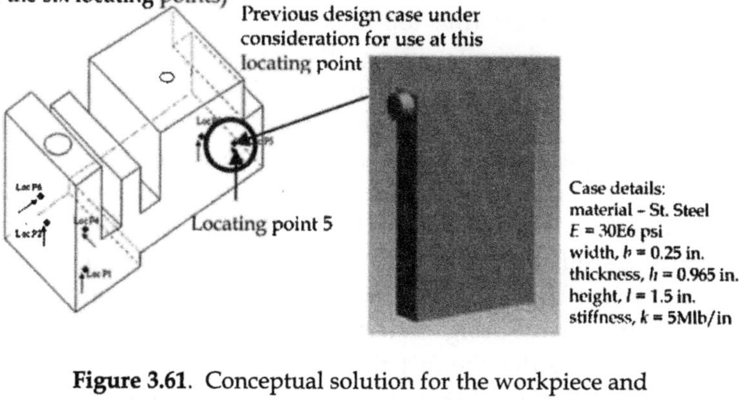

Case details:
material – St. Steel
$E = 30E6$ psi
width, $b = 0.25$ in.
thickness, $h = 0.965$ in.
height, $l = 1.5$ in.
stiffness, $k = 5$Mlb/in

Figure 3.61. Conceptual solution for the workpiece and the fixture unit being evaluated as a solution.

The utility curve for this FR (and its governing equation, which in this case happens to be a quadratic) is shown in Figure 3.62. An individual fixture unit design held in the case base is indexed by an FR having a stiffness value of 5 Mlb/inch (Figure 3.61). Interpolating from the utility curve reveals that in terms of functional similarity this stored case has a utility worth value of 0.5223 in the current design situation. The search through the case base would continue, and for each FR the top three design cases (ranked in term of similarity) would be retrieved for further consideration on the basis of their adaptability.

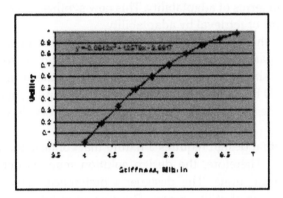

Figure 3.62. The utility curve for $FR_{2.1.2.4}$.

The next stage is to evaluate the top-ranked cases for each FR by determining the effect the changes needed to fix the design have on the ability of the complete fixture design to meet the constraints. In this design, there are only two constraints (fixture weight and cost) whose utility curves are shown in Figure 3.63.

Initially, CAFixD combines all utility values of the design cases that were judged most similar (the top-ranked case from the three cases retrieved for each design), using the multiplicative expression represented by Eq. 3.16, and presents this complete design as the starting point for a solution. CAFixD then proceeds to determine the physical changes that must be made to each case among those that constitute the design in order to make it suitable for the current requirement. It also determines the effect these changes will have on overall utility. The cases that have the most advantageous effect on the utility value will be chosen as the final solution. The second layer of the design matrix is employed to facilitate evaluation of the design in terms of its adaptability (Figure 3.64).

For example, consider the fact that the locator unit detailed in the functional similarity retrieval stage was one of the three top-ranked cases for $FR_{2.1.2.4}$. It has to be altered such that the stiffness (the FR) of the unit (the DP) is increased until it assumes a value of 6.67 Mlb/inch. There are three PVs with which to do this. These are the material stiffness of the locator material and the thickness (or width) of the locator unit. The approach is now to determine the modification of each PV required to achieve the desired FR value. The effect of each change on the design constraints (cost and weight) can then be calculated.

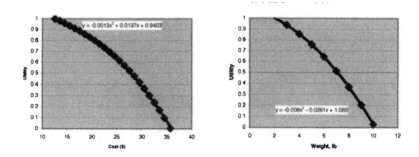

Figure 3.63. The utility curves for cost and weight.

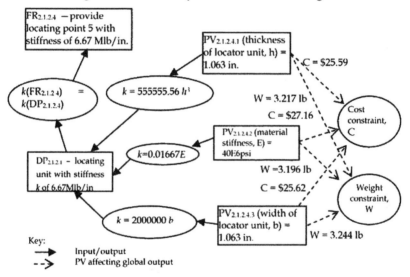

Figure 3.64. The second layer as generated during adaptability-based retrieval.

Initially, the unit thickness can be used to increase the strength. Making the simple assumption that the locator units act in a manner similar to a simple cantilever beam, whereby deflection is due to bending and shear effects are ignored, the thickness of the locator unit (h) can be related to its stiffness (k) by

$$k = \frac{3EI}{l^3} = \frac{3Ebh^3}{12l^3},$$

(3.18)

where E is the modulus of elasticity of the locator material (which in this case is steel), b is the width of the locating unit, l is the height of the locator, and I is the moment of inertia of the locating unit. Substituting the known values for E (30E + 6 psi), b (0.25 in.), and l (0.965 in.), and the desired value for stiffness k (6.67 Mlb/in) yields

$$k = 555555.56h^3 \Rightarrow h = \sqrt[3]{k/555555.56} = 1.063 \text{ in.}$$

Thus, to meet the stiffness requirement this locating unit must have its thickness increased from 0.965 inch to 1.063 inches (an increase of 0.097 over the retrieved case). The cost and weight of the fixture are directly related to the quantity of material used:

Cost = f(material volume) = volume cost of material per unit ✖ volume

Weight = f(material volume) = material density ✖ material volume ✖ gravitational constant

Increasing the thickness of the locator increases the amount of material required for the fixture, which results in a corresponding increase on both cost and weight of the fixture (as detailed in Table 3.5). The utility values of the new weight and cost of the complete fixture can be determined from their respective utility curves, and combined with the single-utility expression (Eq. 3.16) to determine their overall effect on the design. The process is then repeated, but instead of changing the thickness of the locator a different material with an alternative stiffness (E) value is used to increase the locating unit stiffness. From Eq. 3.18, the relationship between stiffness and material stiffness (E) is

$$k = \frac{3Ebh^3}{12l^3} = 0.01667E \Rightarrow E = k/0.01667 = 40Mpsi . \rightarrow$$

Table 3.5. Adaptability-based retrieval results.

	Thickness Modification	Material Stiffness Modification	Width Modification
Required Change	0.965 →1.063 in.	30 → 40 Mpsi	0.25→0.33 in.
Resulting Cost ($)	25.59	27.16	25.62
Resulting Weight (lb)	3.217	3.196	3.244
Resulting Utility (U)	0.7508	0.7078	0.7494

A suitable material identified was chromium (E = 42E6 psi against steel's value of 30E6 psi). Chromium and steel differ in cost and density, and thus this change will have a direct impact on the cost and weight of the fixture unit. These effects can be measured. Similarly, the change (and effect of the change) that has to be made to the width of the unit can also be obtained. All three changes have a negative effect on the utility, but one must be chosen to allow the design case to meet the FR. Altering the thickness of the unit has the least negative impact and would thus appear to be the best candidate. However, before making a final decision the other design cases that were retrieved as potential solutions for this FR need to be evaluated in the same manner. Once all retrieved units for an FR have been evaluated and compared, the one that has the most advantageous effect on the utility of the overall design is selected as the solution for that particular FR. The process is then repeated for all remaining FRs and the complete fixture design is put together in this piecemeal fashion.

3.5.5   Summary

This section has presented the CAFixD methodology, which supports the fixture design process. It seeks to address three main issues. First, it considers the complete operational requirements of a fixturing problem. Second, the approach adopts a rigorous indexing technique in an effort to prevent the problem of inseparability. Third, it uses adaptability to guide retrieval by choosing those cases whose adaptation will have the most desirable effect on the overall design solution, while still using traditional attribute similarity to constrain the search space to prevent control problems.

CAFixD is a software tool that communicates with other CAD/CAM packages necessary to pass or receive information from external sources. The purpose of CAFixD is to process this information, use it to generate a fixture design, and then pass on the details of this fixture design to a CAD package that will create the fixture design drawings. System development focuses on two main areas. The first is that the decomposition-reconstitution approach results in the dynamic creation of constraints during the adaptation stage. These constraint effects must be incorporated into the evaluation of design adaptability. A simple example of this constraint generation is that checks must be performed to ensure that individual units are compatible with each other and can be combined to form a complete fixture. A second issue is the learning mechanism.

This is a key area of development. Due to the high levels of computation involved in CAFixD, work is ongoing to develop strict criteria for managing the growth of the case base.

3.6   SENSOR-BASED FIXTURE DESIGN AND VERIFICATION

3.6.1   Introduction

The main objective of sensor-based fixture design and verification (SFDV) is to confirm the fixture-part compliance in shop floor. Fixture design techniques currently focus on the setup planning, fixture planning, and configuration design and ignore the verification of the design in real time. This approach leads to a system that lacks robustness in the manufacturing environment. The goal of SFDV is to add a new aspect during the fixture design phase to reduce manufacturing ambiguities.

In today's world, everyone is striving hard to improve the quality and productivity of their products. A lot of emphasis is being placed on statistical quality control and six sigma techniques to improve quality. Statistical quality control methods show significant improvements for juvenile processes. However, if the process is mature the main source of defects is not process capability but human error. To achieve zero defect in any manufacturing system, an absolute necessity is implementing poke-yoke/foolproofing/error-proofing techniques. These techniques not only improve quality but also prevent defects before any additional value is added to a product.

Currently, poke-yoke techniques for fixturing are applied based on the experience and knowledge of the fixture designer. This research is focused on automating this process and exploring new possibilities using sensors in the fixtures.

3.6.1.1   Problem Description

The manufacture of precision parts requires accurate fixturing. Currently, in spite of having good manufacturing capabilities parts are often loaded incorrectly or inaccurately into the fixture. The manufacture of parts that involve fixturing complex surfaces especially in aerospace industries for the manufacture of turbine blades, vanes, and other parts need some verification method for location. Figure 3.65 shows a case in which the part has lost contact with locators.

Figure 3.65. A case in which a part has lost contact with locators.

Design verification methods for fixtures include fixture foolproofing (preventing incorrect loading), part location, contact verification, and design of locators with sensors.

Fixture Foolproofing. A part can be loaded into a fixture in a number of incorrect ways (Figure 3.66). The elimination of such occurrences through a robust fixture design is known as fixture foolproofing.

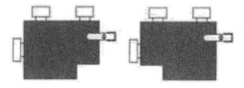

Figure 3.66. Correct (left) and incorrect (right) part loading.

The part CAD geometry is studied and the various features present are used to classify the part based on symmetry/asymmetry. Various rules and algorithms are used to determine solutions. The solutions ranging from a simple foolproofing pin to various sensor-based solutions are explored in order to select the most elegant solution.

Part Location-Contact Verification. Another area of focus is to verify proper contact/location of a part with locating elements after fixturing and during machining. This may be done by placing sensors at various locations to verify the part location. An integrated locator with a built-in sensor is also used to verify the part location. When and where sensor-based systems can be effectively used are also studied based on applications and production requirements.

Design of Locators with Sensors. To accomplish the foolproofing and contact verification tasks, locators are designed to incorporate sensors. Different types of sensors are used for various foolproofing and part verification applications. The integrated locators are studied under various conditions to calculate their optimum performance and limitations.

The major focus in current research is on automatic fixture foolproofing and part location verification. The fixture has to accommodate the sensor systems, and various factors (such as fixture ergonomics and accessibility) have to be considered. Another primary focus is on developing an integrated locator (with sensor) on a modular fixture platform. Such an integrated locator might well be created as a standard component for systems of the future. Sensor-based solutions are expensive and can be justified only by the requirements and economics involved in a particular manufacturing system.

3.6.1.2   Fixture Location Verification

The various published articles in the fixture verification area are studied in this section. Fixture location verification has been a primary research focus for a lot of researchers, with the main focus on locating performance, tolerance analysis, stability analysis, and accessibility. Although research in these areas is not directly related to contact verification, it has high relevance to current research in this area. Notes on important research are included.

The fixture-part relationship has been modeled in 3D space using the Jacobian matrix and kinetic analysis has been performed in regard to deterministic positioning of fixtures and loading/unloading accessibility (Asada 1985).

Extensive research has also been done on tolerance and stability analysis (Rong 1994, 1995; Kang 2003). Locator displacements are map-ped into deviations of locating reference planes. The machining surface deviation is then calculated based on locating reference plane deviations.

A kinetic model of multi-fingered robot hand-grasping problems has been applied to fixture configuration. Based on contact point positions and normal directions, the fixture configuration matrix (grasp matrix, in robotics) is established to model the workpiece-fixture relationship in 3D space (Xiong 1993). This configuration matrix has properties similar to these of the Jacobian matrix. However, the matrix is based on assumptions that apply specifically to robot hand grasping. Applicatbility of the model is suspect because unlike robot hands fixtures have contact points whose positions change with workpiece displace-ment.

3.6.1.3   Sensors in Fixtures

A method has been proposed for optimal sensor placement for automated coordinate checking fixtures (Wang 1999). The decision on how many sensors to use and where to place them is based on Fischer matrix information and statistical analysis. Sensors are placed at locations that maximize the determinant of the Fischer information matrix.

Optimization methods were studied for sensor location in assembly fixtures (Khan 1998, 1999). The main goal of the research apart from sensor location study is fault-type discrimination and manufacturing variation reduction. The model uses three axis measurements at each of the three sensor locations that provide nine variable measures. Optimal sensor location is obtained by maximizing the distance between each dominant eigen vector, obtained for each of the tooling faults. Assembly sequence is decomposed into a sequence of single-fixture subproblems and sensor placement is optimized.

A diagnostic methodology has also been developed for dimensional fault diagnosis of compliant beam structures in automotive and aerospace processes (Rong Q. 2000). Fault variation patterns obtained from measurement data are modeled as eigen-value/eigen-vector pairs using principal component analysis. The mapping of unknown faults against a set of fault pattern models has been developed based on statistical hypothesis tests.

The impact of surface errors on the location and orientation of a cylindrical part in a fixture has been studied by Sanhui (2001). A model is developed using the Newton-Raphson technique to predict the impact of surface errors on the location and orientation of a cylindrical workpiece.

Algorithms were developed for an intelligent fixture system (IFS) to hold a family of cylinder heads for machining operations (Deep 2001). IFS uses a part location system to precisely locate the workpiece relative to a pallet. This system uses a three-axis horizontal coordinate measuring machine (CMM), a wrist with two axes of rotation, and an analog scanning probe. The part location system is used to precisely locate the part with great speed and accuracy by using a micro-positioner to correct any misalignments and to compensate for the difference between the actual and desired position. Two part location algorithms are created and evaluated experimentally.

Although there is a great deal of research involving sensors in fixtures, they are mainly focused on assembly and inspection fixtures for which geometric complexity and accuracy requirements are high.

Several short-term projects have been conducted to study the contact verification problem. A fixture is designed to hold a wide range of abrasive rings and discs, and ultrasonic sensors are used for performing automated inspections of thickness for quality control (Bottino 2001). An existing inspection fixture was redesigned with optic sensors to improve the ergonomics and accuracy of the measurement system (Lippitt 2002). A sensor-based fixture design made for turbine engine vane production where the reliability of part loading was detected has also been studied (Bonczek 1999). A sensor system has been proposed to indicate the correct positioning of a vane within a wire electro discharge machining (WEDM) fixture (Macias 2001). All of these projects focused on specific processes and problems. One of the outcomes from these projects is that the use of sensors in various inspection fixtures has substantially improved the productivity and ergonomics of such fixtures.

3.6.1.4   Fixture Foolproofing

An algorithm has been proposed for fixture foolproofing for polygonal parts in two dimensions (Penev 1995). The algorithm utilizes the results of modular fixture design obtained from the Brost-Goldberg algorithm to find problem positionings of a workpiece in relationship to where it contacts. All candidate grid holes with respect to the identified unwanted configurations are used to place foolproofing pins. However, the algorithm has several limitations. It works only on polygonal parts and does not accept curved edges in the work boundary. In addition, the foolproofing arsenal is limited to fixed size pins and a discrete set of possible locations, and the algorithm does not seem to generalize in 3D dimensional space.

3.6.2   Fixture Foolproofing

Parts can be incorrectly loaded into fixtures due to ambiguities in fixture design (Figure 3.67). The incorrect loading of a part into a fixture can have disastrous consequences including rejection of the manufactured part. Fixture foolproofing aims at detection and prevention of such errors before additional value is added to the parts and before ambiguities creep into the fixturing process.

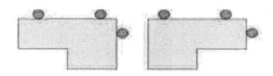

Figure 3.67. Ambiguity in part loading.

The term foolproofing is also known as "mistake proofing" or "poka-yoke." Poka-yoke is a Japanese term meaning "mistake proofing." The term was coined by Japanese industrial engineers in the 1960s (Shigeo 1986). In the United States, the term is seen as poke-yoke. Poke-yoke principles have been widely applied to assembly operations to prevent incorrect assembly by employing various methods.

In any manufacturing environment, safety is a primary concern. Fixtures that accept incorrect part loading pose a direct risk to both the operator and the machine. The consequences can be severe, during machining, machine tool crashes, and the creation of defective parts. Such occurrences are unacceptable. Nevertheless, most manufacturing fixtures are not foolproofed, which is of major concern to the shop floor.

In most cases, the fixtures do not have spare/replacement fixtures in case of fixture failure. When an accident occurs due to the incorrect loading of a part, the loss can be heavy. Such occurrences can be prevented by early foolproofing.

In the automotive industry, the cost of creating a fixture for a cylinder head is about $40,000 to $50,000 (Teresko 2001). The cost involved in foolproofing such a fixture is comparatively low, and the investment is justified in that the reliability of the total manufacturing line is improved. A fixture failure not only adds a replacement burden but brings the total manufacturing setup to a halt, imposing a heavy loss.

3.6.2.1 Fixture Foolproofing for Polygonal Parts

An algorithm has been proposed for foolproofing polygonal parts in 2D (Penev 1995). This algorithm analyzes the solutions for a 2D polygonal part in a modular fixture with three fixed-size locators and a clamp. Incorrect loading potential is analyzed and foolproofing pins are added at required locations in modular fixtures (as illustrated in Figure 3.68) to prevent incorrect loading.

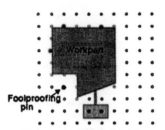

Figure 3.68. Foolproofing pin in modular fixture design.

3.6.2.1   Algorithm for Foolproofing 3D Parts

An independent algorithm for 3D parts that works based on fixture design and part symmetry is proposed here. The main processes involved in this algorithm are as follows.

1. Part classification based on symmetry/asymmetry

2. Determining possible solutions

3. Part peeling/simplification ("banana-peel" algorithm)

4. Checking foolproofing solution

5. Solutions

Solutions are derived via a predefined set of procedures in which the part is classified and later simplified based on part geometry.

Part Symmetric Classification. Symmetry can be defined as a characteristic of certain geometric shapes (2D and 3D) that brings to mind patterns, tiling, and repetitive mirror images. A shape is symmetric if certain motions or rearrangements of its parts leave it unchanged as a whole. These motions or rearrangements are called symmetry transformations. A shape is symmetric if it is congruent to itself in more than one way. For example, if a square is rotated about its center no difference exists between the original figure and its rotated image. They coincide exactly. If a horizontal line is drawn through the center of a square, the top and bottom halves are mirror images of each other. The square is, therefore, symmetrical about this line. Other lines of symmetry are readily identifiable.

In the classification of 3D parts we are concerned only about the mirror transformation about the mid planes in the X, Y, and Z directions. The parts are classified by placing a part's primary locating surface in the Z direction and the secondary and tertiary locating surfaces in the X and Y directions. Parts are classified based on their symmetry across the X-Y-Z planes and on similarity among their respective surfaces. Parts are classified by category as follows:

- *Category I:* Parts are symmetrical on all three planes, as shown in Figure 3.69

- *Category II:* Parts are symetrical on two planes, as shown in Figure 3.70

- *Category III:* Parts are symmetrical along one plane, as shown in Figure 3.71

- *Category IV:* Parts are asymmetrical, as shown in Figure 3.72

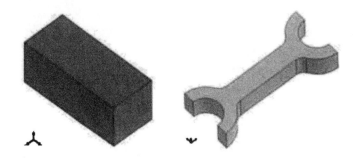

Figure 3.69. Example parts symmetrical on three planes.

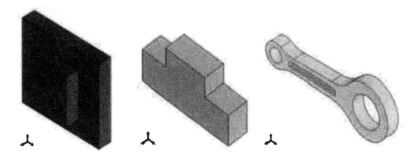

Figure 3.70. Example parts symmetrical on two planes.

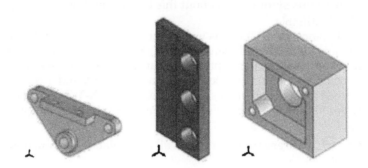

Figure 3.71. Example parts symmetrical on one plane.

Figure 3.72. Examples of asymmetrical parts.

**Determining the Possible Solutions.** The need for such a classification of parts is to determine the various solution spaces that need to be searched. Category I and IV parts do not have any solutions for foolproofing but do allow for error proofing, which is the verification of the proper contact of the part with locators. Category II and III parts are most suitable for foolproofing.

As we are involved in a 3D domain, it is a complex problem to search all solutions in space. This process simplifies and categorizes the problem into individual cases with specific conditions. Once classification of parts is finished, the possible solution spaces can be determined. The solution spaces are classified into two domains: foolproofing and error proofing. Foolproofing is subdivided into pin-based solutions and sensor-based solutions. Table 3.6 shows the possible solution spaces for the various categories of parts.

Table 3.6. Possible solution spaces for symetrical parts

	Foolproofing		Error Proofing
	Pin-based	Sensor-based	Sensor-based
Category I		x	
Category II	x	x	
Category III		x	
Category IV			x

__Part Peeling/Simplification ("Banana-Peel" Algorithm).__ Once it is determined whether there is a possible solution for a given problem, solutions are studied in detail. Our goal here is to simplify the 3D part further into a set of 2D sketches and search for solutions.

In this effort, we might employ what is known as banana-peel algorithm. As the name suggests, the part is peeled like a peel of a banana for part simplification. This is an interesting process; we peel the part based on the part workholding fixture. The fixture we are considering is a traditional 3-2-1 type. The initial surface is the primary locating surface, which forms the base. The peeling process starts with the peeling of the secondary locating surface. All external surfaces in the direction of the secondary locating surface are also peeled. The process is then repeated for the tertiary locating surface. The process is depicted in Figure 3.73.

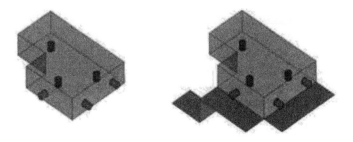

Figure 3.73. Illustration of part peeling process.

As shown in Figure 3.73, we may have more than one surface for every locating direction. In this case, we have multiple faces in the tertiary locating direction. After this process, we generate two simplified representations from the existing peeled part. The first representation is the simplified projection of the peeled part on a plane (Figure 3.74), as a 2D sketch.

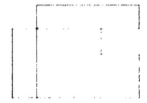

Figure 3.74. Part representation after projection.

The next representation is the same peeled part with the part eliminated (Figure 3.75). This results in a multiple-plane sketch based on the input part.

Figure 3.75. Part representation after eliminating the actual part.

Generating the Foolproofing Solutions. Once we have the simplified part representation we can search the geometry for a solution. This process is divided into several steps. The first step is to utilize the projected part representation and perform symmetry transformations on that part and check for solutions. Later, the same process is repeated for the actual peeled part.

Symmetry transformations are mathematically described motions or rearrangements of a geometric shape or its parts that leave the shape unchanged, including rotation, reflection, and inversion. For any symmetric figure, certain sets of points, lines, or planes are fixed (invariant) under a symmetry transformation.

We perform symmetry transformations on a part's simplified geometry to figure out foolproofing solutions. Although there are a lot of transformations that exist, we are concerned only with a few basic transformations. The most common symmetry transformation is the rotation thansformation. For example, we can rotate a cube 90 degrees about a straight line crossing the middle of two opposite faces. The cube before and after the transformation cannot be distinguished. Another common symmetry operation is the reflection, also called mirror symmetry. Eight basic symmetry transformations can be defined, as shown in Figure 3.76.

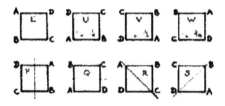

Figure 3.76. Eight basic symmetry transformations for foolproofing solutions.

The first four transformations are rotation transformations as follows.

- **E:** E is called the identity. This is our part without any trans-formations.
- **U:** Rotates the square through a quarter turn (90 degrees).
- **V:** Rotates the square through a half turn (180 degrees).
- **W:** Rotates the square through a three-quarter turn (270 degrees).

The next four transformations are mirror transformations.

- **P:** Reflects about a vertical mirror line through the center.
- **Q:** Reflects about a horizontal mirror line through the center.
- **R:** Reflects about a diagonal mirror line through AC.
- **S:** Reflects about a diagonal mirror line through BD.

These transformations were performed on the part shown in Figure 3.73. The results are outlined in Table 3.7 with 1 representing symmetrical and 0 representing asymmetrical. Row and column have been calculated. The lowest in row total suggests that the probability for a solution is the highest in that plane.

Table 3.7. Symmetry table.

	U	V	W	P	Q	R	S	Total
Primary	0	0	0	0	0	0	0	0
Secondary	0	1	0	1	1	1	1	5
Tertiary	0	1	0	1	1	1	1	5
Column total	0	2	0	2	2	2	2	

In this case we can expect a solution lying in the primary locating plane. We now we need to determine where to place the foolproofing pin/ sensor. The transformations P, Q, R, and S can be used to determine the solution. The area from our initial identity case E is added with the transformed areas and the resulting new area is saved. The solution lies in the region in which all of the transformations interact, which represents the likely location for a foolproofing pin/sensor. An example of this is shown in Figure 3.77. Figure 3.78 shows the placement of the foolproofing pin in a fixture design.

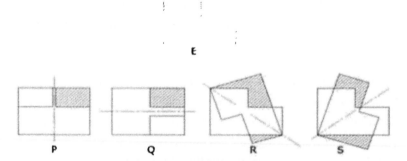

Figure 3.77. Symmetry transformations P, Q, R, and S on the sample part.

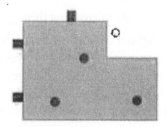

Figure 3.78. The final foolproofing solution.

The same steps can be used to process a multi-plane representation. However, with this representation, we move into 3D space. The 3D case has not been fully investigated, but the same algorithm may be expanded to generalize the solution.

3.6.2.2   Validating the Solutions

The algorithms involved here were verified by applying to sample parts of various categories, as discussed previously. The results from symmetry transformation of the parts shown in Figures 3.70 and 3.71 are outlined in Tables 3.8 through 3.11. The results were verified whether or not the selected surface was actually the preferred surface for foolproofing. In all cases, the minimum row total in the symmetry table represents the most valid solution. Solutions with a tie score are obviously not unique. One strategy in breaking a tie is to give importance to primary locating first, secondary locating next, and tertiary locating last.

Table 3.8. Symmetry table associated with Figure 3.70a.

	U	V	W	P	Q	R	S	Total
Primary	1	1	1	1	1	1	1	7
Secondary	0	0	0	0	1	0	0	1
Tertiary	0	0	0	1	0	0	0	1
Column total	1	1	1	2	2	1	1	

Table 3.9. Symmetry table associated with Figure 3.70b.

	U	V	W	P	Q	R	S	Total
Primary	0	0	0	1	0	0	0	1
Secondary	0	0	0	0	0	0	0	0
Tertiary	0	0	0	1	0	0	0	1
Column total	0	0	0	2	0	0	0	

Table 3.10. Symmetry table associated with Figure 3.71a.

	U	V	W	P	Q	R	S	Total
Primary	0	0	0	1	0	0	0	1
Secondary	0	1	0	1	1	0	0	3
Tertiary	0	0	0	1	0	0	0	1
Column total	0	1	0	3	0	0	0	

Table 3.11. Symmetry table associated with Figure 3.71b.

	U	V	W	P	Q	R	S	Total
Primary	0	0	0	1	0	0	0	1
Secondary	0	0	0	1	1	0	0	2
Tertiary	0	1	0	1	1	0	0	3
Column total	0	1	0	3	2	0	0	

3.6.2.3   Implementation of Algorithms

The various algorithms for foolproofing were implemented in SolidWorks using its API. Figure 3.79 shows various locating planes and symmetrical planes being identified using the system. By applying the algorithm, the foolproofing solution can be obtained.

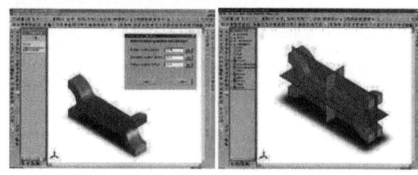

Specifition of locating planes   Identification of symmetrical planes

Figure 3.79. Symetricity identification.

3.6.3   Part Location and Contact Verification

The workholding of parts with free-form surfaces involves complex fixturing. This is especially true in the aerospace industry, which involves many complex surfaces. Turbine blades and vanes are parts that pose a particular problem. Turbine blades are rotating members in a jet engine, and turbine vanes are stationary members that manage gas flow in a jet engine. Turbine blades and vanes are made out of tough alloys. One widely used alloy is Inconel, which is difficult to machine. The blade and vane castings cost about $2,000 to $5,000 each (Purushothaman 2003). Once the blades and vanes are cast, the preferred methods for machining the features are either grinding or electrical discharge machining (EDM). Workholding is critical for these parts because all surface tolerances are very tight.

One fixturing strategy here is to locate and hold using the blade section itself, which involves a complex fixture with unreliable fixturing problems. The most common fixturing problem associated with these fixtures is noncontact of the blade/vane surface with locators in the fixture. Because the surfaces are free-form, locators are spherical in design to achieve best point contact and to minimize fixturing variation. One of the main causes for locating inaccuracy is noncontact of the part with locators, as illustrated in Figure 3.65. This occurs due to misalignment or improper clamping, which are difficult to be detected visually. There are a few two problems. The part may be clamped before it makes contact with all locators or the part may lose contact with one or more locators during clamping. Because free-form surfaces are involved as fixturing surfaces, multiple solutions might present themselves for a given fixture-part design. Multiple solutions mean that the part can be loaded in the fixture in a number of ways and still maintain contact with locators, the contact point on the spherical surface of a locator can change from position to position, as indicated in Figure 3.80. To avoid such a multiple-solution problem, the fixture has to be carefully designed and verified. The sensor-based fixture design solves the problem only in regard to unreliable contact of the part with locators.

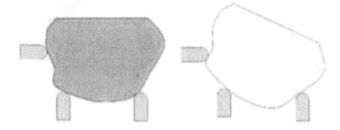

Figure 3.80. Non-unique solution for locating with free-form surfaces.

3.6.3.1   Sensor-based Solutions

With progress in sensing technology in recent years, it has become possible to improve fixture design by placing sensors in fixtures to verify proper part location. One aspect of the application of sensors to fixture design is the determination of sensor location. Although there are different ways to locate sensors in a fixture, the most straightforward way is to place the sensors at locator positions. This is particularly valid in regard to space limitations and high accuracy requirements. The problem is one of how to design a locator with a sensor.

An experimental fixture was designed to demonstrate the effectiveness of using a sensor within a locator. The main functional requirements of a locator with sensor include the following:

- Suitable for parts with different surface geometries and sizes
- Flexible in terms of locator positioning
- Allows for interchangeable locators and sensors in the fixture
- Replicates the current fixture in use in the industry

With these functional requirements in mind, a test fixture was designed. One of the important goals was to design the fixture as closely as possible to the current industry fixture so that solutions could be easily integrated into the existing fixtures. The new locator was designed to replicate the current ceramic locators widely used in the industry. The locator was machined out of aluminum and a hole was drilled on the locator to accommodate an optical fiber sensor. Figure 3.81 shows the locator. Figure 3.82 shows the test fixture used to hold a part involving several locators with sensors.

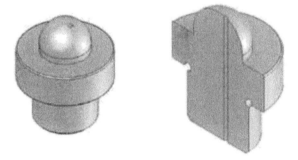

Figure 3.81. Locator with hole for sensor.

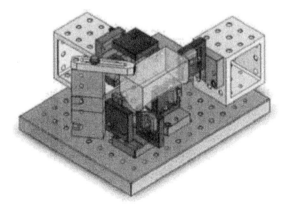

Figure 3.82. Test fixture employing locators with sensors.

3.6.3.2 Experimental Design and Results

An experiment was designed in which tests were conducted to verify theoretical predictions. A part was machined with a free-form surface on one side resembling the profile of turbine blades having a curvature of varying radius. The part was designed with flat surfaces on the other sides to simplify the problem and to study the free-form surfaces. Locators were positioned (based on the 3-2-1 locating scheme) at the locations shown in Figure 3.83. The locators were named A, B, and C in the primary locating surface, 2 and 3 in secondary locating surface, and 1 in tertiary locating surface.

Figure 3.83. Part with locators, viewed upside down.

Locators were placed at three locations of varying radius curvature. Figure 3.84 shows these locator positions and their surface curvature values.

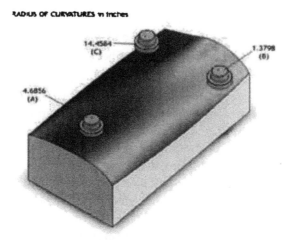

Figure 3.84. Curvature radius at locator positions in primary locating surface.

Figure 3.85 shows curvature radii in a side view. Once the locator positions were decided, the challenge was to position the locators at the required locations accurately. Because the fixture base was made adjustable for functional flexibility, it made accurate positioning of the locators difficult (Purushothaman 2003).

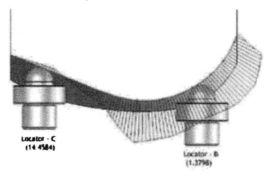

Figure 3.85. Side view of curvature plots and corresponding curvature radii.

Four types of fiber-optic sensors with different specifications were selected to study the performance of each under various experimental conditions. The specifications for the four types of sensors used in the experiments were as follows.

1.   Sensor Xa—Sensor head—ϕ0.06″, core fiber (4 × ϕ0.265 mm (0.01″))

2. Sensor Xb—Sensor head—ϕ0.12″, ϕ0.275 mm (0.11″) lens, beam spot ϕ0.004″

3. Sensor Xc—Sensor head—ϕ0.12″, core fiber (2✕ϕ1.0 mm (0.04″))

4. Sensor Xd—Sensor head—ϕ0.1″, core fiber (2✕ϕ1.0 mm (0.04″))

The test was conducted on all sensors by placing them at different locations. An initial location in which locators were in contact with the workpiece was set as zero. Where the part was not in contact with the locator is indicated as 0 in Tables 3.12 through 3.15. In order to test the sensitivity of the sensors in different positions, a known disturbance (clearance) of 0.0015 inch (0.04 mm) was introduced at the locator positions. An ideal result shows 0 in all spaces of a table except row ZERO.

From the results in Table 3.12, we can see that when the sensor was placed at location B it was very sensitive and detected the disturbances from all locators. The sensors when placed at A and C were least sensitive. They could not read the disturbance at locations 2 and 3. Whenever a disturbance was introduced at location 1, it affected the contacts in other locations. Similar results were obtained in other arrangements of sensors and disturbances, as shown in Tables 3.13 through 3.15.

Table 3.12. Test results for sensor Xa.

Individual Disturbance (at various locators)	Sensor at A κ = 0.21341	Sensor at B κ = 0.72474	Sensor at C κ = 0.06916	Sensor at 1 κ = 0 ($R = \infty$)
ZERO	1	1	1	1
0.0015″ (0.04 mm) at 1	0	0	0	-
0.0015″ (0.04 mm) at 2	1	0	1	0
0.0015″ (0.04 mm) at 3	1	0	1	0
0.0015″ (0.04 mm) at A	-	0	0	0
0.0015″ (0.04 mm) at B	0	-	0	0
0.0015″ (0.04 mm) at C	0	0	-	1

Notee: κ = *curvature,* **R** = *Radius of curvature,* "0" — *contact lost,*
"1" — *maintains contact, and* "-" — *no measurement.*

Table 3.13. Test results for sensor Xb.

Individual Disturbance (at various locators)	Sensor at 1 $\kappa=0$ (R = ∞)	Sensor at 2 $\kappa = 0$ (R = ∞)	Sensor at 3 $\kappa = 0$ (R = ∞)
ZERO	1	1	1
0.0015" (0.04 mm) at 1	-	0	0
0.0015"(0.04 mm) at 2	1	-	1
0.0015" (0.04 mm) at 3	0	0	-
0.0015" (0.04 mm) at A	0	1	0
0.0015" (0.04 mm) at B	0	0	0
0.0015" (0.04 mm) at C	0	1	0

Notee: κ = curvature, R = Radius of curvature, "0" — contact lost, "1" — maintains contact, and "-" — no measurement.

Table 3.14. Test results for sensor Xc.

Individual Disturbance (at various locators)	Sensor at A $\kappa = 0.21341$	Sensor at B $\kappa = 0.72474$	Sensor at C $\kappa = 0.06916$	Sensor at 1 $\kappa = 0$ (R = ∞)
ZERO	1	1	1	1
0.0015" (0.04 mm) at 1	0	0	1	-
0.0015"(0.04 mm) at 2	0	0	1	0
0.0015" (0.04 mm) at 3	0	0	0	0
0.0015" (0.04 mm) at A	-	0	0	0
0.0015" (0.04 mm) at B	0	-	0	0
0.0015" (0.04 mm) at C	0	0	-	0

Notee: κ = curvature, R = Radius of curvature, "0" — contact lost, "1" — maintains contact, and "-" — no measurement.

Individual Disturbance (at various locators)	Sensor at A $\kappa = 0.21341$	Sensor at B $\kappa = 0.72474$	Sensor at C $\kappa = 0.06916$	Sensor at 1 $\kappa = 0$ $(R = \infty)$
ZERO	1	1	1	1
0.0015" (0.04 mm) at 1	0	0	0	-
0.0015"(0.04 mm) at 2	0	0	1	1
0.0015" (0.04 mm) at 3	0	0	1	1
0.0015" (0.04 mm) at A	-	0	1	0
0.0015" (0.04 mm) at B	1	-	1	1
0.0015" (0.04 mm) at C	0	0	-	0

Notee: κ = curvature, R = Radius of curvature, "0" — contact lost, "1" — maintains contact, and "-" — no measurement.

Table 3.15. Test results for sensor Xd.

From the different tests conducted at various locations, inferences on the sensitivity of the sensors include the following:

- The best sensitivity was observed when the sensors were placed at locations with minimum radius of curvature (R) and maximum curvature (κ).

- The worst sensitivity was observed when the sensors were placed at locations with maximum radius of curvature (R) and minimum curvature (κ).

- The sensitivity is also related to the direction of disturbance and the sensing direction.

3.6.4   Locator Design with Sensor

There is a need for an integrated locator/sensor for foolproofing and error-proofing applications. The various foolproofing needs were discussed in previous sections. One of the main requirements of the existing fixtures is the implementation of the new integrated techniques with minimal changes to existing designs and without interfering with the tool paths for existing parts. To realize such needs, an integrated locator with proximity sensors has been designed and tested for various applications.

Besides the ceramic locator with optical fiber sensor (discussed in the last section), proximity sensors were used in the integrated locators for broad range of applications. Proximity sensors can detect metal objects without physical contact with the target. Proximity sensors are classified as three types based on the principles of operation: the electromagnetic inductive type, the magnetic type using a magnet, and the capacitance type using change in capacitance. The electromagnetic high-frequency oscillation type of proximity sensor was selected for the foolproofing applications discussed here, as it best meets the requirements of current applications.

Figure 3.86 indicates the clearances required to flush mount sensors to metal (minimum clearances of ϕD and d are required). In addition, non-shielded sensors interfere with sensors placed close to each other. To simplify the installation process, a self-contained unit with an output indicator built into the unit was employed. The LED built into the unit glows, confirming operation, which eliminates the use of amplifiers and saves space in the fixture.

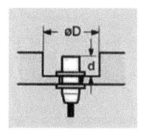

Figure 3.86. The use of a non-shielded proximity sensor.

An integrated locator should be designed based on existing modular fixture components. A new locator might even be included in a modular fixture catalog. Because the sensor in this case is a self-contained type with LED, a slot was machined to enable ease of viewing the LED. The complete design is shown in Figure 3.87.

Various sensors were studied to select the sensor best suited to the fixturing requirements. The type of senors researched were proximity sensors, precision-limit switches, ultrasonic sensors, laser displacement sensors, and photoelectric sensors. One of the promising sensors for the current application was the fiber-optic type. The optical fiber consists of the core and the cladding, which have different indexes. The light beam travels through the core by repeatedly bouncing off the wall of the cladding. The light beam, having passed through the fiber without any loss in light quantity, is dispersed with an angle of approximately 60 degrees and emitted to the target. There are two types of fibers: plastic-fiber and glass-fiber. Optical fibers are also divided as the through-beam type and the reflective type. The reflective type is subdivided according to parallel, coaxial, or separated fibers.

Figure 3.87. Integrated locator with a self-contained proximity sensor.

References

An, Z., S. Huang, J. Li, Y. Rong, and S. Jayaram. "Development of Automated Fixture Design Systems with Predefined Fixture Component Types: Part 1, Basic Design," *International Journal of Flexible Automation and Integrated Manufacturing* 7:3/4, pp. 321–341, 1999.

Asada, H., and A. B. By. "Kinematic Analysis of Workpart Fixturing for Flexible Assembly with Automatically Reconfigurable Fixtures," *IEEE Journal of Robotics and Automation* RA-1:2, pp. 86–94, 1985.

Bi, Z. M., and W. J. Zhang. "Flexible Fixture Design and Automation: Review, Issues and Future Direction," *International Journal of Production Research* 39:13, pp. 2867–2894, 2001.

Bonczek, M. J., C. A. Tolentino, and J. A. Troost. "Continuous Improvement: Manufacturing Process Control of F-100 Turbine Vanes," Senior Project Report, Worcester Polytechnic Institute, Worcester, MA, 1999.

Bottino, J. S. Richards, and B. LeBlanc. "Ultrasonic Sensing of Surface Characteristics of Grinding Wheels," Senior Project Report, Worcester Polytechnic Institute, Worcester, MA, 2001.

Boyle, I., Y. Rong, and D. C. Brown. "Case-based Reasoning in Fixture Design." *SPIE Photonics Technologies for Robotics, Automation, and Manufacturing* 5263, pp. 85–96, 2003.

Bridle, J. S. "Probabilistic Interpretation of Feedforward Classification Network Output, with Relationships to Statistic Pattern Recognition," in *Neurocomputing: Algorithms, Architectures, and Applications*, F. Fogelman-Soulie and J. Herault (eds.), NATO ASI Series, New York: Springer-Verlag, 1991.

Brost, R., and K. Y. Goldberg. "A Complete Algorithm for Synthesizing Modular Fixtures for Polygonal Parts," *IEEE Transactions on Robotics and Automation* RA-12:1, pp. 31–46, 1996.

Chang, H. C., L. Dong, F. X. Liu, and W. F. Lu. "Indexing and Retrieval in Machining Process Planning Using CBR," *Artificial Intelligence in Engineering* 14, pp. 1–13, 2000.

Chou, Y. C. "Automated Fixture Design for Concurrent Manufacturing Planning," *Concurrent Engineering: Research and Applications* 1, pp. 219–229, 1993.

Chou, Y. C., V. Chandru, and M. M. Barash. "A Mathematical Approach to Automatic Configuration of Machining Fixtures: Analysis and Synthesis," *Journal of Engineering for Industry* 111, pp. 299–306, 1989.

DeMeter, E. C. "Fast Support Layout Optimization," *International Journal of Machine Tools and Manufacture* 38:10–11, pp. 1221–1239, 1998.

Fang, B., R. E. DeVor, and S. G. Kapoor. "Influence of Friction Damping on Workpiece-Fixture System Dynamics and Machining Stability," *Journal of Manufacturing Science and Engineering* 124, pp. 226–233, 2002.

Fuh, J.Y.H., and A.Y.C. Nee. "Verification and Optimization of Workholding Schemes for Fixture Design," *Journal of Design and Manufacturing* 4, pp. 307–318, 1994.

Grippo, P. M., M. V. Gandi, and B. S. Thompson. "The Computer-aided Design of Modular Fixturing Systems," *International Journal of Advanced Manufacturing Technology* 2:2, pp. 75–88.

Hoffman, E. G. *Jig and Fixture Design* (3d ed.). Albany, NY: Delmar, 1991.

Hu, W., and Y. Rong. "A Fast Interference Checking Algorithm for Automated Fixture Design Verification," *International Journal of Advanced Manufacturing Technology* 16, pp. 571–581, 2000.

Jang, B.-S., Y.-S. Yang, Y.-S. Yeun, and S.-H. Do. "Axiomatic Design Approach for Marine Design Problems," *Marine Structures* 15, pp. 35–56, 2002.

Kang, Y., Y. Rong, and M. Sun. "Constraint-based Modular Fixture Assembly Modeling and Automated Design," MED-8, pp. 901–908, ASME IMECE, Anaheim, CA, Nov. 15–20, 1998.

Kang, Y., Y. Rong, and J.-C. Yang. "Computer-aided Fixture Design Verification: Part 1, The Framework and Modeling; Part 2, Tolerance Analysis; Part 3, Stability Analysis," *International Journal of Advanced Manufacturing Technology* 21, pp. 827–849, 2003.

Khan, A. M., D. Ceglarek, and J. Ni. "Sensor Location Optimization for Fault Diagnosis in Multi-Fixture Assembly Systems," *ASME Journal of Manufacturing Science and Engineering* 120:4, pp. 781–792, 1998.

Khan, A. M., D. Ceglarek, J. Shi, J. Ni, and T. C. Woo. "Sensor Optimization for Fault Diagnosis in Single Fixture Systems: A Methodology," *ASME Journal of Manufacturing Science and Engineering* 121, pp. 109–117, 1999.

Kong, Z., and D. Ceglarek. "Fixture Configuration Synthesis for Reconfigurable Assembly Using Procrustes-based Pairwise Optimization," *Transactions of NAMRI* 31, pp. 403–410, 2003.

Kow, T. S., A. S. Kumar, and J.Y.H. Fuh. "An Integrated Computer-aided Modular Fixture Design System for Interference Free Design," MED-8, pp. 909–916, ASME IMECE, Anaheim, CA, Nov. 15–20, 1998.

Krishnakumar, K., and S. N. Melkote. "Machining Fixture Layout Optimization Using the Genetic Algorithm," *International Journal of Machine Tools and Manufacture* 40, pp. 579–598, 2000.

Kumar, A. S., and A.Y.C. Nee. "A Framework for a Variant Fixture Design System Using Case-based Reasoning Technique," in *Computer-aided Tooling*, MED 2:1, pp. 763–775, ASME WAM, November 12–17, 1995.

Kumar, A. S., V. Subramaniam, and K. C. Seow. "Conceptual Design of Fixtures Using Genetic Algorithms," *International Journal of Advanced Manufacturing Technology* 15, pp. 79–84, 1999.

Leake, D. B. "CBR in Context: The Present and Future," in *Case-based Reasoning: Experiences, Lessons, and Future*, D. B. Leake (ed.), pp. 3–30, Menlo Park, CA: AAAI/MIT Press, 1996.

Lee, J. D., and L. S. Haynes. "Finite Element Analysis of Flexible Fixturing Systems," *Journal of Engineering for Industry* 109, pp. 134–139, 1987.

Li, B., and S. N. Melkote. "An Elastic Contact Model for Prediction of Workpiece-fixture Contact Forces in Clamping," *Journal of Manufacturing Science and Engineering* 121, pp. 485–493, 1999.

Li, J., W. Ma, and Y. Rong. "Fixturing Surface Accessibility Analysis for Automated Fixture Design," *International Journal of Productivity Research* 37:13, pp. 2997–3016, 1999.

Liao, T. W., Z. M. Zhang, and C. R. Mount. "A Case-Based Reasoning System for Identifying Failure Mechanisms," *Engineering Applications of Artificial Intelligence* 13, pp. 199–213, 2000.

Lippitt, B., and R. Paonessa. "Modularized Design of Inspection Fixtures," Senior Project Report, Worcester Polytechnic Institute, Worcester, MA, 2002.

Ma, W., Z. Lei, and Y. Rong. "FIX-DES: A Computer-aided Modular Fixture Configuration Design System," *International Journal of Advanced Manufacturing Technology* 14, pp. 21–32, 1998.

Ma, W., J. Li, and Y. Rong. "Development of Automated Fixture Planning Systems," *International Journal of Advanced Manufacturing Technology* 15, pp. 171–181, 1999.

Macias, N. J., A. E. Brown, and P. Mouton. "Part Sensing for TSMC Restructured Production Lines," Senior Project Report, Worcester Polytechnic Institute, Worcester, MA, 2001.

McSherry, D. "The Inseparability Problem in Interactive CBR," *Knowledge Based Systems* 15, pp. 293–300, 2002.

Maher, M. L., and A. G. de Garza. "Case-based Reasoning in Design," *IEEE Expert*, Mar./Apr., pp. 34–41, 1997.

Mani, M., and W.R.D. Wilson. "Automated Design of Workholding Fixture Using Kinematic Constraint Synthesis," pp. 437–444, 16th NAMRC, Urbana-Champaign, IL, May 1988.

Marin, R. A., and P. M. Ferreira. "Optimal Placement of Fixture Clamps," *Journal of Manufacturing Science and Engineering* 124, pp. 676–694, 2002.

Markus, A., E. Markusek, J. Farkas, and J. Filemon. "Fixture Design Using Prolog: An Expert System," *Robotics and CIMS* 1:2, pp. 167–172, 1984.

Nee, A.Y.C., and A. S. Kumar. "A Framework for an Object/Rule-based Automated Fixture Design System," *Annual of the CIRP*, pp. 147–151, 1991.

Ngoi, B.K.A., S. H. Yeo, and S. B. Tan. "Tool Collision Detection in Machining Using Spatial Representation Technique," *International Journal of Production Research* 35:7, pp. 1789–1850, 1997.

Nnaji, B. O., S. Alladin, and P. Lyu. "Rules for an Expert Fixturing System on a CAD Screen Using Flexible Fixtures," *Journal of Intelligent Manufacturing* 1, pp. 31–48, 1990.

Ong, S. K., and A.Y.C. Nee. "A Systematic Approach for Analyzing the Fixturability of Parts for Machining," in *Computer-aided Tooling*, ASME WAM, pp. 747–761, 1995.

Pena-Mora, F., and M. Li. "Dynamic Planning and Control Methodology for Design/Build Fast-Track Construction Projects," *Journal of Construction Engineering and Management,* Jan./Feb., pp. 1–17, 2001.

Pham, D. T., and A. de Sam Lazaro. "AUTOFIX: An Expert CAD System for Jigs and Fixtures," *International Journal of Machine Tool and Manufacture* 30:3, pp. 403–411, 1990.

Purushothaman, R. "Sensor-based Fixture Design and Verification," M.S. thesis, Worcester Polytechnic Institute, Worcester, MA, 2003.

Rong, Q., D. Ceglarek, and J. Shi. "Dimensional Fault Diagnosis for Compliant Beam Structure Assemblies," *Journal of Manufacturing Science and Engineering* 122, pp. 773–780, 2000.

Rong, Y., and Y. Bai. "Automated Generation of Modular Fixture Configuration Design," *Journal of Manufacturing Science and Engineering* 119, pp. 208–219, 1997.

Rong, Y., and Y. Bai. "Modular Fixture Element Modeling and Assembly Relationship Analysis for Automated Fixture Configuration Design," *Journal of Engineering Design and Automation* 4:2, pp. 147–162, 1998.

Rong, Y., S. Wu, and T. P. Chu. "Automated Verification of Clamping Stability in CAFD," *Computers in Engineering* 1, pp. 421–426, 1994.

Rong, Y., and Y. Zhu. "An Application of Group Technology in Computer aided Fixture Design," *International Journal of Systems Automation: Research and Applications* 2:4, pp. 395–405, 1992.

Rosenman, M. "Case-Based Evolutionary Design," *AI for Engineering Design, Analysis and Manufacturing* 14, pp. 17–29, 2000.

Roy, U., J. Liao, P.-L. Sun, and M. C. Fields. "Fixture Design Synthesis for a FMS," *Integrated Computer-Aided Engineering* 4:2, pp. 101–113, 1997.

Sangnui, S., and F. Peters. "The Impact of Surface Errors on the Location and Orientation of Cylindrical Workpieces in a Fixture," *ASME Journal of Manufacturing Science and Engineering* 123, pp. 325–330, 2001.

Smyth, B., and M. T. Keane. "Experiments on Adaptation-Guided Retrieval in Case-Based Design," in *First International Conference ICCBR-95: Case-Based Reasoning Research and Development*, M. Veloso and A. Aamodt (eds.), pp. 313–324, Sesimbra, Portugal: Springer-Verlag, 1995.

Smyth, B., and M. T. Keane. "Design a la Déjà vu: Reducing the Adaptation Overhead," Case-based Reasoning, D. B. Leake (ed.), pp. 151–166, Menlo Park, CA: AAAI Press, 1996.

Suh, N. P. *Axiomatic Design Advances and Applications*. New York: Oxford University Press, 2001.

Sun, S. H., and J. L. Chen. "A Modular Fixture Design System Based on Case-based Reasoning," *International Journal of Advanced Manufacturing Technology* 10, pp. 389–395, 1995.

Teresko, J. "Technologies of the Year: Intelligent Fixturing Systems," *Industry Week*, Dec. 2001.

Thompson, B. S., and M. V. Gandhi. "Commentary on Flexible Fixturing," *Applied Mechanics Review* 39:9, pp. 1365–1369, 1986.

Thurston, D. "A Formal Method for Subjective Design Evaluation," *Research in Engineering Design* 3, pp. 105–122, 1991.

Trappey, A.J.C., and C. R. Liu. "An Automatic Workholding Verification System," *Robotics and Computer-Integrated Manufacturing* 9:4/5, pp. 321–326, 1992.

Trappey, A.J.C., C. S. Su, and J. L. Hou. "Computer-aided Fixture Analysis Using Finite Element Analysis and Mathematical Optimization Modeling," MED-2:1, pp. 777–787, ASME IMECE, Nov. 1995.

Umeda, Y., and T. Tomiyama. "Functional Reasoning in Design," *IEEE Expert*, Mar./Apr., pp. 42–48, 1997.

Varma, A., and N. Roddy. "ICARUS: Design and Deployment of a CBR System for Locomotive Diagnostics," *Engineering Applications of AI* 12, pp. 681–690, 1999.

Wang, M. Y. "Automated Fixture Layout Design for 3D Workpieces," in *Proceedings of the IEEE International Conference on Robotics and Automation*, vol. 2, pp. 1577–1582, Detroit, MI: IEEE, 1999.

Wang, M. Y., and S. Nagarkar. "Locator and Sensor Placement for Automated Coordinate Checking Fixtures," *ASME Journal of Manufacturing Science and Engineering* 121, pp. 709–719, 1999.

Wu, Y., Y. Rong, and T. Chu. "Automated Generation of Dedicated Fixture Configuration," *International Journal of Computer Applications in Technology* 10:3/4, pp. 213–235, 1997.

Wu, Y., Y. Rong, W. Ma, and S. LeClair. "Automated Modular Fixture Design: Geometric Analysis," *Robotics and Computer-integrated Manufacturing* 14, pp. 1–15, 1998.

Xiong, C., and Y. L. Xiong. "Stability Index and Contact Configuration Planning for Multifingered Grasp," *Journal of Robotic Systems* 15:4, pp. 183–190, 1998.

Xiong, Y. L. "Theory and Methodology for Concurrent Design and Planning of Reconfiguration Fixture," in *Proceedings of the IEEE International Conference on Robotics and Automation*, vol. 3, pp. 305–311, Atlanta, Georgia: IEEE, 1993.

Yang, H., X. Zhang, J. Zhou, and J. Yu. "A Hierarchy of the OCT-SPHERE Model and Its Application in Collision Detection," *Advances in Design Automation* 20:1, pp. 15–19, 1994.

Yeh, J. H., and F. W. Liou. "Contact Condition Modeling for Machining Fixture Setup Processes," *International Journal of Machine Tools and Manufacture* 39, pp. 787–803, 1999.

Zhu, Y., and Y. Rong. "A Computer-aided Fixture Design System for Modular Fixture Assembly," in *Quality Assurance Through Integration of Manufacturing Processes and Systems*, PED-56, pp. 165–174, ASME WAM, Nov. 8–13, 1992.

Zhu, Y., S. Zhang, and Y. Rong. "Experimental Study on Fixturing Stiffness of T-slot Based Modular Fixtures," in *NAMRI Transactions XXI*, pp. 231–235, Stillwater, OK: NAMRC, 1993.

CHAPTER 4

Computer-aided Fixture Design Verification

4.1 FRAMEWORK AND MODELING

CAFD techniques have advanced to the point that fixture configurations can be generated automatically for both modular fixtures and dedicated fixtures. Computer-aided fixture design verification (CAFDV) is the technique for verifying and improving existing fixture designs. In this chapter, the framework of CAFDV, based on geometric and kinetic models, is introduced. Fixturing tolerance and stability verification are presented in sections following.

4.1.1   Background

Computer technologies have revolutionized the way products are manufactured today. From standalone CAD/CAM applications to enterprise PDM/ERP (product data management/enterprise resource planning) systems that cross borders, computer technologies have fulfilled the dreams of manufacturers: shortened development time, improved product quality, and lowered cost. As part of this revolution, CAFD emerged by integrating fixture design knowledge with CAD platforms (Rong 1999). CAFD empowers engineers with its capabilities for fast prototyping with minimal dependence on human interaction.

The primary users of CAFD had been fixture design engineers, who had used it to generate fixture designs. With the advancement of information technology, supply chain managers joined as new users of CAFD. They outsource fixtures to vendors (usually as part of the production line), and they need tools such as CAFD to inspect and control fixture designs from vendors.

An automated fixture design system typically generates more than one solution, sorted by certain criteria. This leaves the questions to CAFD users: which fixture design is the best, why it is the best, and how it can be improved. Whereas design engineers may have enough expertise to answer such questions, supply chain managers usually do not. They require a tool that enables them to measure and optimize the quality of a fixture design.

This section presents a CAFDV system that verifies the quality of a fixture design. This verification adds a new stage to the earlier division of the three-stage fixture design (Bai 1995). Figure 4.1 shows the four-stage fixture design structure and the role of CAFDV in the structure.

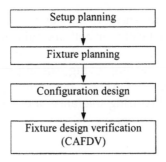

Figure 4.1. Four stages of fixture design.

The requirements of a fixture design are defined by design and manufacturing engineers. Two types of fixture models, geometric and kinetic, are created to describe fixture-workpiece relationships. Four requirements are studied in this work: locating performance analysis, tolerance analysis, stability analysis, and accessibility analysis. Figure 4.2 shows the system structure for CAFDV.

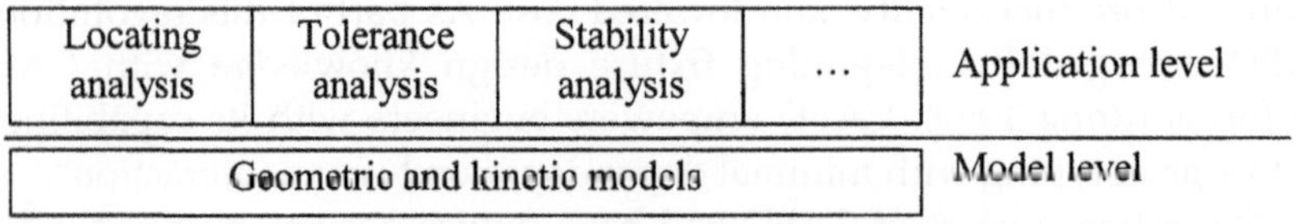

Locating analysis	Tolerance analysis	Stability analysis	...	Application level
Geometric and kinetic models				Model level

Figure 4.2. CAFDV system structure.

4.1.2 Previous Research on Fixture Verification

Earlier researchers had studied several areas of fixture design verification, and each touched on one or more areas. These studies are discussed in the following.

The Jacobian matrix was created to model the fixture-workpiece relationship in 3D space (Asada 1985). Kinetic analysis was used in this model to analyze deterministic positioning, loading/unloading accessibility, bilateral constraint, and total constraint. Screw theory was used for fixture analysis and synthesis (Chou 1989). Deterministic locating, clamping stability, total restraint, clamping point, and clamping force determination were studied, and both kinetics and force analysis was studied for fixture verification (Wu 1995). The contacts between workpiece and fixture were modeled as line and surface contacts, and the stability problem was modeled and solved with screw theory and nonlinear programming technique. A fixture is stable if a solution exists for the nonlinear system. The time-variant stability problem was also discussed, including considerations of fixturing force limits and directions (Trappey 1992). Later, fixture layout was optimized with the finite element analysis (FEA) approach, in which balance is achieved between minimal workpiece deformation and maximal machining accuracy (Trappey 1995).

Table 4.1. Literature overview for fixture verification.

Study	Locating Performance	Tolerance Analysis	Stability Analysis	Accessibility
Asada 1985	X	—	—	X
Chou 1989	X	—	X	—
Lee 1991	X	—	X	—
Trappey 1992	X	—	X	—
Xiong 1998	X	—	X	—
Rong 1994	—	—	X	—
King 1995	X	—	X	—
Rong 1995	—	X	—	—
Wu 1995	X	—	X	—
Rong 1996	—	X	—	—
DeMeter 1998	—	—	X	—
Kashyap 1999	X	—	X	—
Li 1999	—	—	—	X
Wang 1999	X	—	—	—

A number of other studies focused on a single aspect of fixture verification. Table 4.1 outlines the content of these studies based on an in-depth survey of literature relevant to fixture verification.

4.1.3   Fixture Modeling

Two models may be created to formulate fixture-workpiece relationships. A geometric model describes the relationship between workpiece displacement and locator displacements. A kinetic model describes the relationship among external forces, fixture deformation, and workpiece displacement.

4.1.3.1 Geometric Fixture Model

In fixture tolerance analysis, one major task is to find workpiece displacement that has resulted from locating point displacements. Figure 4.3 shows three locating points, each with its own tolerance zone. Given the locating point displacements, we want to determine workpiece displacement or displacement of a specific feature of a workpiece. If we know the workpiece displacement or the feature tolerance, we also want to know if the locating point displacements or the maximum locator displacements allowed tolerances. These questions require a model for determining the relationship between workpiece and locating point displacements.

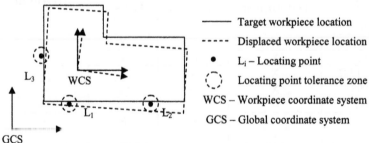

Figure 4.3. Geometric fixture model.

The geometric model is the link between workpiece displacement and locator displacements. It is based on the Jacobian matrix (Asada 1985). The properties of the Jacobian matrix can be used to determine locating performance and locating accuracy. The Jacobian matrix is generally used to formulate the relationship between a 3D object and its locators, as in robotic finger grasping problems (Xiong 1999).

In regard to Figure 4.3, assume the workpiece location is $\{q\}=\{x\ y\ z\ \alpha\ \beta\ \gamma\}^T$. When locating points have displacements $\{\Delta d\}=\{\Delta d_1\ \Delta d_2\ ...\ \Delta d_n\}^T$ along surface normal direction, they will cause the workpiece to be displaced. The displacements between workpiece $\{\Delta q\}$ and locating points $\{\Delta d\}$ can be linked by the Jacobian matrix $[J]$

$$\{\Delta d\} = [J] \cdot \{\Delta q\} \tag{4.1}$$

or $$\{\Delta q\} = [J]^{-1} \cdot \{\Delta d\}, \tag{4.2}$$

where $\{\Delta q\}=\{\Delta x\ \Delta y\ \Delta z\ \Delta\alpha\ \Delta\beta\ \Delta\gamma\}^T$ is the workpiece displacement.

Note that if $[J]$ is singular its pseudo-inverse matrix is used in place of $[J]^{-1}$. From Eqs. 4.1 and 4.2 we can see that once the locating point displacements are known the workpiece displacement can be easily calculated, and vice versa. Eqs. 4.1 and 4.2 serve as the geometric model in CAFDV (its applications include locating performance analysis and tolerance analysis, discussed in material to follow).

4.1.3.2   Kinetic Fixture Model

When external forces (gravity, clamping, machining forces) are applied to a workpiece, the fixture will deform, and the workpiece will be displaced, as indicated in Figure 4.4.

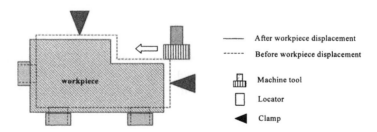

Figure 4.4. Kinetic fixture model.

The kinetic fixture model is used to formulate relationships among external force, workpiece displacement, and fixture reaction forces. Given clamping and machining forces, we are able to calculate fixture reaction forces and workpiece displacement.

In regards to fixturing, we assume that the workpiece is rigid-body (although its actual deformation could be obtained from FEA results) and that fixtures are linear elastic bodies. We also assume that there are friction forces between fixture components and the workpiece. For the workpiece, external forces are balanced by fixture reaction forces.

The workpiece is stable when total external wrench is balanced by total internal wrench, a situation created by fixture reaction forces due to workpiece displacement and fixture deformation. The equilibrium equation for this condition is expressed as

$$\{W_i\} + \{W_e\} = [K] \cdot \{\Delta q\} + \{W_e\} = 0 \qquad (4.3)$$

or $$[K] \cdot \{\Delta q\} = -\{W_e\}; \qquad (4.4)$$

where $\{\Delta q\} = \{\Delta x \;\; \Delta y \;\; \Delta z \;\; \Delta \alpha \;\; \Delta \beta \;\; \Delta \gamma\}^T$ is the workpiece displacement, $\{W_i\} = \{F_{ix}, F_{iy}, F_{iz}, M_{ix}, M_{iy}, M_{iz}\}^T$ is the internal wrench by reaction forces, $\{W_e\} = \{F_{ex}, F_{ey}, F_{ez}, M_{ex}, M_{ey}, M_{ez}\}^T$ is the external wrench, and [K] is the fixture stiff matrix (discussed in Section 4.4).

4.1.4   Locator and Locating Points

There exist many types of locators in fixture design, each with unique geometry and other properties. It is desirable for fixture models to handle various types of locators regardless of their detail geometries.

For this purpose, locators are abstracted in the analysis process as points only. They are abstracted in such a way that a locator and its equivalent locating points constrain the same number of degrees of freedom (DOF) of the workpiece, provide the same level accuracy, and have the same stiffness. This equivalency is achieved through the conversions of locator geometry, tolerance, and stiffness information.

4.1.4.1 Geometry Conversion

The conversion of geometric information (position and surface normal direction) between locators and locating points is shown in Figure 4.5. The number of locating points associated with a locator equals the number of DOF of the workpiece constrained by the locator. It is possible that two locating points share the same position but have different normal directions (Figure 4.5c). Currently, seven commonly used locators are included in the work. More locator types can be added to similar procedures. The conversion of tolerance and stiffness between locators and locating points is similar but more complicated (Kang 2001).

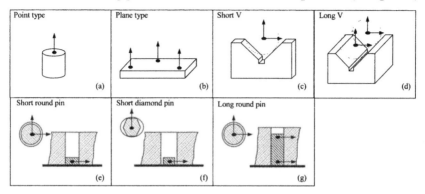

Figure 4.5. Locator types and locating points.

4.1.5 Locating Analysis

Locating analysis involves two major tasks: locator layout evaluation and locator layout optimization. Locator layout is the positioning of locators. A sound layout design is vital to the success of the entire fixture design. Figure 4.6 shows two similar layouts with different bottom-locating positions.

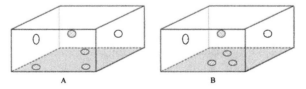

Figure 4.6. Locating performance analysis.

In layout B, three bottom locators are closer to each other than they are in layout A. Intuitively, we can tell that layout A is better because it looks more stable. This stability is defined in this section. Based on the definition, the layout can be optimized.

4.1.5.1   Locator Layout Evaluation

A locator layout is evaluated through two measurements: the number of workpiece DOF constrained by the locators and the locating performance Index (LPI).

Constrained DOF of Workpiece. In Asada's work (1985), it is pointed out that the workpiece DOF constrained by a fixture equal the rank of the Jacobian matrix. This result is further extended so we can assess further details of a locator layout design.

A workpiece is <u>well constrained</u> if the fixture has six locating points and constrains all six DOF of the workpiece. This is the ideal configuration for fixture designs, among which the 3-2-1 setup and its equivalent are the most popular (Figure 4.7a).

A workpiece is <u>under constrained</u> if there exists at least one direction of motion unconstrained. Although there may be six locating points in the fixture design, the workpiece DOF constrained could be less than six. Figure 4.7b shows a bottom-locating surface with three locating points. This layout constrains two DOF with three locating points and is therefore under constrained.

A workpiece is <u>over constrained</u> if there exists a subset of locating points the number of which is greater than workpiece DOF constrained by them. Figure 4.7c shows a bottom-locating surface with four locating points. This layout constrains three DOF with four locating points and is therefore over constrained. An over constrained workpiece is likely to have deflection under clamping and machining forces if locating points are not perfectly aligned.

Via the previous definitions it is possible for a workpiece to be under constrained and over constrained at the same time. Figure 4.7d shows an example of this situation.

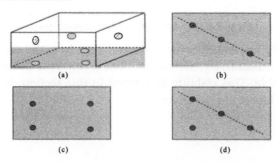

Figure 4.7. Locator layout and constrained DOF.

Locating Performance Index. The LPI measures a fixture ability of tolerating locating errors. In regard to Figure 4.6 between layout designs A and B, although both constrain six DOF of the workpiece, design A obviously has better performance than design B. The reason is simple: design A will have less workpiece overall displacement given the same locator displacements. In other words, it can tolerate more locating errors and achieve higher locating accuracy. This performance index can be applied to more complex locating layouts.

In multi-fingered robot finger-grasping studies, the concept manipulability is used to measure the control of the grasp over the workpiece (Xiong 1998). With given finger movements, the grasp with less workpiece displacement has greater manipulability. In other words, this grasp is able to control the workpiece movement more precisely.

This concept of manipulability is very similar to the locating performance concept discussed previously (i.e., both try to minimize workpiece displacement). Thus, the definition of manipulability can be borrowed to define the LPI as

$$\text{LPI} = \sqrt{\text{gram}([\textsc{j}])} = \sqrt{\left\| [\textsc{j}]^{\text{T}} \cdot [\textsc{j}] \right\|} \text{,} \tag{4.5}$$

where $[\textsc{j}]$ is the Jacobian matrix from the geometric fixture model, $\left\| [\textsc{j}] \right\|$ is the determinant of matrix $[\textsc{j}]$, and $\text{gram}([\textsc{j}]) = \left\| [\textsc{j}]^{\text{T}} \cdot [\textsc{j}] \right\|$ is the grammian of a matrix. LPI is always greater than zero, and its value depends on the size of the workpiece.

4.1.5.2 Locator Layout Optimization

From the last section, we know that a layout design with maximum LPI provides minimum workpiece displacement and therefore maximum locating accuracy. Based on LPI, a locator layout can be optimized. Even if the initial locating positions are unknown, they can be generated and then optimized. The procedure for locator layout optimization (and initial locating position generation) is as follows:

- Find the search space (i.e., all possible surface areas for each locating point).
- Determine the constraints between locating points.
- Generate initial positions for the locating points.
- Search the best positions for locating points (with greatest LPI).

Search Space Representation. The search space for a locating point is the region in which the locating point can be positioned. Because locating points are abstracted from different types of locators, they have different search spaces.

For point and plane locators, locating points are created on surfaces and the searchable areas are the locating surfaces, which are represented by UV parameters ($0 \leq U, V \leq 1$). For short-V locators, locating points are created on the axis of the cylindrical surface and the searchable area is on the axis, which is represented by parameter U ($0 \leq U \leq 1$). For other types of locators (round pin and diamond pin), their respective positions are fixed once the locating surface is known, and thus they do not have a searchable area.

Constraints Between Locating Points. Because locating points are abstracted from locators, there are some constraints between certain locating points. In other words, the locating points are not totally independent of each other.

For short-V locators, the two locating points always share the same position but with directions perpendicular to each other. For pin-hole locating (Figure 4.8) two locating points of the round pin share the same position (one points to the diamond pin and the other is perpendicular with the first). The direction of the locating point from the diamond pin is the same as that of the second locating point of the round pin. All these constraints need to be satisfied in optimizing locating point positions.

Short round pin   Short diamond pin

Figure 4.8. Locating points for pin-hole locating.

Initial Position Generation. Theoretically, initial locating points could be anywhere in the searchable areas. These are easily determined based on experience. For point-type and plane-type locators, locating points are generated around the center position of the locator on the surface of the workpiece. For "short-V" locators, the initial locating points are generated around the center of the axis.

The reason they are generated around center instead of at center is that if two or more points have exactly the same position the determinant of the Jacobian matrix will be zero (det[J]=0), which will prevent coincident points from being optimized. A random-number generator is used to generate the initial locating positions around the surface (axis) center. The difference between two initial point coordinates resides within a specified range.

Position Optimization. Locating positions are optimized by searching for the best position for each locating point. A "better" position is defined as increasing overall fixture LPI.

As illustrated in Figure 4.9, the locating surface is discretized into grids. The locating point was at position 0, and the LPI of the total locator layout is calculated as LPI(0). Then the LPIs are calculated when this locating point is at positions 1, 2, 3, and 4, which translate respectively as LPI(1), LPI(2), LPI(3), and LPI(4). Compare LPI(0) through LPI(4). The position with maximum LPI is the best position among these five points. If it is not position 0, then this locating point is moved to the new position with maximum LPI.

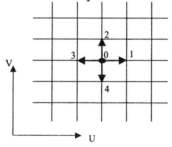

Figure 4.9. Layout optimization on surface.

For each iteration, each locating point position is optimized as described previously. The iteration stops when none of the locating points needs further optimization (none moved) or user-specified maximum iteration number is reached.

For a searchable area that is an axis, the procedure is similar to that for a surface, except it is 1D instead of 2D. Given the surface, it is very possible that the final locating point position is on the edge of the outer loop of the surface. In addition, because the parameter of a surface does not have associated information on surface details, such as a hole or a boss on the surface, it is possible that the point is located in an inaccessible area. For such types of problems, post processing is needed to adjust the locating point positions to generate a feasible result.

A margin percentage can be set to define the minimal distance allowed for a locating point to be close to the surface outer loop, based on fixture component size and workpiece geometry. Figure 4.10 shows a flowchart of procedures for locator layout optimization.

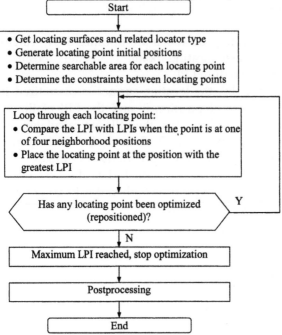

Figure 4.10. Layout optimization flowchart.

4.1.6   Implementation and Integration with CAD

A realistic and nontrivial problem facing CAFDV software design is the variety of today's CAD systems and operating systems. To maximize the portability of CAFDV among different CAD systems and operating systems and to minimize maintenance cost, the CAFDV software is divided and capsulated into modules so that common modules can be reused as much as possible.

Figure 4.11 shows a diagram of the software architecture. The CAFDV software contains four modules (shaded in the figure), and each module is functionally self-contained. An arrow from module A to B indicates the dependency of module B on A. Table 4.2 outlines the descriptions of all modules.

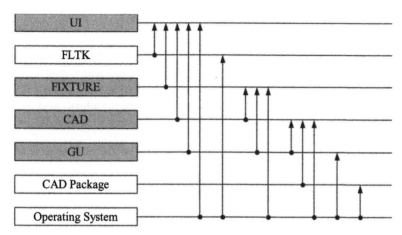

Figure 4.11. CAFDV software architecture.

Table 4.2. CAFDV software modules.

Module Name	Description
UI	Functions related to user interface, such as the windows, menu bars, dialog boxes and so on.
FLTK	FLTK is a third-party multiplatform user interface library.
FIXTURE	Functions related to fixture design algorithms, such as finding the Jacobian matrix, tolerance assignment, and so on.
CAD	Functions related to CAD functions, such as selecting a surface interactively, getting the part name, mass center, and so on.
GU	General utilities. This is a set of functions as a utility library. It includes data structure, matrix, geometry, and other utilities.
CAD Package	The API (application program interface) provided by the CAD package to allow access to its geometry data.
Operating System	We used IRIX.

The CAFDV system has been developed and integrated with CAD. Figure 4.12 shows the startup screen of the software. The following is setup information for a real world case in production.

Workpiece: The workpiece is a simplified engine block (Figure 4.13).

Machining surfaces: The machining surfaces under this setup are the bottom surfaces of the four lugs on both sides of the engine block (as circled in Figure 4.13a). Each machining surface has two tolerances — a surface profile of 0.08 and a parallelism of 0.04, with the bottom-locating surface as datum.

Locating surfaces and locators: There are two point-type locators on surface A, and one on surface B. A short round pin locator and a short diamond pin locator are placed in the holes on surface B (Figure 4.13b).

Users can choose locator type, set locator parameters, and select locator position interactively using a CAD system (Figure 4.14). There are five locators in this case.

- Locator 1: Point-type locator on surface A.
- Locator 2: Point-type locator on surface A.
- Locator 3: Point-type locator on surface B.
- Locator 4: Short-round pin locator inside hole I on surface B.
- Locator 5: Short-diamond pin locator inside hole II on surface B.

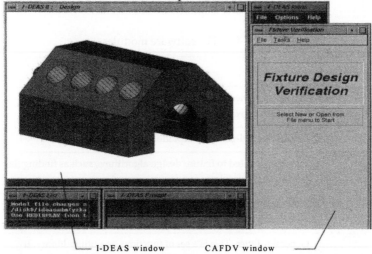

I-DEAS window   CAFDV window

Figure 4.12. CAFDV CAD integration: startup screen.

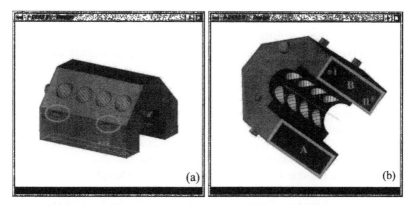

Figure 4.13. Workpiece model with locating specifications.

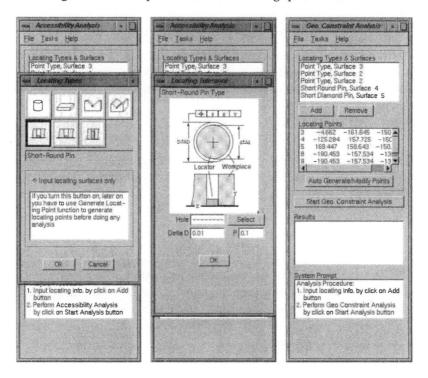

Figure 4.14. Locator selection and positioning.

After locators are selected and positioned, they are converted into equivalent locating points. The following is data for the six locating points converted from locators, including their positions and surface normal direction.

Point 1 – from Locator 1 Position: -1.2528e+02 1.5772e+02 -1.5000e+02
    Normal: 0.0000e+00 0.0000e+00 -1.0000e+00
Point 2 – from Locator 2 Position: 1.6945e+02 1.5864e+02 -1.5000e+02
    Normal: 0.0000e+00 0.0000e+00 -1.0000e+00
Point 3 – from Locator 3 Position: -4.6620e+00 -1.6164e+02 -1.5000e+02
    Normal: 0.0000e+00 0.0000e+00 -1.0000e+00
Point 4 – from Locator 4 Position: -1.9045e+02 -1.5753e+02 -1.3000e+02
    Normal: -1.0000e+00 0.0000e+00 0.0000e+00
Point 5 – from Locator 5 Position: -1.9045e+02 -1.5753e+02 -1.3000e+02
    Normal: 0.0000e+00 -1.0000e+00 0.0000e+00
Point 6 – from Locator 6 Position: 1.8977e+02 -1.5753e+02 -1.3000e+02
    Normal: 0.0000e+00 -1.0000e+00 0.0000e+00

Based on workpiece location (in this case a 4×4 identity matrix), and locating point positions and normal directions, the Jacobian Matrix can be constructed (Kang 2001). In this case, the Jacobian matrix looks as foolows:

```
0.0000e+00  0.0000e+00  1.0000e+00   1.5772e+02   1.2528e+02   0.0000e+00

0.0000e+00  0.0000e+00  1.0000e+00   1.5864e+02  -1.6945e+02   0.0000e+00

0.0000e+00  0.0000e+00  1.0000e+00  -1.6164e+02   4.6620e+00   0.0000e+00

1.0000e+00  0.0000e+00  0.0000e+00   0.0000e+00  -1.3000e+02   1.5753e+02

0.0000e+00  1.0000e+00  0.0000e+00   1.3000e+02   0.0000e+00  -1.9045e+02

0.0000e+00  1.0000e+00  0.0000e+00   1.3000e+02   0.0000e+00   1.8977e+02
```

The rank of this Jacobian matrix is 6, indicating the workpiece is well constrained. The LPI can be calculated once the Jacobian matrix is available, following the procedure outlined in Section 4.1.5: LPI(1) = 3.58304e+007. If the locator layout changes so that locator 2 is closer to locator 1 at new position via

Position: 0.0000e+00 1.5864e+02 -1.5000e+02, and
Normal: 0.0000e+00 0.0000e+00 -1.0000e+00,
New LPI is computed as LPI(2) = 1.52546e+007.

Comparing LPI(1) with LPI(2), we can see the previous layout has a greater LPI, and thus better overall locating accuracy.

4.1.7 Summary

A framework for CAFDV has been presented. In this work, two fixture models (geometric and kinetic) are established. In the geometric model, the Jacobian matrix links the workpiece displacement with locator displacements. In the kinetic model, the fixture stiffness matrix links the external force with workpiece displacement and fixture deformation. Locator layout verification and optimization were also presented in this section, as well as theie implementation in a CAD system.

4.2 FIXTURING TOLERANCE ANALYSIS

Tolerance analysis is the most important issue in CAFD, and it is an important implementation in CAFDV. This study presents a new approach to fixture tolerance analysis that is more generalized and that can be used to assign locator tolerances based on machining surface tolerance requirements. Such tolerance analysis is also capable of handling various types of fixture designs, workpieces, datum features, and machining feature tolerances. Locator tolerance assignment distributes tolerances to locators based on sensitivity analysis.

Automated fixture design technique has been developed in recent years. Once a fixture is designed with a CAFD system, design performance needs to be verified before the fixture is constructed and applied in production. As part of CAFDV, fixture tolerance analysis has been explored extensively. Studies on fixture tolerance analysis can generally be classified into two categories: inter-setup and intra-setup tolerance analysis. Inter-setup fixture tolerance analysis deals with tolerance stack-up and allocation for setup planning (Zhang 2001; Rong 1996; Huang 1994; Lee 1991), whereas intra-setup tolerance analysis deals with tolerance calculation and assignment within a single setup.

This section focuses on tolerance analysis within a single setup in regards to two objectives: machining surface accuracy analysis and locator tolerance assignment. Accuracy analysis predicts machining surface errors based on given locator errors, and tolerance assignment allocates locator tolerances so that they can ensure machining accuracy.

Three perpendicular locating reference planes were established based on locator types and positions. Locator displacements are mapped into the deviations of locating reference planes (Rong 1995). The machining surface deviation is then calculated based on the locating reference plane deviations. A model has been developed that relates to datum error measurement to locator geometric variability (Choudhuri 1999). This model is limited to dimensional and profile tolerances applied to spherical tip locators, planar workpiece datum features, and linear machined features bounded by planar workpiece surfaces.

In this research, the Jacobian matrix is used to formulate the relationship between locating point displacements and workpiece displacement (Asada 1985). It takes into account error caused by both locator position inaccuracy and locator deformation. Given locator tolerance and displacement, this model can predict the deviation for any machining surface. Given machining surface tolerances, it can assign the tolerances to locators. There is no limitation as to which types of locator or tolerance can be included in this model.

The rest of this section is organized as follows. First, the geometric fixture model is briefly introduced, along with its role in tolerance analysis. Then the tolerances are defined based on surface sample points, followed by detailed discussion on accuracy analysis and tolerance assignment. An example and conclusions follow.

4.2.1   Tolerance Analysis Overview

In CAFDV, tolerance analysis has two tasks: machining surface accuracy analysis and locator tolerance assignment. The former calculates machining surface accuracy with given locator tolerances, and the latter finds optimal locator tolerances based on machining surface tolerance requirements. Figure 4.15 illustrates the relationship between accuracy analysis and tolerance assignment.

The relationship between locator displacements and machining surface deviation is an n-to-1 relationship (i.e., the machining surface displacement can be uniquely determined with a given set of locator displacements, but not the other way around). There is no single solution to tolerance assignment, and to find the optimal solution locator sensitivity analysis needs to be performed.

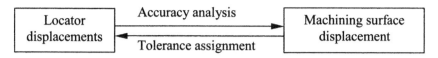

Figure 4.15. Accuracy analysis and tolerance assignment.

The tolerance analysis in this study is based on the geometric fixture model, which uses the Jacobian matrix to formulate the relationship between fixture and workpiece displacements (Asada 1985; Kang 2003). This relationship can be expressed as

$$\{\Delta d\} = [J] \cdot \{\Delta q\} \tag{4.6}$$

or $$\{\Delta q\} = [J]^{-1} \cdot \{\Delta d\}, \tag{4.7}$$

where $\{\Delta d\} = \{\Delta d_1 \ \Delta d_2 \ ... \ \Delta d_n\}^T$ is the locating point displacements and $\{\Delta q\} = \{\Delta x \ \Delta y \ \Delta z \ \Delta \alpha \ \Delta \beta \ \Delta \gamma\}^T$ is the workpiece displacement.

From Eqs. 4.6 and 4.7, we can see that once the locating point displacements are known the workpiece displacement can be calculated, and vice versa (if the Jacobian matrix is a known constant). Once the workpiece displacement is known, the machining surface deviation can be obtained. This chain relationship is the foundation for fixture tolerance analysis. Figure 4.16 shows the relationship among locating point displacements, workpiece displacement, and machining surface deviation.

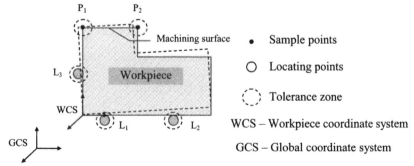

Figure 4.16. Tolerance analysis.

4.2.2   Definition of Surface Deviation and Accuracy

Because maximum deviation always occurs at points on a surface contour, sample points can be used to represent the machining surface in tolerance analysis to enhance computer implementation simplification and ease of use. Figure 4.17 shows an example of surface sample points.

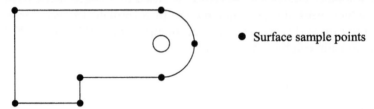

Figure 4.17. Surface sample points.

Once surface sample point displacements are known, deviation can be calculated relative to the ideal datum based on the tolerance type specified on the surface. For each of the 14 tolerance types defined in ANSI Y14.5M (ANSI 1994), the deviation is calculated differently. Figure 4.18 shows the target surface and the deviated surface, along with their sample points.

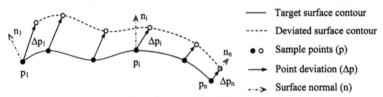

Figure 4.18. Surface deviation.

4.2.2.1 Surface Profile and Line Profile Deviation

For surface and line profiles, tolerance zones are defined as double the maximum sample point deviation in the direction normal to the surface or perpendicular to the line axis. They can be calculated as

$$\text{dev} = 2 \times \max\{\Delta p_1^n \quad \Delta p_2^n \quad \cdots \quad \Delta p_n^n\}, \tag{4.8}$$

where $\Delta p_i^n = \Delta p_i \cdot n_i$ is the sample point deviation along surface normal direction.

4.2.2.2 Parallelism, Perpendicularity, and Angularity Deviation

For parallelism, perpendicularity, and angularity, their surface deviations are calculated as the difference between maximum and minimum sample point deviations in the direction normal to the surface according to

$$\text{dev} = \max\{\Delta p_1^n \quad \Delta p_2^n \quad \cdots \quad \Delta p_n^n\} - \min\{\Delta p_1^n \quad \Delta p_2^n \quad \cdots \quad \Delta p_n^n\}. \tag{4.9}$$

4.2.2.3 Position Deviation

The deviation calculation for position type is a little different from other types. The sample points are derived from the cylinder axis instead of from the surface contour. The deviation is defined as double the maximum deviation from the target axis (Figure 4.19) according to

$$\text{dev} = 2 \times \max\{\Delta d_1^n \quad \Delta d_2^n \quad \cdots \quad \Delta d_n^n\}. \tag{4.10}$$

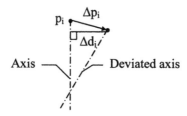

Figure 4.19. Position deviation

Other types of deviation, such as plane surface flatness, cylindrical surface runout, and symmetry, are not incorporated here because they are not affected by locator displacements.

4.2.3 Machining Surface Accuracy Analysis

Machining surface accuracy is defined by the worst case of all possible surface deviations. The task is to find the largest machining surface deviation under a given set of locating point deviations.

When locating points entail deviations, they will cause deviation of the workpiece. Sample points will also have deviations, which can be used to calculate surface deviation.

As shown by the geometric fixture model, once the locating point deviations $\{\Delta d\}$ are known the workpiece location deviation $\{\Delta q\}$ can be calculated as

$$\{\Delta q\} = [J]^{-1} \cdot \{\Delta d\}. \tag{4.11}$$

Let $\{q_0\}$ be the ideal workpiece location, $T_G^W(q)$ be the 4 x 4 workpiece transformation matrix based on location $\{q\}$, and $\{P_i^W\}$ be the surface sample point coordinates in WCS. We can then express sample point deviations in GCS $\{\Delta P_i^G\}$ as

$$
\begin{aligned}
\Delta P_i^G &= P_{i2}^G - P_{i1}^G = \left[T_G^W(q_0 + \Delta q)\right] \cdot P_i^W - \left[T_G^W(q_0)\right] \cdot P_i^W \\
\Rightarrow \Delta P_i^G &= \left[T_G^W(q_0 + \Delta q) - T_G^W(q_0)\right] \cdot P_i^W \\
\Rightarrow \Delta P_i^G &= \left[T_G^W\left(q_0 + [J]^{-1} \cdot \{\Delta d\}\right) - T_G^W(q_0)\right] \cdot P_i^W.
\end{aligned}
\tag{4.12}
$$

WCS and GCS have been established (Figure 4.16). For a given set of locating point deviations $\{\Delta d\}$, machining surface deviation can be calculated via the "definition of surface deviation" (given previously) as

$$
\begin{aligned}
\Delta p_i^n &= \Delta p_i \cdot n_i, \\
\mathrm{dev} &= \mathrm{dev}\{\Delta p_1^n \quad \Delta p_2^n \quad \cdots \quad \Delta p_n^n\}
\end{aligned}
\tag{4.13}
$$

By varying the locating point displacements in the locating point tolerance zone, we can obtain a set of machining surface deviations. Machining surface accuracy is determined by the worst case of all surface deviations.

$$\mathrm{acc} = \max\{\mathrm{dev}_1 \quad \mathrm{dev}_2 \quad \cdots \quad \mathrm{dev}_m\} \tag{4.14}$$

Figure 4.20 shows a block diagram of the machining surface accuracy check. It should be mentioned that although the worst-case scenario is used in this research the method presented here can also be applied to statistical tolerance analysis (e.g., the application of Monte Carlo simulation methods).

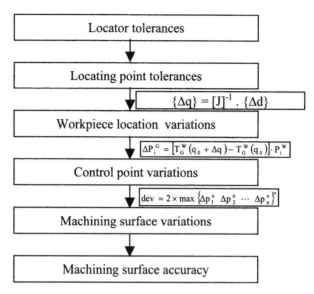

Figure 4.20. Machining surface accuracy analysis for profile tolerance deviation.

4.2.4   Locator Tolerance Assignment

Locator tolerance assignment is used to find the tolerance specifications of locators so that all machining surface tolerance requirements can be satisfied. When machining surface accuracy is assured under relatively low locating tolerances, production cost is reduced. To reasonably distribute tolerances to each locator, we first need to find out how sensitive the machining surface is to each locator. More sensitive locators should require tighter tolerance specifications.

4.2.4.1 Surface Sensitivity on Locators

Sensitivity analysis is used to evaluate the relationship between surface deviation and locating-point deviation. Such analysis measures the impact individual locating point deviations have on machining surface accuracy and on tolerance distribution to locating points according to respective sensitivity measurements.

For machining surface tolerance T_j ($j = 1, ..., m$), let P_i ($i = 1, ..., n$) be the locating point and $\{\Delta d\} = \{0 \, ... \, 1 \, ... \, 0\}$ (only the ith element is 1) be the locating point deviation in normal direction. The surface deviation resulting from unit deviation of the locating point is then

$$\text{dev}_{ij} = \text{dev}(\Delta d).$$
(4.15)

The sensitivity of the surface on the locating point S_{ij} can be found by normalizing the deviations for all locating points as following.

$$S_{ij} = \frac{\text{dev}_{ij}}{\text{dev}_{1j} + \text{dev}_{2j} + \cdots + \text{dev}_{nj}} \qquad \left(\sum_{i=1}^{n} S_{ij} = 1 \right)$$
(4.16)

A sensitivity matrix can then be constructed for all surface tolerances and locating points.

4.2.4.2 Tolerance Distribution

For each machining surface tolerance, locating point tolerances are assigned based on their sensitivities. In the case of multiple machining surface tolerances, the tightest tolerance is selected as the final tolerance for each locating point. This procedure is outlined in the following.

For machining surface tolerance T_j ($j = 1, ..., m$), an initial tolerance t_o is used to assign the locating point tolerances t_{ij} ($i = 1, ..., n$), based on their sensitivities. This is done through a weighting factor w_{ij}:

$$t_{ij} = w_{ij} \cdot t_o.$$
(4.17)

The initial tolerance t_{ij} can be determined according to the economic manufacturing accuracy in general production. A point with larger sensitivity should have tighter tolerance, and thus w_{ij} is defined as

$$w_{ij} = 1 - k \cdot S_{ij}.$$
(4.18)

The factor k is to prevents zero tolerance when the sensitivity $S_{ij} = 1$. It can be tuned to achieve an optimal result. In our implementation of locator tolerance assignment, k = 0.9 is assumed. Combining the forgoing equations, locator tolerances are assigned as

$$t_{ij} = t_o \cdot (1 - k \cdot S_{ij}).$$
(4.19)

In the case of multiple tolerances on a machining surface, first the locating point tolerance is assigned for all surface tolerances, and then the tightest tolerance among them is selected as the final locating point tolerance. This is shown in Table 4.3.

Table 4.3. Tolerance assignment for multiple surface tolerances.

Locating Point Tolerance t_{ij}	**Machining Surface Tolerances** T_j (j=1, …, m)				**Final Locating Point Tolerance** t_i
Locating Points P_i (i=1, …, n)	t_{11}	t_{12}	$\cdots$	t_{1m}	$t_1 = \min\{t_{11} \ \cdots \ t_{1m}\}$
	t_{21}	t_{22}	$\cdots$	t_{2m}	$t_2 = \min\{t_{21} \ \cdots \ t_{2m}\}$
	$\vdots$	$\vdots$		$\vdots$	$\vdots$
	t_{n1}	t_{n2}	$\cdots$	t_{nm}	$t_n = \min\{t_{n1} \ \cdots \ t_{nm}\}$

With tolerance assigned, positional accuracies of locators ensure that all machining surface tolerances will be satisfied. Figure 4.21 shows the procedure for tolerance assignment.

4.2.5 Implementation

The tolerance analysis functions in CAFDV have been integrated with a CAD system. The material that follows describes a simple case study that demonstrates the effectiveness of tolerance analysis functions.

Figure 4.22 shows the workpiece and setup information under study. The workpiece is a simplified V-8 engine block from a real case. The machining surfaces under this setup are the bottom surfaces of the four lugs on both sides of the engine block (as circled in Figure 4.22a). Each machining surface has two tolerances (a surface profile of 0.08 mm and a parallelism of 0.04 mm) with the bottom-locating surface serving as datum. The bottom surface (the combination of surfaces A and B shown in Figure 4.22b) is the primary locating surface and constrains three DOF. A short round pin locator and a short diamond pin locator are placed in the holes on surface B (Figure 4.22b).

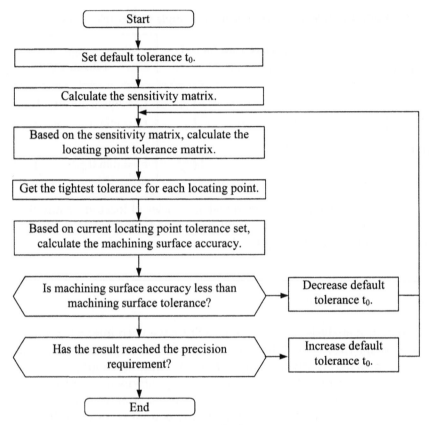

Figure 4.21. Procedure of tolerance assignment.

After conversion of locators to locating points, the Jacobian matrix is constructed. When the locating plan has been verified through geometric constraint analysis, locating position optimization (Kang 2003), and accessibility analysis (Li 1999), tolerance analysis is conducted.

4.2.5.1 Locator Tolerance Assignment

To satisfy the two tolerance specifications of the machining surface, locator tolerances are assigned following the procedures discussed previously. The following are locator tolerances assigned. The notations are illustrated in Figure 4.23. More details can be found in (Kang 2001).

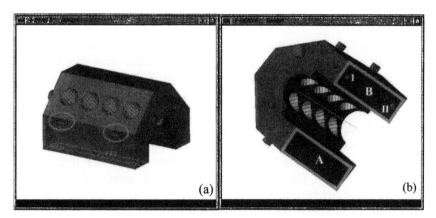

Figure 4.22. Workpiece and setup information.

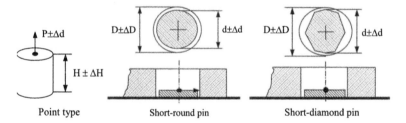

Figure 4.23. Locator tolerances.

- Locator 1, point-type locator: $\Delta h = 0.0193046$
- Locator 2, point-type locator: $\Delta h = 0.0225628$
- Locator 3, point-type locator: $\Delta h = 0.0128038$
- Locator 4, short-round pin locator:
  - $D = 34.0137$ (from workpiece)
  - $d = 33.9827$
  - $\Delta d = 0.0210339$
  - $\Delta D = 0.01$
- Locator 5, short-diamond pin locator:
  - $D = 33.9791$ (from workpiece)
  - $d = 33.948$
  - $\Delta d = 0.0210339$
  - $\Delta D = 0.01$

4.2.5.2 Machining Surface Accuracy Analysis

If locator tolerances are known, we can also analyze machining surface accuracy. If we use the tolerances assigned in the previous step to check the surface profile, we will get the profile as 0.0770921, which is slightly less than the specified 0.08. This result also validates the correctness of tolerance assignment.

If we change the tolerance of locator 1 from 0.0193046 to 0.15 and perform the machining surface accuracy check again, the surface profile accuracy is calculated to be 0.0710263 (Figure 4.24). The result is in agreement with the fact that when the locator tolerance tightens it provides higher machining accuracy.

It should be noted that the locating error is not the only source resulting in the machining surface error. When other process errors are considered, the machining surface accuracy needs to be higher than the product design specifications. It could be 1/5-1/3 of the overall tolerance specifications in final design requirements, based on general knowledge and experience in manufacturing.

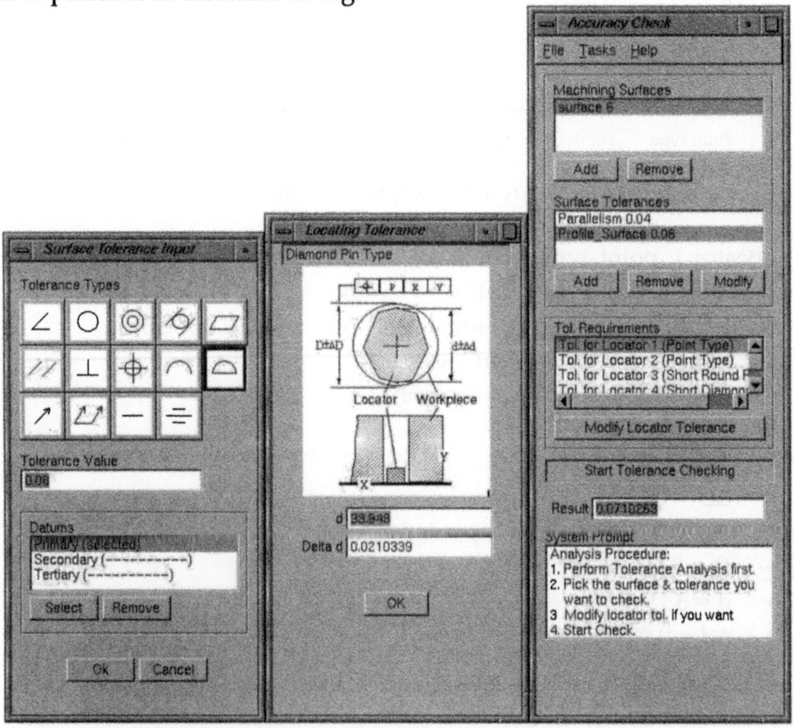

Figure 4.24. Tolerance assignment and accuracy analysis.

4.2.6   Summary

In CAFDV, when locator tolerances are given machining surface accuracy can be predicted. On the other hand, given machining surface tolerance specifications, locator tolerances can be determined to satisfy requirements. To achieve generality, the Jacobian matrix is adopted to formulate the fixture-workpiece relationship. The locators are represented with equivalent locating points and machining surfaces are represented with surface sample points. For computer implementation, machining surfaces are represented by their sample points. Six fixture-related tolerances are then defined with the surface sample points. In locator tolerance assignment, surface sensitivities on locating points are measured to best distribute tolerances among locating points.

4.3 FIXTURING STABILITY ANALYSIS

In fixture design, workpieces need to remain stable throughout fixturing and machining processes to ensure safety and achieve machining accuracy. This requirement is verified by one function of the CAFDV system. This section presents a methodology of fixturing stability analysis in CAFDV. A kinetic fixture model is created to formulate the stability problem, and a fixture stiffness matrix (FSM) is derived to solve the problem. This approach not only verifies fixturing stability but finds minimum clamping forces, fixture deformation, and fixture reaction forces. Clamping sequences can also be verified with this approach.

4.3.1   Overview

The stability analysis module of CAFDV verifies workpiece stability during clamping and machining. These stabilities cannot be easily verified when friction forces are taken into consideration. To achieve this, a kinetic fixture model is created to formulate the stability problem mathematically, and an FSM is derived to solve the problem.

Using the kinetic fixture model, related problems can be solved. The first is to find the minimum clamping forces required to stabilize the workpiece during the machining process. The second is to measure fixture deformation and its impact on machining accuracy. And the third is to measure the influence of the clamping sequence on fixturing stability.

There is a great deal of literature on stability analysis regarding both fixturing and robotic grasping. There are many different assumptions, approaches, and applications including stability analysis, including the consideration of friction force, workpiece and fixture deformation, clamping sequence, and the like. Table 4.4 compares various studies.

Table 4.4. Comparison of studies on fixturing stability.

Study	Friction force	Minimum clamping force	Clamping sequence	Workpiece deformation	Fixture deformation	FEA method	Optimization
Chou 1989	–	X	–	–	–	–	–
Lee 1991	X	X	–	–	–	–	–
Cogun 1992	–	–	X	–	–	–	–
Trappey 1992	X	–	–	–	–	–	–
Xiong 1998	–	–	–	–	–	–	X
Rong 1994	X	–	–	–	–	–	–
Chen 1995	–	–	X	–	–	–	–
King 1995	–	–	–	–	–	–	X
Wu 1995	X	–	–	–	–	–	–
DeMeter 1998	–	–	–	X	–	X	X
Kashyap 1999	–	–	–	X	–	X	X

4.3.2   Kinetic Fixture Model

When a workpiece remains stable during machining process (Figure 4.25), it is balanced by two types of forces: external and internal forces. External forces are active forces (including clamping and machining forces) and internal forces are reactive forces, including reaction forces (with friction) from locators.

To isolate the problem, fixture components are assumed to be linear elastic bodies, and the workpiece is assumed to be a rigid body. Note that integration with workpiece deformation is a separate effort and is outside the scope of this study presented here. When under external forces, locators may deform, and the workpiece may be displaced. In a stable situation, the workpiece may eventually be balanced between external forces and reaction forces generated from locator deformations.

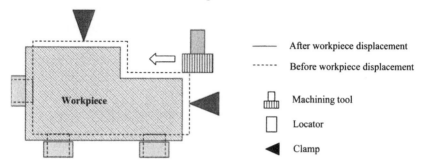

Figure 4.25. Kinetic fixture model.

To establish the workpiece equilibrium equation, we use the wrench from the screw-theory (Ohwovoriole 1981) to represent the force and moment generated at a point. A wrench in 3D space $\{W\}=\{F_x, F_y, F_z, M_x, M_y, M_z\}^T$ consists of three force elements and three torque elements. The workpiece is stable when the total external wrench is balanced by the total internal wrench, which is generated by fixture reaction forces due to workpiece displacement and fixture deformation. This equilibrium equation is

$$\{W_i\}+\{W_e\}=[K]\cdot\{\Delta q\}+\{W_e\}=0 \qquad (4.20)$$

or $\qquad [K]\cdot\{\Delta q\}=-\{W_e\}, \qquad (4.21)$

where $\{\Delta q\}=\{\Delta x\ \Delta y\ \Delta z\ \Delta\alpha\ \Delta\beta\ \Delta\gamma\}^T$ is the workpiece displacement, $\{W_i\}=\{F_{ix}, F_{iy}, F_{iz}, M_{ix}, M_{iy}, M_{iz}\}^T$ is the internal wrench by reaction forces, $\{W_e\}=\{F_{ex}, F_{ey}, F_{ez}, M_{ex}, M_{ey}, M_{ez}\}^T$ is the external wrench, and $[K]$ is the fixture stiff matrix (discussed in material to follow).

4.3.2.1 Derivation of the Fixture Stiffness Matrix

This section explores the derivation of the FSM in detail. First, three types of coordinate systems used in this section are introduced, followed by the concept of contact point stiffness. Then the derivation of the FSM is listed in five steps.

Three Coordinate Systems. The following three types of coordinate systems (CSs) are used (Figure 4.26).

- Global coordinate system (GCS): CS fixed in 3D space. It serves as the ultimate reference frame for all other coordinate systems
- Workpiece coordinate system (WCS): CS attached to each workpiece. In CAD packages, this is determined by the user at the workpiece modeling creation.
- Local coordinate system (LCS): CS attached to each contact point. It is generated based on locating position and locator orientation.

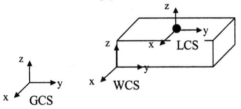

Figure 4.26. Global, workpiece, and local coordinate systems.

Contact Point Stiffness. The contact (locating/clamping) point is modeled as a linear elastic element with its stiffness in three directions, $\{k_x, k_y, k_z\}$, and in touch with the workpiece surface (Figure 4.27).

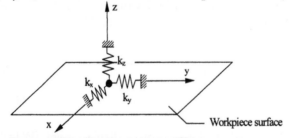

Figure 4.27. Contact point stiffness.

A contact point represents either a locating point or a clamping point, depending on the circumstance. When external forces are applied, the workpiece displaces and the contact point displaces with the surface. The displacement of the contact point indicates the displacement of the locator/clamp. Therefore the reaction force applied on the workpiece under the LCS, $\{f^L\}$, is

$$\begin{Bmatrix} f_x^L \\ f_y^L \\ f_z^L \end{Bmatrix} = - \begin{bmatrix} k_x & 0 & 0 \\ 0 & k_y & 0 \\ 0 & 0 & k_z \end{bmatrix} \cdot \begin{Bmatrix} \Delta d_x^L \\ \Delta d_y^L \\ \Delta d_z^L \end{Bmatrix} \tag{4.22}$$

or $\quad \{f^L\} = -[k] \cdot \{\Delta d^L\}.$ $\hspace{3cm}$ (4.23)

Locator stiffness is estimated offline using the FEA method. This stiffness can be converted into equivalent locating point stiffness (Kang 2001).

Formulation Outline. The following are the steps involved in establishing the equilibrium equation. Details of each step are discussed in the sections that follow.

1. Assume that the workpiece displacement in GCS is $\{\Delta q\} = \{\Delta x \ \Delta y \ \Delta z \ \Delta \alpha \ \Delta \beta \ \Delta \gamma\}^T$.
2. Find the contact point displacement $\{\Delta d^G\}$ in GCS using $\{\Delta q\} \rightarrow \{\Delta d^G\}$.
3. Transform the contact point displacement from GCS into LCS suing $\{\Delta d^G\} \rightarrow \{\Delta d^L\}$.
4. Calculate the elastic contact force in LCS using $\{\Delta d^L\} \rightarrow \{f^L\}$.
5. Transform the contact force from LCS into GCS using $\{f^L\} \rightarrow \{f^G\}$.
6. Combine all contact forces into the internal wrench using $\{f^G\} \rightarrow \{W_i\}$.
7. The result of this combination is $\{\Delta q\} \rightarrow \{W_i\} = [K] \cdot \{\Delta q\}$.

Contact Point Displacement in GCS. When the workpiece displaces, the contact point on the workpiece surface displaces too. Because the WCS is attached to the workpiece, the contact point coordinates change in GCS but remain the same in WCS. The displacement of a contact point in GCS is found by the following procedure. First, the contact point is transformed from WCS, $\{p^G\}$, to GCS, $\{p^w\}$ using

$$\{p^G\} = [T_G^W] \cdot \{p^w\}, \tag{4.24}$$

where $[T_G^W]$ is the transformation matrix from WCS to GCS, which is a function of workpiece location $\{q_w\} = \{x_w \ y_w \ z_w \ \alpha_w \ \beta_w \ \gamma_w\}^T$.

We then take the derivative of $\{q_w\}$ on both sides of Eq. 4.24 to obtain

$$\left\{ d(p^{G}) \right\} = \left[\frac{\partial \left(\left[T_{G}^{w} \right] \cdot \left\{ p^{w} \right\} \right)}{\partial q} \right] \cdot \left\{ dq \right\} = [G] \cdot \left\{ dq \right\} \tag{4.25}$$

$$[G] = \left[\left[\frac{\partial T}{\partial x_{w}} \right] \left\{ p^{w} \right\} \quad \left[\frac{\partial T}{\partial y_{w}} \right] \left\{ p^{w} \right\} \quad \left[\frac{\partial T}{\partial z_{w}} \right] \left\{ p^{w} \right\} \quad \left[\frac{\partial T}{\partial \alpha_{w}} \right] \left\{ p^{w} \right\} \quad \left[\frac{\partial T}{\partial \beta_{w}} \right] \left\{ p^{w} \right\} \quad \left[\frac{\partial T}{\partial \gamma_{w}} \right] \left\{ p^{w} \right\} \right],$$

$$\tag{4.26}$$

where $[G]$ is a 3 x 6 matrix. Finding this matrix is similar to finding the Jacobian matrix (Kang 2001). For small displacement (as in fixture deformation), we use the approximation

$$\left\{ \Delta d^{G} \right\} = \left\{ \begin{matrix} \Delta d_{x}^{G} \\ \Delta d_{y}^{G} \\ \Delta d_{z}^{G} \end{matrix} \right\} = [G] \cdot \left\{ \begin{matrix} \Delta x_{w} \\ \Delta y_{w} \\ \Delta z_{w} \\ \Delta \alpha_{w} \\ \Delta \beta_{w} \\ \Delta \gamma_{w} \end{matrix} \right\} = [G] \cdot \left\{ \Delta q \right\}. \tag{4.27}$$

From this equation, we get the relationship between the contact point displacement in GCS, $\{\Delta d^{G}\}$, and the workpiece displacement, $\{\Delta q\}$.

Contact Point Displacement in LCS. If the contact point displacement in GCS, $\{\Delta d^{G}\}$, is known, this displacement in LCS, $\{\Delta d^{L}\}$, can be calculated by transforming it from GCS to LCS as

$$\left\{ \Delta d^{L} \right\} = \left[T_{G}^{L} \right]^{-1} \cdot \left\{ \Delta d^{G} \right\}. \tag{4.28}$$

Contacting Force in LCS. At each contact point, the contact force in LCS, $\{f^{L}\}$, is generated by point displacement. Because we know that the stiffness matrix of the contact point is $[k_{i}]$ and that local displacement is $\{\Delta d_{i}^{L}\} = \{\Delta d_{ix}^{L} \quad \Delta d_{iy}^{L} \quad \Delta d_{iz}^{L}\}^{T}$, the contact force in LCS, $\{f^{L}\}$, can be expressed as

$$\left\{ f_{i}^{L} \right\} = \left\{ \begin{matrix} f_{ix}^{L} \\ f_{iy}^{L} \\ f_{iz}^{L} \end{matrix} \right\} = - \left[\begin{matrix} k_{ix} & 0 & 0 \\ 0 & k_{iy} & 0 \\ 0 & 0 & k_{iz} \end{matrix} \right] \cdot \left\{ \begin{matrix} \Delta d_{ix}^{L} \\ \Delta d_{iy}^{L} \\ \Delta d_{iz}^{L} \end{matrix} \right\} = - [k_{i}] \cdot \left\{ \Delta d_{i}^{L} \right\}. \tag{4.29}$$

The contacting forces for all points are

$$
\{f^L\} = \begin{Bmatrix} \{f_1^L\} \\ \{f_2^L\} \\ \vdots \\ \{f_3^L\} \end{Bmatrix} = - \begin{Bmatrix} [k_1] & & & \\ & [k_2] & & \\ & & \ddots & \\ & & & [k_m] \end{Bmatrix} \cdot \begin{Bmatrix} \{\Delta d_1^L\} \\ \{\Delta d_2^L\} \\ \vdots \\ \{\Delta d_m^L\} \end{Bmatrix} = -[k^L] \cdot \{\Delta d^L\} . \quad (4.30)
$$

Contacting Force in GCS. The contacting force in GCS can be calculated once the forces in LCS are known. For each contact point, the relationship between global contacting force, $\{f_i^G\}$, and local contacting force, $\{f_i^L\}$, is

$$
\{f_i^G\} = [T_{Gi}^L] \cdot \{f_i^L\} ,
$$

where $[T_{Gi}^L]$ is the transformation matrix from LCS to GCS. The contacting force in GCS for all points can be expressed as

$$
\{f^G\} = \begin{Bmatrix} \{f_1^G\} \\ \{f_2^G\} \\ \vdots \\ \{f_m^G\} \end{Bmatrix} = \begin{Bmatrix} [T_{G1}^L] & & & \\ & [T_{G2}^L] & & \\ & & \ddots & \\ & & & [T_{Gm}^L] \end{Bmatrix} \cdot \begin{Bmatrix} \{f_1^L\} \\ \{f_2^L\} \\ \vdots \\ \{f_m^L\} \end{Bmatrix} = [T_G^L] \cdot \{f^L\} . \quad (4.31)
$$

Internal Wrench. A wrench generated by external force is an external wrench, and a wrench generated by reaction force at a contact point is an internal wrench. Let the ith contact point in GCS be $\{p_i^G\} = \{p_{ix}^G \quad p_{iy}^G \quad p_{iz}^G\}$. The torque generated by contacting force $\{f_i^G\}$ is then

$$
\begin{cases} M_{ix} = f_{iz}^G \cdot p_{iy}^G - f_{iy}^G \cdot p_{iz}^G \\ M_{iy} = f_{ix}^G \cdot p_{iz}^G - f_{iz}^G \cdot p_{ix}^G \\ M_{iz} = f_{iy}^G \cdot p_{ix}^G - f_{ix}^G \cdot p_{iy}^G \end{cases} . \quad (4.32)
$$

The ith internal wrench at this point can be written as

$$\{W_{ii}\} = \begin{Bmatrix} F_{ix} \\ F_{iy} \\ F_{iz} \\ M_{ix} \\ M_{iy} \\ M_{iz} \end{Bmatrix} = \begin{bmatrix} 1 & 0 & 0 \\ 0 & 1 & 0 \\ 0 & 0 & 1 \\ 0 & -p_{iz}^G & p_{iy}^G \\ p_{iz}^G & 0 & -p_{ix}^G \\ -p_{iy}^G & p_{ix}^G & 0 \end{bmatrix} \cdot \begin{Bmatrix} f_{ix}^G \\ f_{iy}^G \\ f_{iz}^G \end{Bmatrix} = [\Sigma_i] \cdot \{f_i^G\} \tag{4.33}$$

By combining wrenches at all m contact points, we get the total internal wrench:

$$\{W_i\} = \sum \{W_{ii}\} = \begin{bmatrix} [\Sigma_1] & [\Sigma_2] & \cdots & [\Sigma_m] \end{bmatrix} \cdot \begin{Bmatrix} \{f_1^G\} \\ \{f_w^G\} \\ \vdots \\ \{f_m^G\} \end{Bmatrix} = [\Sigma] \cdot \{f^G\} . \tag{4.34}$$

__Internal Wrench Final Calculation.__ By combining the previous steps, we can now get the internal wrench as a function of workpiece deviation,

$$
\begin{aligned}
\{W_i\} \\
&= [\Sigma] \cdot \{f^G\} \\
&= [\Sigma] \cdot [T_G^L] \cdot \{f^L\} \\
&= -[\Sigma] \cdot [T_G^L] \cdot [k^L] \cdot \{\Delta d^L\} \\
&= -[\Sigma] \cdot [T_G^L] \cdot [k^L] \cdot [T_G^L]^{-1} \cdot \{\Delta d\} \\
&= -[\Sigma] \cdot [T_G^L] \cdot [k^L] \cdot [T_G^L]^{-1} \cdot [G] \cdot \{\Delta q\}.
\end{aligned}
\tag{4.35}
$$

4.3.2.2   The Fixture Stiffness Matrix

To see the relationship between total external wrench and workpiece displacement we can now restate the stability equilibrium equation as

$$[K] \cdot \{\Delta q\} = -\{W_e\} \tag{4.36}$$

$$[K] = -[\Sigma] \cdot [T_G^L] \cdot [k^L] \cdot [T_G^L]^{-1} \cdot [G], \tag{4.37}$$

where [K] is the 6 ×6 fixture stiffness matrix.

4.3.3 Stability Criteria

With the kinetic fixture model, we can solve the linear equation system (Eq. 4.36) and calculate the reaction forces at all locating points. By using this result, the workpiece stability can be verified. Workpiece stability is defined so that there is no slippage between any locating point and the workpiece surface. This criterion is based on the concept of the friction cone and the definition of the contact stability index (CSI).

4.3.3.1 Friction Cone

As we know, the condition for a point in contact with a surface is that the contact force falls within the friction cone, as shown in Figure 4.28.

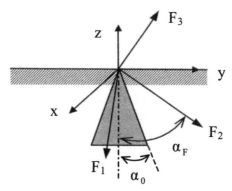

Figure 4.28. Friction cone and contact stability index.

In Figure 4.28, three cases are shown with different forces exerted onto the surface at the contact point. F_1 falls within the friction cone (shaded area) and will remain in contact with the surface. It is a stable situation. F_2 falls outside the cone but still points toward the inside of the surface. It will cause slippery displacement. F_3 points toward the outside of the surface. It will cause the separation of the workpiece from the locator. The friction cone is defined by the maximum friction force limitation

$$-\infty \le \frac{f}{\mu \cdot N} \le -1,$$
(4.38)

where f is the friction force, μ is the static friction coefficient, and N is the normal force.

4.3.3.2 Contact Stability Index (CSI)

To evaluate the stability at a contact point, it is desirable to have a quantitative measurement. It is also desirable that the measurement be normalized, so that the stability index can be read directly from the value. To fulfill these objectives, the CSI is defined to measure the stability of a contact point. It will have the following properties:

- $-1 \leq CSI < 0$:   Outside the friction cone, unstable
- $CSI = 0$:   On the friction cone, marginally stable.
- $0 < CSI \leq 1$:   Inside the friction cone, stable

To satisfy these conditions, the CSI is formulated as

$$CSI = \begin{cases} 1.0 - \dfrac{\alpha_F}{\alpha_0} & \alpha_F \leq \alpha_0 \\ -\dfrac{\alpha_F - \alpha_0}{\pi - \alpha_0} & \alpha_F > \alpha_0 \end{cases}, \tag{4.39}$$

where α_0 is the angle of the friction cone and α_F is the angle between the force vector and the -z axis, as illustrated in Figure 4.28.

4.3.3.3 Workpiece Stability

After obtaining the CSI, it is easy to check the workpiece stability. It requires that every locating point remain in contact with the workpiece surface. That is, at all locating points $CSI \geq 0$.

4.3.4   Minimum Clamping Forces

Some fixture designs rely on friction forces to stabilize the workpiece. In such cases, the clamping forces require a minimum amplitude to ensure the stability and should not be exceeded to the value that may cause excessive forces and unnecessary workpiece deformation.

The clamping forces are optimized by the following rule: if certain contact points are found to need larger normal force to maintain stable, all clamping forces are searched and the most helpful clamping forces will be adjusted. This is accomplished via the CSI sensitivity matrix.

4.3.4.1 CSI Sensitivity Matrix

Assume a fixture with m locating points (L_1, ..., L_i, ..., L_m) and n clamping points (C_1, ..., C_j, ..., C_n). To evaluate the effect at locating point L_i by clamping force at point C_j, we set a unit clamping force at C_j, and find out the CSI at L_i, α_{ij}, by Eq. 4.39. After finding the CSI at all locating points by assessing all clamping points, we get the CSI matrix as follows,

$$[C] = \begin{bmatrix} \alpha_{11} & \cdots & \alpha_{1j} & \cdots & \alpha_{1n} \\ \vdots & \ddots & & & \\ \alpha_{i1} & & \alpha_{ij} & & \\ \vdots & & & \ddots & \\ \alpha_{m1} & & & & \alpha_{mn} \end{bmatrix}, \tag{4.40}$$

where α_{ij} shows how the jth clamp affects the ith locator stability. $\alpha > 0$ means that the clamp is stabilizing the contact at the locating point, whereas $\alpha < 0$ means that the clamp is causing slippage at the locating point. For example, Figure 4.29 shows three locators and two clamps in a fixture design. The CSI for this would be

$$[C] = \begin{bmatrix} -0.25 & 1.0 \\ 1.0 & -0.5 \\ 1.0 & -0.5 \end{bmatrix}.$$

From this CSI sensitivity matrix we can see that the clamping force at point C_1 decreases the contact stability at locating point L_1 ($\alpha_{11} = -0.25$) but increases it at L_2 ($\alpha_{21} = 1.0$) and L_3 ($\alpha_{31} = 1.0$). The clamping force at point C_2 increases contact stability at L_1 ($\alpha_{12} = 1.0$) but decreases it at L_2 ($\alpha_{22} = -0.5$) and L_3 ($\alpha_{23} = -0.5$).

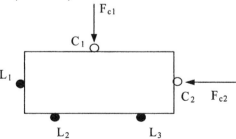

Figure 4.29. Example of CSI sensitivity matrix.

4.3.4.2 Search for Minimal Clamping Forces

From the CSI sensitivity matrix presented in the last section, we can determine the extent to which a clamping force affects contact stability at each locating point. After solving the stability equation, if we find (for example) a case of instability at locating point L_i with the CSI less than zero, we should adjust the clamping forces according the CSI sensitivity matrix at point L_i (ith row) and then resolve the stability equation with adjusted clamping forces. This procedure is repeated until the workpiece is stable or the maximal number of iterations has been reached, which means there is no solution for the case. If the ith locating point is not stable, for each clamping point its clamping force is adjusted by

$$f = f_0 \cdot [1 + pos(C_{ij})] \quad (j = 1 \cdots n), \tag{4.41}$$

where f_0 is the force before adjustment, f is the force after adjustment, $pos(x) = \begin{cases} x & x \geq 0 \\ 0 & x < 0 \end{cases}$, and C_{ij} is the element in the CSI sensitivity matrix at ith row and jth column. Figure 4.30 shows a block diagram of the determination of minimum clamping forces. If there were three locating points and 3 clamping points, the CSI sensitivity matrix would look as follows:

$$[C] = \begin{bmatrix} 0.25 & 1.0 & -0.5 \\ 1.0 & 0.0 & -0.5 \\ 1.0 & -0.5 & -0.25 \end{bmatrix}$$

If we found that the CSI at locating point L_1 is negative, we would want to adjust the clamping forces based on the CSI sensitivity values at this point (first row of the matrix) and we would

- increase clamping force at C_1 by 25%,
- increase clamping force at C_2 by 100%, and/or
- keep clamping force at C_3 unchanged.

The interval size could then be adjusted to achieve the best result.

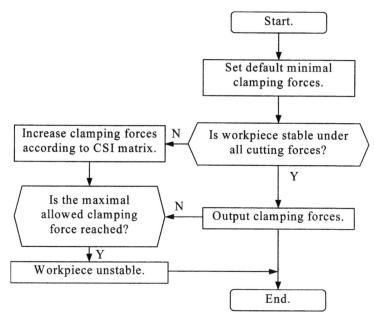

Figure 4.30. The determination of minimum clamping forces.

4.3.5 Clamping Sequence and Stability

To solve a stability problem, we can treat all locating points as contact points in the kinetic model, and combine the gravity force, clamping forces, and machining forces as a single external wrench.

However, this scheme is only true if all clamps are applied at one time. When friction forces are taken into consideration, the stability problem becomes clamping sequence dependent. This is because when the clamps are applied one by one the previously applied clamp also serves as a new contact point as the next clamp is applied.

From the kinetic model, we know the stability problem is a linear system. Thus, a multi-load stability problem can be decomposed into several independent stability problems. Each step contains one more contact point from the previous step, and the final solution is the combination of all solutions from sub steps. For example, the stability problem shown in Figure 4.31 can be decomposed into four sub problems (Figure 4.32), each with its own contact points and external forces.

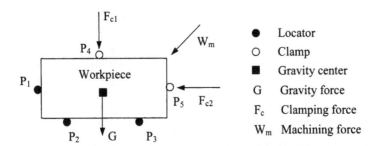

Figure 4.31. Multi-load stability problem.

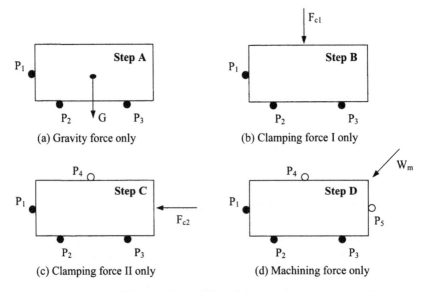

Figure 4.32. Stability decomposition.

In step A (Figure 4.32a), the workpiece is placed on the three locating points. In this step, these three locating points serve as the contact points, and the gravity force serves as the external force. In step B (Figure 4.32b), clamping force F_{c1} is applied. The contact points are still the three locating points, and the external force is the clamping force F_{c1}. In step C (Figure 4.32c), the clamping point from the previous step becomes a new contact point. Thus, there are a total of four contact points, and the external force is clamping force F_{c2}. In step D (Figure 4.32d), there is a total of five contact points: three locating points and two clamping points. The external force in this step is the machining force.

When checking the stability at each stage, the contact point displacements and reaction forces in LCS are the sum of those in current and all previous steps. In Figure 4.32, if the contact point displacement and reaction forces in each stage are $\{d_i^L\}$ and $\{f_i^L\}$ ($i = A, B, C, D$), then we have the displacements and locating reaction forces calculated, as shown in Table 4.5. For a workpiece to be stable throughout the loading/fixturing/machining processes, it must be stable under each of the steps described previously.

Table 4.5. Stability decomposition.

Step	Displacements in LCS	Reaction Forces in LCS
A	$\{d^L\} = \{d_A^L\}$	$\{f^L\} = \{f_A^L\}$
B	$\{d^L\} = \{d_A^L\} + \{d_B^L\}$	$\{f^L\} = \{f_A^L\} + \{f_B^L\}$
C	$\{d^L\} = \{d_A^L\} + \{d_B^L\} + \{d_C^L\}$	$\{f^L\} = \{f_A^L\} + \{f_B^L\} + \{f_C^L\}$
D	$\{d^L\} = \{d_A^L\} + \{d_B^L\} + \{d_C^L\} + \{d_D^L\}$	$\{f^L\} = \{f_A^L\} + \{f_B^L\} + \{f_C^L\} + \{f_D^L\}$

4.3.6   Implementation

The fixturing stability function in CAFDV is implemented in a CAD system. Figure 4.33 shows the CAD interface.

One case study conducted with the CAFDV system includes a simplified V-8 engine block from a real case. In the specific setup, the machining surfaces are the bottom surfaces of the four lugs on both sides of the engine block. Each machining surface has two tolerances (a surface profile of 0.08 and a parallelism of 0.04) with the bottom-locating surface as datum. The bottom surface is the primary locating surface, which restricts three DOF. A short round pin locator and a short diamond pin locator are placed in the holes on the bottom surface.

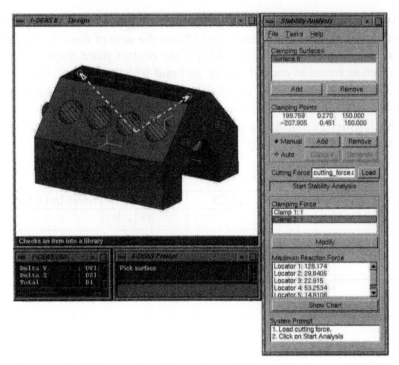

Figure 4.33. CAD interface of fixturing stability verification for CAFDV.

First, the locators are converted into locating points, and the stiffness measures are converted into contact spring constants based on the structural information in fixture design (Kang 2001). With locating point stiffness measures calculated, the fixture stiffness matrix can be constructed. In this case, it looks as follows.

```
-1.0000e+06   0.0000e+00   0.0000e+00   0.0000e+00   1.3000e+08  -1.5753e+08
 0.0000e+00  -2.0000e+06   0.0000e+00  -2.6000e+08   0.0000e+00   6.8000e+05
 0.0000e+00   0.0000e+00  -3.0000e+06  -1.5472e+08   3.9508e+07   0.0000e+00
 0.0000e+00  -2.6000e+08  -1.5472e+08  -1.0997e+11   7.8760e+09   8.8400e+07
 1.3000e+08   0.0000e+00   3.9508e+07   7.8760e+09  -6.1330e+10   2.0479e+10
-1.5753e+08   6.8000e+05   0.0000e+00   8.8400e+07   2.0479e+10  -9.7100e+10
```

The input cutting force is a series of static forces with spatial parameters (position, direction, and magnitude) in time steps. The workpiece stability is solved at each time step. At each time step, the external forces include gravity force, clamping forces, and cutting forces. The calculation result is illustrated in five steps. The reaction forces are calculated following the procedure discussed previously. The output for each locating point is shown in Figure 4.34.

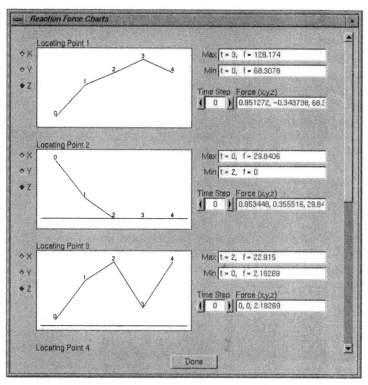

Figure 4.34. Reaction force at locating points.

4.3.7   Conclusion

This section explored the kinetic fixture model established to formulate the frictional stability problem in fixture design. A fixture stiffness matrix is derived to link the external forces with the workpiece displacement. By solving a linear equation system, workpiece displacement, fixture deformation, and reaction forces can be obtained. Workpiece stability is defined through the contact stability index (CSI). With the kinetic fixture model, we are also able to find the minimum clamping forces required in machining operations, and the impact of clamping sequence on fixturing stability. You also learned that the fixturing stability verification function has been integrated into CAFDV and implemented in the CAD environment.

References

ANSI (American National Standards Institute). *Dimensioning and Tolerancing*, ANSI Y14.5M, New York: The American Society of Mechanical Engineers, 1995.

Asada, H., and A. By. "Kinematic Analysis of Workpart Fixturing for Flexible Assembly with Automatically Reconfigurable Fixtures," *IEEE Journal of Robotics and Automation* 1, pp. 86–94, 1985.

Bai, Y., and Y. Rong. "Establishment of Modular Fixture Component Assembly Relationship for Automated Fixture Design," *Symposium on Computer-aided Tooling* MED-2:1, pp. 805–816, San Francisco: ASME IMECE, 1995.

Chen, Y. C. "Proper Clamping Sequences in the Fixturing of Prismatic Workpieces," *Proceedings of the IEEE International Symposium on Assembly and Task Planning*, pp. 368–373, Pittsburgh, PA: IEEE, 1995.

Chou, Y. C., V. Chandru, and M. M. Barash. "A Mathematical Approach to Automatic Configuration of Machining Fixtures: Analysis and Synthesis," *Journal of Engineering for Industry* 111, pp. 299–306, 1989.

Choudhuri, S. A., and E. C. DeMeter. "Tolerance Analysis of Machining Fixture Locators," *Journal of Manufacturing Science and Engineering* 121:2, pp. 273–281, 1999.

Cogun, C. "Importance of the Application Sequence of Clamping Forces on Workpiece Accuracy," *Transactions of the ASME, Journal of Engineering for Industry* 114:4, pp. 539–543, 1992.

DeMeter, E. C. "Fast Support Layout Optimization," *International Journal of Machine Tools and Manufacture* 38:10–11, pp. 1221–1239, 1998.

Huang, X., and P. Gu. "Tolerance Analysis in Setup and Fixture Planning for Precision Machining," in *Proceedings of The Fourth International Conference on Computer Integrated Manufacturing and Automation Technology*, pp. 298–305, Troy, NY: IEEE Computer Society Press, 1994.

Kang, Y. "Computer-aided Fixture Design Verification," Ph.D. dissertation, Worcester Polytechnic Institute, Worcester, MA, 2001.

Kang, Y., Y. Rong, and A. Yang. "Geometric and Kinetic Model Based Computer-Aided Fixture Design Verification," *Journal of Computing and Information Science in Engineering* 3:3, pp. 187–199, 2003.

Kashyap, S., and W. R. DeVries. "Finite Element Analysis and Optimization in Fixture Design," *Structural Optimization* 18:2–3, pp. 193–203, 1999.

King, L. S., and F. Ling. "Force Analysis Based Analytical Framework for Automatic Fixture Configuration," MED-2:1, *Manufacturing Science and Engineering*, pp. 789–800, American Society of Mechanical Engineers, 1995.

Lee, S. H., and M. R. Cutkosky. "Fixture Planning with Friction," *Journal of Engineering for Industry* 113, pp. 320–327, 1991.

Li, J., W. Ma, and Y. Rong. "Fixturing Surface Accessibility Analysis for Automated Fixture Design," *International Journal of Production Research* 37:13, pp. 2997–3016, 1999.

Ohwovoriole, M. S., and B. Roth. "An Extension of Screw Theory," *ASME Journal of Mechanical Design* 103, pp. 725–735, 1981.

Rong, Y., and Y. Bai. "Machining Accuracy Analysis for Coputer-aided Fixture Design," *Journal of Manufacturing Science and Engineering* 118, pp. 289–300, 1996.

Rong, Y., W. Li, and Y. Bai. "Locator Error Analysis for Fixturing Accuracy Verification," in *Computers in Engineering: Proceedings of the International Conference and Exhibit*, Boston, MA: ASME, pp. 825–832, 1995.

Rong, Y., S. Wu, and T. P. Chu. "Automated Verification of Clamping Stability in Computer-aided Fixture Design," in *Computers in Engineering: Proceedings of the International Conference and Exhibit*, vol. 1, pp. 421–426, Minneapolis, MN: ASME, 1994.

Rong, Y., and Y. Zhu. *Computer-aided Fixture Design*. New York: Marcel Dekker, 1999.

Trappey, A.J.C., and C. R. Liu. "Automatic Workholding Verification System," *Robotics and Computer-Integrated Manufacturing* 9:4–5, pp. 321–326, 1992.

Wang, M. Y. "Automated Fixture Layout Design for 3D Workpieces," in *IEEE International Conference on Robotics and Automation*, vol. 2, Detroit, MI: IEEE, pp. 1577–1582, 1999.

Wu, N. H., K. C. Chan, and S. S. Leong. "Fixturing Verification Based on the Analysis of Multi-discipline Frictional Contacts," MED-2:1, *Manufacturing Science and Engineering*, pp. 735–744, American Society of Mechanical Engineers, 1995.

Xiong, C., and Y. Xiong. "Stability Index and Contact Configuration Planning for Multifingered Grasp," *Journal of Robotic Systems* 15:4, pp. 183–190, 1998.

Xiong, C., Y. Li, Y. Xiong, H. Ding, and Q. Huang. "Grasp Capability Analysis of Multifingered Robot Hands," *Robotics and Autonomous Systems* 27, pp. 211–224, 1999.

Zhang, Y., W. Hu, Y. Rong, and D. W. Yen. "Graph-based Setup Planning and Tolerance Decomposition for Computer-aided Fixture Design," *International Journal of Production Research* 39:14, pp. 3109–3126, 2001.

CHAPTER 5

Fixturing Stiffness Analysis

Fixturing-induced deformation contributes to locating accuracy and machining dynamics. This chapter examines the deformation of fixtured workpieces and the fixture stiffness.

5.1 DEFORMATION OF FIXTURED WORKPIECE

Machining fixtures are used to locate and constrain a workpiece during a machining operation. To ensure that the workpiece is manufactured according to specified dimensions and tolerances, it must be appropriately located and clamped. Minimizing workpiece and fixture tooling deflections due to clamping and cutting forces in machining is critical to machining accuracy. An ideal fixture design maximizes locating accuracy and workpiece stability while minimizing displacements.

This section introduces a method for modeling workpiece boundary conditions and applied loads during a machining process, analyzing the workpiece deformation, and optimizing support locations using FEA. Workpiece boundary conditions are defined through locators and clamps that are in contact with the workpiece. The locators are placed in a 3-2-1 fixture configuration, constraining all six DOF of the workpiece, and are modeled using linear spring-gap elements. Clamps are modeled as point loads. The cutting force in drilling and milling operations are modeled as quasi-static load acting on the workpiece.

This research verifies fixture design integrity. The ANSYS Parametric Design Language (APDL) code is used to develop an algorithm to automatically optimize fixture support and clamp locations, as well as the clamping forces, in order to minimize workpiece deformation, and subsequently increasing machining accuracy. By implementing FEA in CAFD environment, unnecessary and uneconomical trial-and-error experimentation in the machine shop can be eventually eliminated.

5.1.1   Introduction

Machining fixtures are used to locate and constrain a workpiece during a machining operation. To ensure that the workpiece is manufactured according to specified dimensions and tolerances, it must be located and clamped. Production quality depends considerably on the relative position of the workpiece and machine tools. Minimizing workpiece and fixture tooling deflections due to clamping and cutting forces in machining is critical to machining accuracy. The workpiece deformation during machining is directly related to the workpiece-fixture system stiffness. An ideal fixture design maximizes locating accuracy, workpiece stability, and stiffness, while minimizing displacements.

Traditionally, fixtures were designed by trial and error, which is expensive and time consuming. Research in flexible fixturing and CAFD has significantly reduced manufacturing lead time and cost. A computer-aided tool is developed for modeling workpiece boundary conditions and applied loads in machining.

The majority of FEA research conducted in fixture design considers workpiece boundary conditions to be rigid and applied loads to be concentrated. In all cases where friction is considered, Coulomb friction is assumed. Cutting tool torque, which results in a trend of workpiece rotation, is not considered. Clamping forces are considered to be constant point loads.

This study acknowledges that workpiece boundary conditions are deformable and influence the global stiffness of the workpiece-fixture system. The boundary conditions of the workpiece (i.e., the locators) are modeled as multiple springs (in parallel) attached to the actual workpiece-fixture contact area on the surface of the workpiece. In addition, tangential and normal stiffness components in the boundary conditions are assumed not to be equal as in Coulomb friction, but are assigned independently. In applying loads representing the machining operation, torque, axial, and transverse loads due to feeding are considered.

In this study, both the FEA and optimization are conducted in ANSYS. In the analysis, the general procedure is as follows. A workpiece is imported in IGES (initial graphics exchange specifications) format from a CAD model. Material properties, element type, and real constants are specified. The workpiece is meshed and boundary conditions and loads are applied. The model is then solved. Finally results are retrieved parametrically and support locations, clamp locations, and clamping forces are optimized to minimize workpiece deflection (Amaral 2001).

Principles of fixture design and precedent FEA research in fixture design have been well studied. Although some research has been conducted in fixture design, a comprehensive finite element model that accurately represents applied boundary conditions and loads has not been developed. Tables 5.1 and 5.2 summarize the precedent research conducted on FEA in fixture design.

Table 5.1. Literature survey of workpiece models.

Reference	Workpiece Model				
	Material				Element Type
	Type	E (psi)	ν	μ	
Lee 1987	Steel Homogeneous Isotropic linear elastic	1.0×10^5	0.3	U/A*	3D solid 8-node brick
Pong 1993	Aluminum Homogeneous Isotropic linear elastic	1.0×10^7	0.3	U/A*	3D solid 10-node tetrahedral; ANSYS SOLID92
Trappey 1995	Aluminum Homogeneous Isotropic linear elastic	1.0×10^7	0.3	0.3	U/A*
Cai 1996	Steel Isotropic linear elastic	3.0×10^7	0.3	U/A*	2D 4-node rectangular element; MSC NASTRAN QUAD4
Kashyap 1999	Aluminum Homogeneous Isotropic linear elastic	1.0×10^7	0.3	U/A*	3D solid tetrahedral elements

* U/A stands for "unavailable"; N/A stands for "not applicable."

Table 5.2. Literature survey of boundary conditions and loading.

Reference	Fixture Component Model		Steady-State Load Model	
	Locators	Clamps	Drilling	Milling
Lee 1987	Rigid area constraints Rigid Coulomb friction	U/A*	U/A*	Normal and shear point loads
Pong 1993	3D spring-gap interface element, Rigid Coulomb friction	N/A*	Normal point loads	N/A*
Trappey 1995	3D solid deformable constraints	Point loads	Normal point loads	Normal and shear point loads
Cai 1996	Rigid point constraints	N/A*	Normal point loads	Normal and shear point loads
Kashyap 1999	Rigid point constraints	Point loads	Normal point loads	Normal and shear point loads

* U/A stands for "unavailable"; N/A stands for "not applicable."

FEA was used to minimize workpiece deflection by Lee (1987). Lee's workpiece was modeled as linear elastic. However, fixture tooling was modeled as rigid. The objective function included the maximum work done by clamping and machining forces, the deformation index, and the maximum stress on the workpiece. The study considers the importance of part deformation with respect to the necessary number of fixturing elements and the magnitude of clamping forces (Pong 1993). Coulomb's law of friction was used to calculate frictional forces attendant to the workpiece-fixture contact points. Machining forces were applied at nodal points. Manassa (1991) conducted similar research to that of Lee (1987), but modeled fixturing elements as linear elastic springs.

Pong (1993) used spring-gap elements with stiffness, separation, and friction capabilities to model elastic workpiece boundary conditions. 3D tetrahedral elements were used to mesh the finite element model of the solid workpiece. All contacts between the workpiece and fixture were considered to be point contacts, and machining forces were applied sequentially as point loads. The positions of locators and clamps, and clamping forces, were considered as design variables for optimization. Trappey (1995) developed a procedure for the verification of fixture performance. FEA was used to analyze the stress-strain behavior of the workpiece when machining and clamping forces were applied. A mathematical optimization model was formulated to minimize workpiece deformation with a feasible fixture configuration.

Cai (1996) used FEA to analyze sheet metal deformation and optimized support locations to minimize resultant displacements of workpiece. Kashyap (1999) used FEA to model workpiece and fixture tool deformation, and developed an optimization algorithm to minimize deflections at selected nodal points by considering support and tool locations as design variables.

A summary of research on FEA and fixture design optimization is shown in Table 5.3. The majority of research conducted in FEA and fixture design optimization resulted in the development of a computer algorithm for optimization. Pong (1993) used the ellipsoid method to optimize support locations and minimize nodal deflection. Trappey (1995) used an external software package, GINO, to optimize support locations and clamping forces. Cai (1996) used a sequential quadratic programming algorithm in an external FORTRAN-based software package, VMCON, to perform a quasi-Newton nonlinear constrained optimization of N-2-1 support locations to minimize sheet metal deflection. Kashyap (1999) developed a discrete computer algorithm for optimization.

Table 5.3. Literature survey of optimization analysis.

Reference	Optimization Analysis		
	Method	Objective Function	Software Package
Pong 1993	Ellipsoid method	Nodal deflection	N/A*
Trappey 1995	Nonlinear mathematical algorithm	Nodal deflection	GINO
Cai 1996	Sequential quadratic programming algorithm	Nodal deflection normal to sheet metal surface	VMCON
Kashyap 1999	Discrete mathematical algorithm	Nodal deflection	N/A*

* U/A stands for "unavailable"; N/A stands for "not applicable"

5.1.2   Fixture Design Analysis Methodology

The flowchart shown in Figure 5.1 is a summary of the fixture design analysis methodology developed and used in this work. In summary, workpiece IGES geometry is imported from the solid modeling package, the workpiece model is meshed, boundary conditions are applied, the model is loaded (representative of a machining operation) the model is solved, and then boundary conditions are optimized to minimize workpiece deflections.

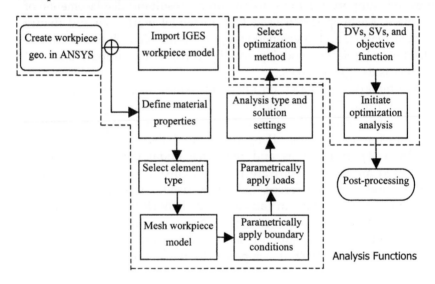

Figure 5.1. Fixture design analysis methodology.

5.1.3   Workpiece Model

The workpiece model is the starting point of the analysis. This research currently limits the workpiece geometry to solids with planar locating surfaces. Some workpiece geometry may contain thin walls and non-planar locating surfaces, which are not considered in this study.

In regard to geometry, the workpiece model is created in a CAD package and exported to ANSYS in IGES format. IGES is a neutral standard format used to exchange models among CAD/CAM/CAE systems. ANSYS provides two options for importing IGES files:

DEFAULT and ALTERNATE. The DEFAULT option allows for file conversion without user intervention. The conversion includes automatic merging and creation of volumes to prepare the model for meshing. The ALTERNATE option uses the standard ANSYS geometry database and provides backward compatibility with the previous ANSYS import option. The ALTERNATE option has no capabilities for automatically creating volumes, and modes imported through this translator require manual repair through the PREP7 geometry tools.

In regard to materials properties, the workpiece material in this study is homogeneous, isotropic, linear elastic, and ductile. This is consistent with the material properties of most metal workpieces. The material selected is SAE/AISI 1212 free-machining grade(a) carbon steel with Young's Modulus, $E = 30 \times 10^6$ psi, Poisson's ratio ($\nu = 0.295$) and density ($\rho = 0.283$ lb/in^3), and hardness of 175 HB. Although SAE1212 steel was selected for use in this study (it is commonly used and is a benchmark material for machinability), any other material could also be used for the workpiece by simply changing the isotropic material properties in ANSYS.

5.1.3.1 Meshed Workpiece Model

Three ANSYS 3D solid elements are selected to potentially mesh the workpiece, including (1) a four-node tetrahedral element (SOLID72) with six DOF at each node and linear displacement behavior; (2) a ten-node tetrahedral element (SOLID92) with three DOF at each node and quadratic displacement behavior; and (3) an eight-node hexahedral element (SOLID45) with three DOF at each node and linear displacement behavior. The linear SOLID45 element in the hex configuration is more desirable but less accurate than the quadratic SOLID92 element. SOLID45 is used for 3D modeling of solid structures. The element is defined by eight nodes with three translation-DOF at each node. The SOLID45 element can be degenerated to a four-node tetrahedral configuration with three DOF per node. The tetrahedral configuration is more suitable for meshing non-prismatic geometry, but is less accurate than the hex configuration.

5.1.3.2 Boundary Conditions

Locators and clamps define the boundary conditions of the workpiece model. The locators can be modeled as point or area contact and clamps are modeled as point forces.

Locators. In regard to point contact, the simplest boundary condition is a point constraint on a single node. An LCS, referenced from the global coordinate system origin, is created at the center of each locator contact area such that the Z-axis is normal to the workpiece locating surface. The node closest to the center of the local coordinate system origin is selected and all three translational degrees of freedom (μ_x, μ_y, and μ_z) are constrained. The point constraint models a rigid locator with an infinitesimally small contact area.

To model locator stiffness and friction at the contact point, a 3D interface spring-gap element is placed at the center of the LCS. The element is connected to existing nodes on the surface of the workpiece and to a fully constrained copied node offset from the workpiece surface in the z direction of the local coordinate system (i.e., perpendicular to the surface). Figure 5.2 is a model of the CONTAC52 element used to represent a linear elastic locator.

$$\mu_x = \mu_y = \mu_z = 0$$

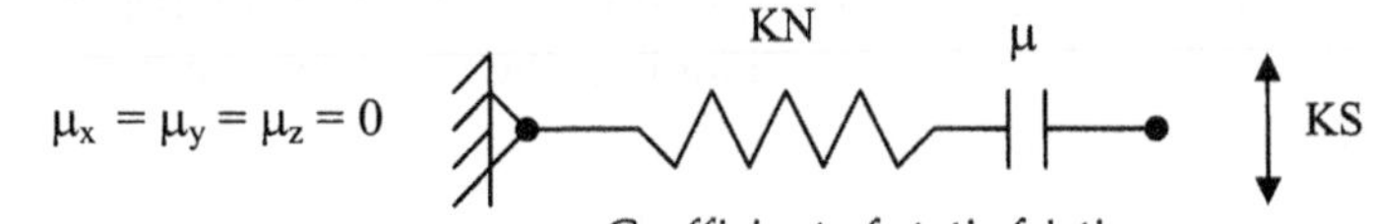

μ = Coefficient of static friction
KN = Total normal stiffness of the locator
KS = Total tangential stiffness of the locator

Figure 5.2. CONTAC52 element used to model point contact for locators.

In regard to area contact, to model a rigid locator with a contact _area_, multiple nodes are specified within the contact area. An LCS is created on the workpiece surface at the center of the locator contact area. For a circular contact area, a cylindrical LCS is created and nodes are selected at $0 < R < R_L$. For a rectangular contact area, a Cartesian LCS is created and nodes are selected at $0 < x < x_L$ and $0 < y < y_L$. All three translational degrees of freedom (μ_x, μ_y, and μ_z) of each of the nodes are constrained. This model assumes rigid constraints. In reality, however, locators are elastic.

A more accurate representation of the elastic locators consists of multiple ANSYS CONTAC52 elements in parallel. Nodes are selected within the locator contact area and are offset perpendicular to the locating surface. Each selected node is connected to the node sequentially with the CONTAC52 element. Figure 5.3 shows the contact area model with multiple spring-gap elements in parallel used to represent a linear elastic locator. It is important to note that the user is constrained to the number of nodes within the specified contact area when attaching the CONTAC52 elements.

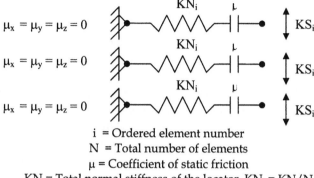

$$\mu_x = \mu_y = \mu_z = 0$$

$$\mu_x = \mu_y = \mu_z = 0$$

$$\mu_x = \mu_y = \mu_z = 0$$

i = Ordered element number
N = Total number of elements
μ = Coefficient of static friction
KN = Total normal stiffness of the locator, KN_i = KN/N
KS = Total tangential stiffness of the locator, KS_i = KS/N

Figure 5.3. CONTAC52 elements in parallel, used to model contact area for locators.

The method for obtaining the normal and tangential stiffness for a locator is shown in Figure 5.4. The stiffness divided by the total number of springs is assigned accordingly to each spring-gap element, in the real constant set. Similarly, a point load is applied to the 3D finite element model of the real locator, normal to the contact area and the normal stiffness can be obtained by the applied force divided by the resulted normal displacement. A point load is applied tangent to the contact area of the real locator to determine the tangential or sticking stiffness of the locator. The stiffness values are then assigned to the CONTAC52 elements.

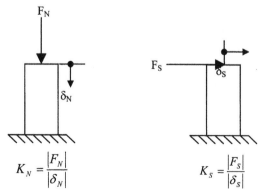

$$K_N = \frac{|F_N|}{|\delta_N|}$$

$$K_S = \frac{|F_S|}{|\delta_S|}$$

Figure 5.4. Normal and tangential stiffness for locator.

5.1.3.3 Clamps

Clamps are used to fully constrain the workpiece, once it is located. It is common to use multiple clamps and clamping forces that are generally constant for each clamp. The clamping force, F_{cl} is applied through either a toggle mechanism or a bolt mechanism, which lowers a strap that comes into contact with the workpiece. Although friction is just as important in clamping as it is in locating, it is not modeled at the clamp contact area due to limitations in ANSYS. To model friction, a comprehensive 3D model of the entire workpiece-fixture system is required, with contact and target surfaces defined at the workpiece-fixture contact areas. The clamping forces are modeled in ANSYS as point loads on nodes selected either within a rectangular area for a clamp strap or a circular area on the workpiece surface for a toggle clamp. Both clamps may also be modeled with a single point load at the center of the clamp contact area.

5.1.3.4 Loading

This section examines the machining operations (milling and drilling). The purpose of this research is not to accurately model the machining process but to apply the torque and forces that are transferred through the workpiece in machining to determine the reactions at the boundary conditions of the workpiece. The desired result of the load model is the trend of rotation from the applied torque of the cutting tool and translation due to axial feeding of the workpiece and transverse motion of the table in milling.

Drilling. The forces in a drilling operation include torque, T, to generate tool rotation, a shear force, V, created by tool rotation at the cutting edge contact for chip removal, and an axial load, P, due to feeding. The forces in drilling are time and position dependent because of the dynamic nature of machining. Fluctuations in the cutting force are also due to cutting tool tooth distribution during rotation. In this study, the torque and thrust forces in feeding are applied as steady state loads because the initial tool entry is not considered. In previous FEA fixture design research, loads were applied as steady state. Previous studies also neglected cutting tool torque and workpiece deflections due to the trend of rotation in the fixture. The model consisted of placing key points on an LCS created on the machining surface of the workpiece. The key points were located at exact R, θ, and Z positions on the cutting tool perimeter. At each key point forces were applied to model a drilling operation. The torque was modeled with tangential forces placed at the outer radius of the cutting tool contact area. The tangential couple forces were decomposed into global X and Y components. The axial load was modeled by applying forces at each key point in global Z direction. Figure 5.5 shows the load model for drilling. Note that node i may be slightly offset from the cutting tool perimeter. Because a node may not exist in the exact location specified by R, θ, and Z, the node closest to that location in the LCS is selected and forces are applied as point loads with global X, Y, and Z components. The user may minimize the distance between a specified coordinate location and an existing node by increasing the mesh density. The nodes are selected at equivalent θ intervals on or near the cutting tool perimeter. At each selected node, global X and Y components of the tangential couple force (F_{ti}) and axial load component (F_{ci}) are applied. The applied torque is equal to the sum of the tangential forces multiplied by the cutting tool radius, r. F_{tiX} and F_{tiY} are the global X and Y components respectively of the tangential force, F_{ti}. F_{ci} is equal to the total axial load, F_c, divided by the number of nodes over which it is applied.

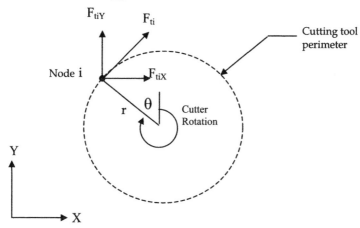

Figure 5.5. Drilling load model.

A simplified model entails the use of a single point force normal to the surface of the workpiece to model the cutting tool axial load, and a two-point couple to model the applied torque. A study was conducted to determine whether multiple point forces applied along the cutting tool perimeter are actually necessary to model the axial load and assess the validity of the simplified model.

Milling. The loading in a milling operation involves an axial load of a transverse load due to the linear feeding of the workpiece, a torque to generate tool rotation (which is transmitted through the workpiece) and shear force in the cutting area. Figure 5.6 is the loading model for end milling. The end milling model is the same as the drilling model, with the transverse load added. Because the objective of the analysis is to determine the maximum resultant displacements and equivalent stresses in the workpiece during the operation (and because tool entry is not considered), only the average steady state load magnitude is addressed. In this study, the cutting forces are applied as steady-state loads. In previous FEA research, forces in milling were traditionally modeled as steady state single-point loads and torque was neglected. The axial load due to feeding can be applied as multiple point loads on the cutting tool perimeter or as a single point load. The transverse load, F_{tri}, is applied as a single point load at the center of the cutting tool.

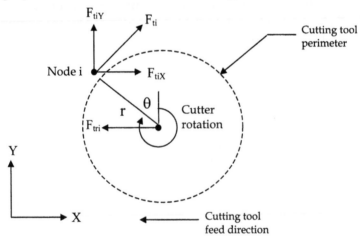

Figure 5.6. Milling load model.

5.1.3.5 Boundary Condition Study

A study was conducted to determine whether multiple spring-gap elements in parallel and distributed over an area are necessary to accurately model workpiece supports (locators). The maximum resultant displacement results of an elastically simply supported shear beam were compared for commercially available locators of different contact areas, modeled both with a single CONTAC52 element and multiple CONTAC52 elements in parallel. The purpose of the study was to determine whether there is a threshold ratio of workpiece surface area to support contact area for which there is a significant change in the maximum resultant displacement results due to the size effect of the constraints. If there is no significant change in results as the locator contact area decreases, it is simpler and just as appropriate to model the locator as an elastic point constraint on the workpiece, with a single CONTAC52 element. In past studies, supports were modeled as single point constraints.

In this study, each of the three locators supporting the beam was modeled with both a single CONTAC52 element and multiple CONTAC52 elements in parallel, attached to nodes selected within the support contact area on a 10-inch×2-inch×2-inch shear beam. A 1,000-pound point load was applied at the top center of the beam. Figure 5.7 shows the simply supported shear beam with elastic constraints used in this study. The material properties assigned to the locators and beam (workpiece) are listed in Table 5.4.

Table 5.4. Workpiece and locator material properties.

	Material	E (psi)	ρ (lb/in³)	ν	σ_y (psi)
Workpiece	AISI 1212	2.9×10^7	0.284	0.295	3.3×10^4
Locators	AISI 1144	2.9×10^7	0.284	0.295	9.7×10^4

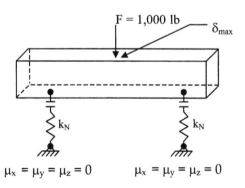

Figure 5.7. Simply supported shear beam with elastic constraints.

Five locators were selected that traverse the range of commercially available sizes of Carr Lane modular fixtures (Carr Lane 2001). In this study, the largest commercially available locator (which had a contact area of 0.785 in²) was used. Their part numbers and dimensions are listed in Table 5.5 and diagramed in Figure 5.8. The normal and tangential stiffness values were determined numerically for each locator through FEA of the locator models. The locator model in ANSYS did not include small geometric features, such grooves on the shaft, chamfers, or fillets due to their insignificance in the stiffness calculations. SOLID45, four-node tetrahedral elements were used to mesh the locator model. All translational degrees of freedom on the bottom surface of each locator are fixed. A 1,000-pound point load was applied to the top center of the locator, normal to the contact area, to determine the normal stiffness. A 1,000-bound point load was applied tangent to the contact area of the locator to determine the tangential or sticking stiffness of the locator. The stiffness values were then assigned to the CONTAC52 elements. The locator stiffness results are listed in Table 5.6.

Table 5.5. Locator dimensions (Carr Lane 2001) in Figure 5.8.

Locator Part Number	A (in.)	B (in.)	C (in.)	D (in.)
CL-12-RB	0.1855	0.3750	0.2500	0.3750
CL-2-RB	0.2500	0.5000	0.2500	0.5000
CL-6-RB	0.3750	0.6250	0.5000	0.6250
CL-9-RB	0.5000	0.6250	0.6250	0.8750
CL-18-RB	0.6250	0.7500	0.7500	1.0000

Table 5.6. Locator stiffness.

Locator Part Number	Number of Nodes	Number of Elements	δ_N 10^{-3} (in.)	k_N 10^6 (lb/in.)	δ_S 10^{-3} (in.)	k_S 10^5 (lb/in.)
CL-12-RB	1858	8905	1.200	0.833	4.810	0.208
CL-2-RB	1756	8449	0.890	1.130	2.460	0.407
CL 6 RB	1506	7010	0.640	1.560	1.580	0.633
CL-9-RB	1167	5193	0.410	2.430	0.068	1.470
CL-18-RB	1242	5769	0.350	2.900	0.050	2.000

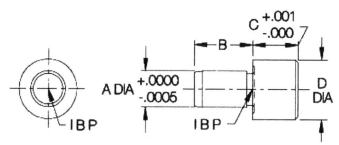

Figure 5.8. Locator dimensions (Carr Lane 2001) in Table 5.5.

Table 5.7 lists the ratio of workpiece surface area to contact area of each locator, and the maximum resultant displacement in the workpiece for both the elastic point constraint and elastic area constraint models. The maximum resultant displacement occurs at the load point, as shown in Figure 5.7. δ_{Pmax} is the maximum resultant displacement in the point constraint model and δ_{Amax} is the maximum resultant displacement in the area constraint model. A_S is the workpiece surface area and A_C is the locator contact area.

Table 5.7. Shear beam maximum resultant displacement results.

Locator Part Number	As /Ac	δ_{Pmax} 10^{-3} (in.)	δ_{Amax} 10^{-3} (in.)	$\dfrac{\delta_{Pmax} - \delta_{Amax}}{\delta_{Pmax}}$
CL-12-RB	181.12	1.36	1.280	5.54%
CL-2-RB	101.86	1.24	1.150	6.96%
CL-6-RB	65.22	1.14	1.030	9.89%
CL-9-RB	33.26	1.06	0.093	11.93%
CL-18-RB	25.46	1.03	0.091	11.74%

As seen in Figure 5.9, the maximum resultant displacements do not vary significantly between the single-contact element and multiple-contact element models. A difference on the order of ten thousandths of an inch does not contribute significantly to the overall results in determining machining accuracy. Percentages of difference in maximum resultant displacement between the two models are listed in Table 5.7. The point constraint model is more conservative because it results in larger displacements and is less likely to give a false positive solution in fixturing stability evaluation for an optimum fixture design. Therefore, it is appropriate to use a single contact element to model an elastic locator when the workpiece surface area to locator contact area ratio is greater than or equal to 25.46.

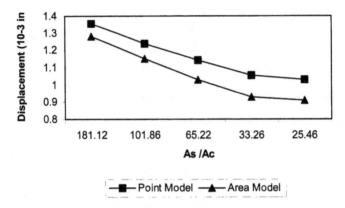

Figure 5.9. Maximum resultant displacement versus workpiece surface area to locator contact area ratio.

5.1.4   Loading Study

The two machining operations considered in this study are drilling and milling. As discussed in the fixture design analysis methodology section, drilling consists of an axial load and a torque. Milling consists of an axial load, a torque, and a transverse load. The axial load in each case may be applied as multiple point forces distributed along the cutter perimeter or as a single point load in the center of the cutter perimeter. The torque may also be applied as multiple point loads tangent to the cutter perimeter or as a simple couple. In this loading study, the effect of applying the torque as a simple couple versus multiple tangential forces on the cutter perimeter is not addressed, because the resultant torque around the center of the cutter perimeter is the same. Although it is not practical, it is possible that the torque may be zero. The transverse load is applied to model linear feeding of the cutter in the milling operation. The two limiting cases are (1) the axial load, $F_c = 0$, because there is no axial feeding, and (2) the transverse load, $F_{tr} = 0$, because there is no linear feeding. Therefore, the drilling model developed in this study may be considered a limited case of milling, where $F_{tr} = 0$.

Two studies were conducted independently to analyze the components of the load model common to both drilling and milling: the axial load and the torque. The first study was conducted to determine whether multiple point loads distributed along the cutting tool perimeter are necessary to accurately model the cutting tool axial load applied normal to the workpiece surface for both machining operations (i.e., milling and drilling). The second study was conducted to determine the effect of neglecting torque in the load model. The significance of neglecting torque depends on the magnitude of the applied torque, and therefore several torque magnitudes are considered.

5.1.4.1 Axial Load Study

The loading in this study may be applied as a single point load or multiple point loads in parallel. However, there may be a threshold ratio where a large cutting tool cannot be appropriately modeled with a single point load due to significant differences in the maximum resultant displacement. If there is a significant difference in the maximum resultant displacement, it is evident that accurately applying loads is dependent on which model is used, based on the size of the cutting tool. The purpose of this study is to determine whether that threshold exists. The geometry, mesh, boundary conditions, and number of nodes used to distribute the load were held unchanged. The only variable was the radius of the cutting tool perimeter. The torque was neglected in this study, in that it is applied the same way regardless of the size of the cutting tool contact area. There are several commercially available cutting tool diameters. Because the purpose of this study is to determine whether there is a threshold where the size effect will be observed, cutting tool diameters that cover the entire range of available sizes were selected. A minimum diameter of zero and a maximum diameter of two inches are used to ensure that the common range of commercially available cutting tool diameters used in machining is considered. The single point load model has the zero-diameter and is the basis for comparison of the maximum displacement results between the two models. Because the selection of cutting tool diameters is relatively coarse, should there be a significant difference in the maximum resultant displacements between any of the cutting tool diameters the study could be refined to a smaller range of values, encompassing those where a difference was observed. The cutting tool diameters, radii, and ratios of workpiece surface area to cutting tool contact area are listed in Table 5.8. Figure 5.10 shows the cutting tool perimeter and distribution of point loads for both models. The axial load models shown are common to both the drilling and milling load models. Figure 5.10a shows the multiple point load model discussed in the fixture design analysis methodology section. Figure 5.10b shows the simplified single-point load model. As seen from the maximum resultant displacement results in Table 5.7, there is no significant change in displacement as a result of applying the axial load over a large perimeter as opposed to using a single point load.

Table 5.8. Maximum resultant displacement with different cutter size.

Cutter Diameter, ϕ_c (in.)	A_s/A_C	$\delta_{max} 10^{-3}$ (in.)	% Difference
0	∞	1.355	0
0.5	101.86	1.100	18.8
1.0	25.46	1.100	18.8
2.0	6.37	1.186	12.5

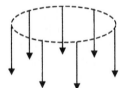   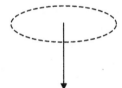

(a) Multiple point load model   (b) Single point load model

Figure 5.10. Axial load models.

The percent difference between the single point load model ($\phi_c = 0$) and multiple point load models is listed in Table 5.8. Although the percent difference between the two models appears significant, with a mean of 16.7% the displacement differences (which are on the order of only 10^{-4} inches) are not.

In addition to being easier to apply, the single point load is more conservative because it results in slightly larger displacements and is less likely to give a false positive solution for an optimum fixture design (i.e., a viable configuration obtained in the optimization analysis that would fail in application). In addition, because the local state of stress is not of concern, the point load is as appropriate as a distributed load for the purpose of workpiece deflection analysis.

5.1.4.2 Torque Study

The purpose of this study is to determine the effect of neglecting torque when modeling the applied loads in a machining operation. In the axial load study, it was concluded that the single-point axial load model is valid. In addition, the axial load and torque models are common to both drilling and milling. Therefore, two comparisons need to be made in torque model: (1) drilling with a single-point axial load and torque versus drilling with a single-point axial load and no torque, and (2) milling with a single-point axial load, transverse load, and torque versus milling with a single-point axial load, transverse load, and no torque. In this study, a block (workpiece) $2 \times 2 \times 1$ inches is used because a 3-2-1 fixture configuration is used in order to assess the effect of applied torque.

The block is meshed with SOLID45 8-node brick elements. The boundary conditions are applied as elastic point constraints, with the properties of the CL-12-RB locator placed in a 3-2-1 configuration. Three locators are placed on the primary reference plane, which is the largest locating surface. The locators are initially placed as far apart as possible, to maximize workpiece stability. The two remaining mutually perpendicular locating surfaces are the same size. Therefore, it is irrelevant which surface is selected as the secondary reference plane and which is selected as the tertiary reference plane. There are two locators on the secondary reference plane, placed as far apart as possible at the mid-plane of the surface. One locator is placed at the center of the tertiary reference plane at the mid-plane of the surface.

The drill diameter is arbitrarily selected as 1 inch. A constant 1,000-pound axial load is applied at the top center of the block. For milling, a constant 500-pound transverse load is applied at the top center of the block.

Four different torque values were considered: 0 lb-in., 50 lb-in., 100 lb-in., and 200 lb-in. The maximum resultant displacement and von Mises effective stress at each of the supports were analyzed.

The results in Table 5.9 show that the torque contributes significantly to the maximum resultant displacement and local von Mises effective stress in the workpiece. Note the increase in maximum resultant displacement and von Mises stress magnitude as the applied torque increases. The torque primarily affects the workpiece stress results at locators 1, 2, and 3, which are placed on reference planes perpendicular to the loaded surface. Although the entire range of possible torque magnitudes was not covered in this study, it can be concluded that by neglecting the applied torque the workpiece deformation results would be inaccurate and the fixture design optimization problem would be misleading.

Table 5.9. Torque study results.

Load Case		δ_{max} 10^{-3} (in.)	von Mises Stress σ_{vm} 10^4 (psi) at Locator					
F_c (lbf) F_{tr} (lbf)	T (lb-in.)		Loc 1	Loc 2	Loc 3	Loc 4	Loc 5	Loc 6
Drilling $F_c = 1,000$ $F_{tr} = 0$	0	1.23	0.84	0.45	0.37	2.31	2.42	4.80
	50	1.23	0.87	0.74	0.08	2.31	2.42	4.80
	100	1.47	1.00	0.93	0.05	2.28	2.49	4.84
	200	3.30	1.31	1.27	0.05	2.18	2.48	4.89
Milling $F_c = 1,000$ $F_{tr} = 500$	0	1.90	5.29	0.55	0.27	4.07	0.69	4.74
	50	2.00	5.29	0.85	0.01	4.00	0.67	4.77
	100	2.44	5.45	1.00	0.02	3.96	0.67	4.84
	200	6.76	5.77	1.16	0.02	3.96	0.64	4.95

5.1.5   Fixture Design Optimization

To minimize workpiece deformation and maximize locating accuracy, the boundary conditions (support locations and clamp location, and clamping force magnitude) of the model are optimized. The object of optimization is to maximize machining accuracy by minimizing workpiece deformation. The locators satisfy two functional requirements: (1) locate and stabilize the workpiece and (2) serve as supports to minimize workpiece deflections. The optimization analysis attempts to satisfy both functional requirements with a single design parameter: the position of the locators on the workpiece surface.

The optimization analysis is performed in ANSYS 5.6.2. The ANSYS program offers two optimization methods to accommodate a wide range of optimization problems. The subproblem approximation method is an advanced zero-order method that can be efficiently applied to most engineering problems. The first-order method is based on design sensitivities and is more suitable for problems that require high accuracy. For both the subproblem approximation and first-order methods, the program performs a series of analysis-evaluation-modification cycles. That is, an analysis of the initial design is performed, the results are evaluated against specified design criteria, and the design is modified if necessary. This process is repeated until all specified criteria are met. In addition to the two optimization techniques available, the ANSYS program offers a set of strategic tools that can be used to enhance the efficiency of the design process. For example, a number of design iterations can be performed. The initial data points from the random design calculations can serve as starting points to feed the optimization methods mentioned previously.

The design variables (DVs) are state variables (SVs) and are referred to as the optimization variables against the objective function. In an ANSYS optimization, these variables are represented by user-named variables called parameters. The user must identify which parameters in the model are DVs, which are SVs, and which is the objective function.

The analysis file is an ANSYS input file that contains a complete analysis sequence (preprocessing, solution, and postprocessing). It must contain a parametrically defined model, using parameters to represent all inputs and outputs, which will be used as DVs, SVs, and the objective function. The loop file resides in the working directory and is used by the control file to build the model. The control file initializes the design variables; defines the feasible design space, optimization analysis method, and looping controls; and executes the optimization analysis (Looman 2001).

A loop is one pass through the analysis cycle. Output for the last loop performed is saved on file *Jobname*.OPO. An optimization iteration is one or more analysis loops that result in a new design set. Typically, an iteration equates to one loop. However, for the first-order method one iteration represents more than one loop. The optimization database contains the current optimization environment, which includes optimization variable definitions, parameters, all optimization specifications, and accumulated design sets. This database can be saved to *Jobname*.OPT or resumed at any time in the optimizer (ANSYS 2001).

DVs are independent quantities that are varied in order to achieve the optimum design. Upper and lower limits are specified to serve as constraints on the DVs. The DVs in the optimization are the positions of locators and clamps as well as clamping force. SVs are quantities that constrain the design. They are also known as dependent variables, which are functions of the DVs. A SV may have a maximum and minimum limit, or it may be single sided. The SV is the von Mises effective stress in this analysis. The objective function is the dependent variable you are attempting to minimize. It should be a function of the DVs. That is, changing the values of the DVs should change the value of the objective function. In our beam example, the total weight of the beam could be the objective function. The objective function in this study is the maximum resultant displacement in the model. Table 5.10 lists all optimization variables used in this study.

Table 5.10. Optimization variables.

Design variables	Position of locators: Locator 1 (X_1, Y_1, Z_1) Locator 2 (X_2, Y_2, Z_2) Locator 3 (X_3, Y_3, Z_3) Locator 4 (X_4, Y_4, Z_4) Locator 5 (X_5, Y_5, Z_5) Locator 6 (X_6, Y_6, Z_6)
	Position of clamps : Clamp 1 (X_1, Y_1, Z_1) Clamp 2 (X_2, Y_2, Z_2)
	Clamping force magnitude Clamp 1 (Fcl_1) Clamp 2 (Fcl_2)
State variables	Von Mises effective stress (VON MISES)
Objective function	Maximum resultant displacement (DMAX)

A design set is simply a unique set of parameter values that represents a particular model configuration. Typically, a design set is characterized by the optimization variable values. However, all model parameters are included in the set. A feasible design is one that satisfies all specified constraints on the SVs, as well as constraints on the DVs. The best design is the one that satisfies all constraints and produces the minimum objective function value.

Because there are a finite number of positions where the modular tooling can be fastened to the base plate, the optimization algorithm is discrete. There are also geometric constraints on the locators and clamps. For example, although it would be ideal to position the primary reference plane supports directly under the applied load during machining because the forces would be transferred directly through the support and the bending moment would be zero, it is impractical in some instances (such as in drilling of a through hole because of interference with the support). For maximum workpiece stability and locating accuracy, the supports on the primary reference plane should be placed as far apart as possible. However, to minimize workpiece deformation the supports should be placed as close to the loads normal to the primary surface as possible. The support locations are optimized where workpiece deflections are minimized and locating accuracy is highest. Locating accuracy, workpiece stability, and workpiece deformations are all affected by the support locations and contribute to the overall fixture stiffness and subsequently the machining accuracy (Pong 1993).

5.1.5.1 Benchmark Optimization Study

A sample optimization analysis shown in Figure 5.11 was conducted to demonstrate the validity of the APDL batch code in ANSYS. As mentioned previously, the optimization analysis is used to minimize the maximum resultant displacement in the workpiece by optimizing support locations, clamp locations, and clamping force magnitudes. The same 3-2-1 fixture configuration used for the workpiece in the loading study was used as the initial configuration in the optimization analysis. The algorithm for selecting initial support locations is explicitly described in the loading study. Three feasible design sets resulted from the optimization analysis. The results are listed in Table 5.11. Design set 1 is the initial fixture configuration. Design set 2 is the optimized configuration given a limited design space, as shown in Figure 5.11. Design set 3 is the optimized configuration given an extended design space. The design space for the optimization analysis resulting in design set 2 is shown in Figure 5.11 as a dashed square. The design space for the optimization analysis resulting in design set 3 was extended to include the entire surface on each reference plane. The von Mises stress at each support location is compared to the yield stress of the workpiece material (AISI 1212 Steel with σ_y = 58,015 psi) to ensure that the material does not exhibit plastic deformation during machining. The von Mises stress is treated as a state variable.

The von Mises stresses at the locators on the secondary and tertiary reference planes (SEQV1, SEQV2, and SEQV3) vary between design sets due to their position and the magnitude of the clamping forces. Note that on the primary reference plane the von Mises stresses (SEQV4, SEQV5, and SEQV6) remain relatively constant, in that the axial thrust force magnitude is constant. The clamping force is increased to 249 lbf in design set 2 from 100 lbf in design set 1. In design set 3, it is only increased to 112 lbf. The maximum resultant displacement was subsequently reduced by 8.4%, from 1.47×10^{-3} inches (design set 1) to 1.34×10^{-3} inches (design set 2). In design set 3, the optimized fixture configuration did not vary significantly from the initial configuration. The maximum resultant displacement was only reduced by 0.75% from 1.47×10^{-3} inches to 1.46×10^{-3} inches.

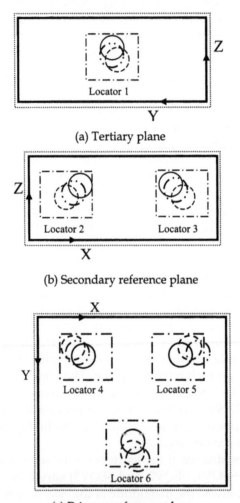

(a) Tertiary plane

(b) Secondary reference plane

(c) Primary reference plane

Figure 5.11. Benchmark fixture design configurations.

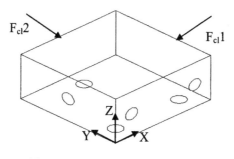

(d) Isometric view of 2 x 2 x 1 block

○ Design set 1   ⸤⸥ Feasible design space (set 2)
○ Design set 2   ⸤⸥ Feasible design space (set 3)
⸑ Design set 3

Figure 5.11. Benchmark fixture design configurations (Continued).

Table 5.11. Benchmark problem optimization analysis results.

Optimization Variable	Variable Type	Design Set 1 (Feasible)	Design Set 2 (Feasible)	Design Set 3 (Feasible)
SEQV1	SV	10043 psi	21832 psi	3248 psi
SEQV2	SV	9263 psi	16853 psi	10459 psi
SEQV3	SV	507 psi	4259 psi	297 psi
SEQV4	SV	22767 psi	27531 psi	26366 psi
SEQV5	SV	24855 psi	20742 psi	24675 psi
SEQV6	SV	48435 psi	47182 psi	44871 psi
FCL1	DV	100 lbf	249 lbf	112 lbf
FCL2	DV	100 lbf	249 lbf	112 lbf
DMAX	OBJ	1.467×10^3 in.	1.344×10^3 in.	1.456×10^3 in.

In design set 2, note that the locators on the primary reference plane (4, 5, and 6) were moved closer to the center of the plane to minimize deflections due to the applied axial load. The locators on the secondary and tertiary reference planes were moved up to minimize deflections due to the applied torque.

It is obvious that without some knowledge base in fixture design the optimization analysis is meaningless. An initial fixture configuration must be provided. If all of the supports are initially placed at the global coordinate system origin, for example, the optimization analysis will not result in a feasible design set. The user must also specify the design space by selecting the range of values for the design variables. It is more appropriate to declare the entire surface on each reference plane as feasible design space, but the analysis is more time intensive than if the design space is limited to a smaller range of values.

5.1.5.2 Industrial Case Study of Fixture Layout Design Optimization

An industrial case study was conducted to validate the fixture design analysis method developed in this study. The workpiece model is a simplified die-cast aluminum brake caliper taken from Delphi Automotive Systems. The model is simplified to protect proprietary features and dimensions. The locators are placed in a 3-2-1 configuration. Three locators are placed on the primary reference plane (one on the bottom of the caliper and two directly below the slide bushing holes). Two locators are placed on the secondary reference plane (which is on the side of the caliper) and one locator is placed on the tertiary reference plane, directly behind the cylinder bore at the center of the cylinder. The configuration is shown in Figure 5.12. The clamps are placed directly opposite the locators on each reference plane, so that the clamping force is transferred directly through the workpiece to the locator without generating any bending moments. Because the tertiary reference plane is perpendicular to the direction of applied loading, it is not necessary to place a clamp against the locator. Brake caliper model parameters and results are listed in Table 5.12. Table 5.13 lists the locator and clamp positions in millimeters relative to the origin of the global coordinate system. Delphi provided the initial fixture configuration, clamping force magnitude, machining forces, and locator stiffness values. The locators were modeled with multiple ANSYS CONTAC52 spring-gap elements in parallel, attached to a circular contact area at specified fixturing points on the brake caliper. The loading is representative of a boring operation.

The maximum resultant displacement in the preloaded workpiece model is 0.0032 mm, and increases slightly to 0.0036 mm in the fully loaded workpiece model. Thus, it is evident that the preloading due to clamping is the major contribution to the resultant displacement throughout the machining operation. The displacement near the cylinder bore increases significantly, by as much as 100%, but does not exceed the maximum resultant displacement in the preloaded workpiece model.

Table 5.12. Brake caliper model parameters and results.

Element type	ANSYS SOLID45 4-node tetrahedral
Mesh type	Free tetrahedral
Workpiece material type	6061-T6 aluminum
Locator material type	AISI 1144 steel
Locator normal stiffness	1.75×10^5 N/mm
Locator tangential stiffness	1.75×10^4 N/mm
Young's modulus, E	70 Gpa
Workpiece material yield strength, σ_y	0.17 GPa
Poisson's ratio, ν	0.35
Coefficient of static friction, μ	0.61
Thrust force, F_c	249.1 N
Torque, T	18,865 N-mm
SEQV1	7.67×10^{-4} GPa
SEQV2	5.95×10^{-4} GPa
SEQV3	7.40×10^{-4} GPa
SEQV4	1.31×10^{-4} GPa
SEQV5	2.66×10^{-4} GPa
SEQV6	4.11×10^{-4} GPa
Clamping force, FCL1	200 N
Clamping force, FCL2	200 N
Clamping force, FCL3	200 N
Clamping force, FCL4	200 N
DMAX	0.0036 mm

Figures 5.13 and 5.14 are the resultant displacement and von Mises stress plots, respectively, for the preloaded model (clamping loads, no machining loads). Figures 5.15 and 5.16 are the resultant displacement and von Mises stress plots, respectively, for the loaded model.

There is a stress concentration at the bottom of the cylinder bore, as shown in Figure 5.16, during machining due to bending moments generated by the thrust force. The maximum von Mises stress occurs at the contact area of clamp 3, located opposite locator 3 on the primary reference plane.

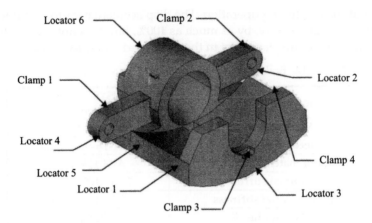

Figure 5.12. A simplified brake caliper model.

Table 5.13. Optimized brake caliper locator and clamp positions.

Locator	Initial configuration (mm)			Optimized configuration (mm)		
	X	Y	Z	X	Y	Z
1	37.95	17.00	-89.50	37.95	17.00	-89.50
2	37.95	17.00	89.50	37.95	17.00	89.50
3	133.85	48.00	0.00	133.85	48.00	0.00
4	78.42	17.51	-76.00	78.42	17.51	-76.00
5	126.84	17.51	-76.00	126.84	17.51	-76.00
6	0.00	0.00	0.00	0.00	0.00	0.00
Clamp	X	Y	Z	X	Y	Z
1	37.95	-10.00	-89.50	37.95	-10.00	-89.50
2	37.95	-10.00	89.50	37.95	-10.00	89.50
3	141.85	27.00	0.00	141.85	27.00	0.00
4	102.61	17.51	76.00	102.61	17.51	76.00

An optimization analysis was conducted to validate the optimization tool developed in ANSYS. Because the fixture configuration for the caliper has been optimized experimentally, the desired result of the optimization analysis is that ANSYS will produce the same fixture configuration. As expected, the support location optimization resulted in the same fixture configuration. However, ANSYS further reduced the maximum resultant displacement in the workpiece by minimizing the clamping force magnitude. The clamping force was reduced to 100 N, subsequently reducing the maximum resultant displacement by 31% to 0.0025 mm. The von Mises stresses at the supports, which are located directly opposite the clamps, were also reduced significantly (as outlined in Table 5.14). The von Mises stress at locator 6 (SEQV6) remained the same in that locator 6 is not reacting to the clamping forces but rather to the applied machining loads, which remained constant.

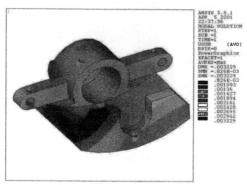

Figure 5.13. A contour plot of preloaded brake caliper displacement (mm).

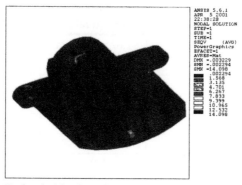

Figure 5.14. Preloaded brake caliper von Mises stress (MPa) contour plot.

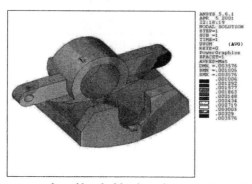

Figure 5.15. A contour plot of loaded brake caliper resultant displacement (mm).

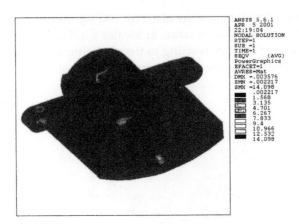

Figure 5.16. A contour plot of loaded brake caliper von Mises stress (MPa).

Table 5.14. Optimized brake caliper results.

Optimization variable	Variable type	Initial configuration	Optimized configuration
SEQV1	SV	7.67×10^{-4} GPa	4.72×10^{-4} GPa
SEQV2	SV	5.95×10^{-4} GPa	1.98×10^{-4} GPa
SEQV3	SV	7.40×10^{-4} GPa	3.68×10^{-4} GPa
SEQV4	SV	1.31×10^{-4} GPa	0.68×10^{-4} GPa
SEQV5	SV	2.66×10^{-4} GPa	1.41×10^{-4} GPa
SEQV6	SV	4.11×10^{-4} GPa	4.11×10^{-4} GPa
FCL1	DV	200 N	100 N
FCL2	DV	200 N	100 N
FCL3	DV	200 N	100 N
FCL4	DV	200 N	100 N
DMAX	OBJ	0.0036 mm	0.0025 mm

5.1.6   Conclusions

In this study a finite element model was developed for fixtured work-piece boundary conditions and applied loads in machining using ANSYS 5.6.2. As opposed to precedent FEA research in fixture design, boundary conditions modeled as both area and point constraints were considered in this study to determine whether a single-point constraint model is appropriate. Pong (1993) modeled boundary conditions to be elastic and deformable, but only considered elastic point constraints. In his research, it was not specified whether an elastic area constraint model was considered.

A more accurate representation of machining loads was also developed. The load model developed in this study includes torque, which is neglected in all precedent research. Distributed and concentrated loading is considered in this study, whereas in previous research all machining forces are applied as single point loads.

Because the model boundary conditions and loads are applied parametrically, the APDL code can be used for solid models with planar locating surfaces and user-defined (1) support locations, (2) clamp locations, (3) clamping force magnitude, (4) cutting tool location, (5) axial load, (6) transverse load, and (7) torque magnitude. The following conclusions are realized based on the research conducted throughout this study:

- *Workpiece elements*: The SOLID45 eight-node brick element is suitable for meshing prismatic geometry. The SOLID45 four-node tetrahedral element may not be as accurate as the brick element, but is suitable for displacement analysis of non-prismatic geometry.

- *Locator model*: It is appropriate to model locators with a single elastic point constraint for a large ratio of workpiece surface area to locator contact area. If the surface area to locator contact area ratio is small, the multiple spring-gap element model using ANSYS CONTAC52 elements must be used.

- *Load model*: It is appropriate to model the cutting tool axial load with a single point load for large workpiece surface area to cutting tool contact area ratios. In addition to being easier to apply, the single point load is more conservative because it results in slightly larger displacements. Because the local state of stress is not of concern, the point load is as appropriate as a distributed load for the purpose of workpiece deflection analysis. The torque component of the load model is critical to workpiece deformation.

- *Optimization*: In this study, a method of fixture design optimization was developed. The method is valid for solid workpieces with planar locating surfaces and may be used to optimize support locations, clamp locations, and clamping force magnitude. The user must have some basic fixture design knowledge to define the initial fixture configuration and design space. The method is capable of minimizing the maximum resultant displacement and assessing workpiece stability. If the workpiece is not stable, it will enter a state of rigid body motion and will not be solved by ANSYS.

This study focused on the minimization of the maximum resultant displacement in the workpiece as a result of applied machining loads to demonstrate the capabilities of the modeling methods developed. The displacement results can be retrieved parametrically at any user-specified location in the workpiece, critical to the quality of the finished part. The displacements in the workpiece are elastic and the concern is local displacements occurring during a machining operation, which are critical to the accuracy of the machined feature. The total machining error, which should be within specified workpiece design tolerances, is the sum of the locating error of the fixture, the machine tool resolution, machine error, cutting tool deflection, fixture component deflection, and the workpiece deflections due to machining loads. Cutting tool resolution and machine error for milling centers can vary significantly depending on machine component quality.

5.2 FINITE ELEMENT ANALYSIS OF FIXTURE UNIT STIFFNESS

This section presents a systematic finite element model to predict fixture unit stiffness by introducing nonlinear contact elements on the contact surface between fixture components. The contact element includes three independent springs: two in tangential directions and one in the normal direction of the contact surface. Strong nonlinearity is caused by possible separation and sliding between two fixture components. The problem is formulated by the penalty function method and is solved by the Newton-Raphson procedure. The model was validated by two cases of analysis of a linear cantilever beam and a simple fixture unit with two components. Results are in agreement with the corresponding analytical solution of beams and the previous experimental results for fixtures in the literature.

5.2.1   Introduction

Manufacturing involves tooling-intensive operations. As an important aspect of tooling, fixturing significantly contributes to the quality, cost, and cycle time of production. In machining processes, fixtures are used to accurately position and adequately constrain workpieces relative to the cutting tool. Fixturing accuracy and reliability are crucial to the success of machining operations. Product quality is often sacrificed when the workholding capability of the fixture is not predictable. Approximately 40% of rejected parts are due to dimensioning errors attributed to poor fixture design (Wardak 2001).

CAFD with verifications has become a means of providing solutions in production operation improvement. Although fixtures can be designed using CAD functions, a lack of scientific tools and a systematic approach for evaluating design performance leads to a trial-and-error design and results in several problems: (1) over design in functions, which is very common and sometimes depredates the performance (e.g., unnecessary heavy design); (2) the quality of design cannot be ensured before production; (3) the long cycle time of fixture design, fabrication, and testing, which may take weeks if not months; and (4) a lack of technical evaluation of fixture design in the quoting processes in the business cycle. In the past 15 years, CAFD has been recognized as an important area and has been studied from fixture planning to fixture design to fixturing analysis/verification. Fixture planning seeks to determine the locating datum surfaces and locating/clamping positions on workpiece surfaces for totally constrained locating and reliable clamping. Fixture design seeks to generate a design of fixture structure as an assembly, according to different production requirements such as production volume and machining conditions. Fixture design verification seeks to evaluate fixture design performances for satisfying production requirements, such as completeness of locating, tolerance stack-up, accessibility, fixturing stability, and ease of operation.

For many years, fixture planning has been the focus of academic research with significant progress in both theoretical (Chou 1989; Xiong 1998; Wu 1998; Brost 1996; Asada 1985; Marin 2002; DeMeter 1998; Wang 1999; Whitney 1999; Roy 2002) and practical (Ma 1999; Fuh 1994) studies. Most analyses are based on strong assumptions, such as frictionless smooth surfaces in contact, rigid fixture body, and single objective function for optimization. Fixture design is a complex problem with considerations of many operational requirements. Four generations of

CAFD techniques and systems have been developed: group technology (GT)-based part classification for fixture design and on-screen editing (Grippo 1987, Rong 1992), automated modular fixture design (Rong 1997; Kow 1998), permanent fixture design with predefined fixture components types (Wu 1997; An 2000; Chou 1993), and variation fixture design for part families (Han 2003). The study on a new generation of CAFD just started to consider operational requirements (Rong 2003). Geometric reasoning (Ma 1998; An 2000), knowledge-based reasoning (Markus 1988; Nee 1991; Nnaji 1990; Pham 1990), and case-based reasoning (CBR) (Kumar 1995; Sun 1995) techniques have been intensively studied for CAFD. How to make use of best-practice knowledge in fixture design and verify fixture design quality under different conditions has become a challenge in the CAFD study.

In fixture design verification, it was proved that when fixture stiffness and machining force are known as input information the fixturing stability problem can be completely solved (Kang 2003). However, most of the studies are focused on the fixtured part model (i.e., how to configure positions of locators and clamps for an accurate and secured fixturing). FEA methods have been extensively used to develop fixture-workpiece models (e.g., Fang 2002; Lee 1987; Trappey 1995) with an assumption of rigid or linear elastic fixture stiffness as a boundary condition (as shown in Figure 5.17, where KN is a normal contact stiffness and KS is a tangential contact stiffness). The models and computational results cannot represent the nonlinear deformation in fixture connections identified in previous experiments (Zhu 1993). As Beards (1986) pointed out, up to 60% of the deformation and 90% of the damping in a fabricated structure can arise from various connections. The determination of fixture contact stiffness is the key barrier in the analysis of fixture stiffness. The existing work is very preliminary, by either simply applying the Hertzian contact model or considering the effective contact area (Yeh 1999; Li 1999).

The development of CAFD tools will enhance both the flexibility and performance of workholding systems by providing a more systematic and analytic approach to fixture design. Fixture functional elements, such as locating pads, buttons, and pins, immediately contact with the workpiece when the workpiece is loaded. Subsequent clamping (by movable elements) creates preloaded joints between the workpiece and each fixture component. In addition, there may be supporting components and a fixture base in a fixture. In fixture design, a thoughtful and economic fixture-workpiece system maintains uniform maximum joint stiffness throughout machining while providing the fewest fixture components, open workpiece cutting access, and shortest setup and unloading cycles. Both static and dynamic stiffness in this fixture-workpiece system rely on the component number, layout, and static stiffness of the fixture structure. These affect fixture performance and must be addressed through appropriate design solutions integrating the fixture with other process elements to produce a highly rigid system. This requires a fundamental understanding of fixture stiffness in order to develop an accurate model for the fixture-workpiece system.

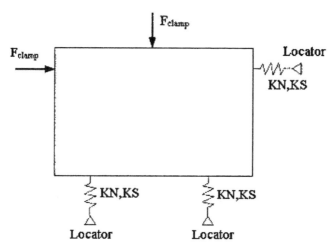

Figure 5.17. A fixture-workpiece model.

5.2.2   CAFD with Predictable Fixture Stiffness

The final goal is to develop a fixture design technique with predictable fixture stiffness. First, the stiffness of typical fixture units is studied with considerations of contact friction conditions. The results of the fixture unit stiffness analysis are integrated in fixture design as a database, with variation capability driven by parametric representations of fixture units. When a fixture is designed with CAFD, the fixture stiffness at the contact locations (locating and clamping positions) to the workpiece can be estimated and/or designed based on machining operation constraints (e.g., fixture deformation and dynamic constraints). Figure 5.18 shows a diagram of the integrated fixture design system.

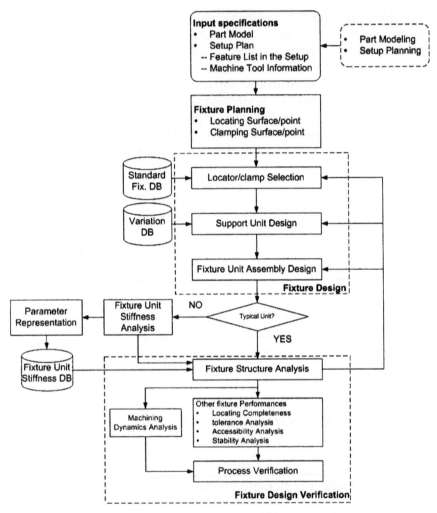

Figure 5.18. Integrated fixture design system.

    To study the fixture stiffness in a general manner, fixture structure is decomposed into functional units which consist of fixture components and functional surfaces (Rong 1999). In a fixture unit, all components are connected one to another where only one is in contact directly with the fixture base and one or more in contact with the workpiece serving as the locator, clamp, or support. Figure 5.19 shows a sketch of the fixture units in a fixture design. When a workpiece is located and clamped in the fixture, fixture units are subjected to the external loads, which pass through the workpiece. If the external load is known and is acting on a fixture unit, and the displacement of the fixture unit at the contact position is measured or calculated based on an FEA model, the fixture unit stiffness can be determined.

Fixture unit stiffness is defined as the force required for a unit deformation of the fixture unit in normal and tangential directions at the contact position with the workpiece. The stiffness can be static if the external load is static (such as clamping force), and dynamic if the external load is dynamic (such as machining force). It is a key parameter to analyze the relative performance of various fixture designs and in optimizing the fixture configuration.

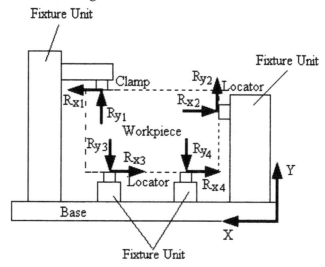

Figure 5.19. Sketch of fixture units.

Analysis of fixture unit stiffness may be divided into three categories: analytical, experimental, and numerical such as FEA. Conventional structural analysis methods may not work well in estimating fixture unit stiffness. A preliminary experimental study has shown the nature of fixture deformation in T-slot-based modular fixtures (Zhu 1993). An integrated model of a fixture-workpiece system was established for surface quality prediction (Liao 2001) based on the experiment results in Zhu (1993), but combining Zhu's experimental work and FEA. Hurtado (2002) used one torsional spring and two linear springs, one in the normal direction and the other in the tangential direction, to model the stiffness of the workpiece, contact, and fixture element. There is no work on FEA of fixture unit stiffness due to the complexity of the contact condition and the great computation effort for many fixture components involved.

This section presents a systematic FEA model to predict fixture unit stiffness based on the elastic body's contact method. The method used here for dealing with contact problems is based on the penalty function method. The contact elements are presented to solve the highly nonlinear problem. The friction condition is also taken into account. To illustrate the validity and accuracy of the model, ANSYS-based FEA code is used to simulate two cases that are compared with analytical solution of beams and previous experimental results in Zhu (1993).

5.2.3   FEA Model With Nonlinear Contact Conditions

In this section, the basic concepts of contact problems in fixture unit stiffness analysis are presented and the concepts are formulated in an FEA solution for the fixture stiffness analysis.

5.2.3.1   FEA Formulation

Consider a general fixture unit with two components I and J (as shown in Figure 5.20). For multicomponent fixture units, a model can be easily expanded. The fixture unit is discretized into an FEA model via the standard procedure. An exception is for the contact surface, for which any node on the finite element mesh for the contact surface is modeled by a pair of nodes at the same location belonging to components I and J, respectively, which are connected by a set of contact springs. The basic assumptions include that material is linearly elastic, displacements and strains are small in both components I and J, and the frictional force acting on the contact surface follows the Coulomb Law of friction.

The total potential energy (Π_p) of the structure element is expressed as the sum of the internal strain energy U and the potential energy Ω of the nodal force; that is,

$$\Pi_p = U + \Omega. \tag{5.1}$$

In the standard procedure of FEA, the strain energy is

$$U = \frac{1}{2}\{q\}^T[K]\{q\}, \tag{5.2}$$

where [K] is the element stiffness matrix and {q} is the element nodal displacement vector. The potential energy of the nodal force is

$$\Omega = -\{q\}^T \{R\},$$ (5.3)

where {R} is the vector of the nodal force.

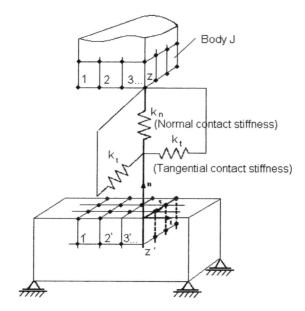

Figure 5.20. Contact model of two fixture components.

When the two components I and J are in contact, a number of 3D contact elements between each contact or potential contact surface are in effect on the contact surfaces. Note that the problem is strongly nonlinear, partially due to the fact that the number of contact elements may vary (i.e., depending on the deformation condition on the contact surface the originally contacting nodes might separate or recontact after separation. The contact elements are capable of supporting a compressive load in the normal direction and a tangential force in the tangential direction. When the two components are in contact and the displacements in the tangential direction and normal direction are assumed independent, the element itself can be treated as three independent linear springs: two having stiffness k_t and k_τ in the tangential directions of the contact surface at the contact point and one having stiffness k_n in the normal direction.

There are two methods typically used to include the contact condition in the energy equation: Lagrange multiplier and penalty function methods (Cook 1989). To understand the Lagrange multiplier and penalty function methods, a physical model of the contact conditions is shown in Figure 5.21. When two contact surfaces of fixture components (i.e., body I and body J) are loaded together, they will contact at a few asperities, such as those shown in Figure 5.21b. The contact condition can be written as:

$$\eta \leq 0; \quad f_{nj} \geq 0; \quad \eta f_{nj} = 0,$$

where η is distance from a contact point i in fixture component I to a contact point j on the fixture component J in the normal direction of contact, and f_{nj} is the contact force acting on point j of body J in the normal direction.

To prevent interpenetration, for each contact points the separation distance η must be greater than or equal to zero. If $\eta > 0$, the contact force is $f_{nj} = 0$. When $\eta = 0$, the points are in contact and $f_{nj} < 0$. If $\eta < 0$, it means that the penetration occurs. To prevent the penetration, the actual contact area increases and the contact stiffness is enhanced when the load increases. Thus the contact stiffness is nonlinear as a function of the preload, as shown in Figure 5.21e. In the penalty function method, an artificial penalty parameter is used to prevent the penetration between contact pairs. In the Lagrange multiplier method, the function w (η, f_n) represents the constraint, which prevents penetration between contact pairs. In the penalty function method, the contact condition is considered as the constraint equation

$$\{t\} = [K_c]\{q\} - \{Q\}, \tag{5.4}$$

where $\{t\}$ is the resultant force of the internal reaction force and the external force. When the system is in equilibrium condition, $\{t\} = 0$, $[K_c]$ is the contact element stiffness matrix, and $\{Q\}$ is the force vector of the active contact node pairs. It includes the applied nodal force $\{F\}$ and contact load $\{f\}$. Thus when $\{t\} = 0$ the constraint equation becomes

$$[K_c]\{q\} = \{F\} + \{f\}. \tag{5.5}$$

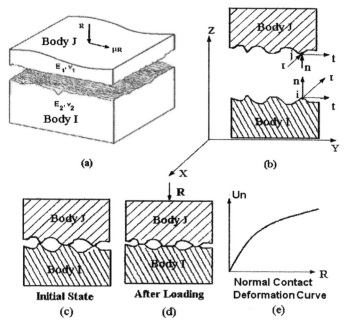

Figure 5.21. Physical model of the contact conditions.

When $\{t\} = \{0\}$, it means that the constraints have been satisfied. The usual potential Π_p in Eq.5.1 can be augmented by a penalty function $\frac{1}{2}\{t\}^T[\alpha]\{t\}$ where $[\alpha]$ is a diagonal matrix of penalty value α_i. The potential energy with the penalty function method becomes

$$\Pi_{pP} = \frac{1}{2}\{q\}^T[K]\{q\} - \{q\}^T\{R\} + \frac{1}{2}\{t\}^T[\alpha]\{t\}.$$

(5.6)

The minimization of Π_{pP} with respect to $\{q\}$ requires that $\left\{ \partial\Pi_{pP} \big/ \partial q \right\} = \{0\}$, which reads to

$$\left([K] + [K_c]^T[\alpha][K_c]\right)\{q\} = \{R\} + [K_c]^T[\alpha]\{Q\},$$

(5.7)

where $[K_c]^T[\alpha][K_c]$ is the penalty matrix. If $[\alpha] = [0]$, the constraints are ignored. At the limit of $[\alpha] \to \infty$, the constraint Eq. 5.4 is satisfied exactly.

In the Lagrange multiplier method, the contact constraint equation can be written as

$$w = \{\gamma\}^{T} ([K_c]\{q\} - \{Q\}),$$ (5.8)

where components of the row vector $\{\gamma\}^{T}$, γ_i ($i = 1,2,...,N$), also often referred as Lagrange multipliers. Adding Eq. 5.8 to the potential energy in Eq. 5.1, we have the total energy in the Lagrange multiplier method:

$$\Pi_{PL} = \frac{1}{2}\{q\}^{T}[K]\{q\} - \{q\}^{T}\{R\} + \{\gamma\}^{T} ([K_c]\{q\} - \{Q\})$$ (5.9)

The minimization of Π_{PL} with respect to $\{q\}$ and $\{\gamma\}$ requires that $\left\{\partial\Pi_{PL}/\partial q\right\} = \{0\}$ and $\left\{\partial\Pi_{PL}/\partial\gamma\right\} = \{0\}$, which leads to

$$\left\{\partial\Pi_{PL}/\partial q\right\} = [K]\{q\} + [K_c]^{T}\{\gamma\} - \{R\} = \{0\}$$ (5.10)

$$\left\{\partial\Pi_{PL}/\partial\gamma\right\} = [K_c]\{q\} - \{Q\} = \{0\}.$$ (5.11)

In a matrix form, Eqs. 5.10 and 5.11 can be expressed as

$$\begin{bmatrix} [K] & [K_c]^{T} \\ [K_c] & [0] \end{bmatrix} \begin{Bmatrix} \{q\} \\ \{\gamma\} \end{Bmatrix} = \begin{Bmatrix} \{R\} \\ \{Q\} \end{Bmatrix}.$$ (5.12)

Although the constraints in Eq. 5.8 can be satisfied, the Lagrange multiplier method has disadvantages. Because the stiffness matrix in Eq. 5.12 may contain zero components in its diagonal, there is no guarantee for the existence of the saddle point. In this situation, the computational stability problem may occur. To overcome the difficulty, the following perturbed Lagrange multiplier method has been introduced (Aliabadi 1993).

$$\Pi^{P}_{PL} = \Pi_{PL} - \frac{1}{2\alpha'}\{\gamma\}^{T}\{\gamma\}$$ (5.13)

$$= \frac{1}{2}\{q\}^{T}[K]\{q\} - \{q\}^{T}\{R\} + \{\gamma\}^{T} ([K_c]\{q\} - \{Q\}) - \frac{1}{2\alpha'}\{\gamma\}^{T}\{\gamma\}.$$

where α' is an arbitrary positive number. At the limit $\alpha' \rightarrow \infty$, the perturbed solutions converge to the original solutions. The introduction of α' will maintain a small force across and along the interface. This will not only maintain stability but avoid the stiffness matrix being singular due to rigid-body motion. Similarly, the minimization of $\Pi^p{}_{PL}$ with respect to $\{q\}$ and $\{\gamma\}$ results in the following matrix:

$$\begin{bmatrix} [K] & [K_c]^T \\ [K_c] & -\dfrac{1}{\alpha'}[I] \end{bmatrix} \begin{Bmatrix} \{q\} \\ \{\gamma\} \end{Bmatrix} = \begin{Bmatrix} \{R\} \\ \{Q\} \end{Bmatrix} \tag{5.14}$$

From Eq. 5.14, the relationships can be established as

$$[K]\{q\} = \{R\} - [K_c]^T\{\gamma\} \tag{5.15}$$

$$\{\gamma\} = \alpha'([K_c]\{q\} - \{Q\}). \tag{5.16}$$

Substituting Eq. 5.16 into Eq. 5.15, it becomes

$$([K] + [K_c]^T \alpha'[K_c])\{q\} = \{R\} + [K_c]^T \alpha'\{Q\}$$

For simplicity, if α' is written as a diagonal matrix and let $\alpha_i = \alpha'$, thus the perturbed Lagrange multiplier will be equivalent to the penalty function method.

In the Lagrange multiplier method, both displacement and contact force are regarded as independent variables. The constraint (contact) conditions can thus be satisfied and the contact force can be calculated. It has disadvantages such that the stiffness matrix contains zero components in its diagonal. In addition, the Lagrange multiplier terms must be treated as addition variables and this leads to the construction of an augmented stiffness matrix, the order of which may significantly exceed the size of the original problem in the absence of constraint equations (Aliabadi 1993). In comparison to the Lagrange multipliers method, the implementation of the penalty function method is relatively simple and does not require additional independent variables. It is often adopted in practical analysis because of its easy implementation.

5.2.3.2 Contact Conditions

Based on an iterative scheme (Mazurkiewicz 1983), the contact conditions are classified into the following three cases:

- Open condition: gap remains open
- Stick condition: gap remains closed and no sliding motion occurs in shear direction
- Sliding condition: gap remains closed and the sliding occurs

Let f_{ji} and u_{ji} be the contact nodal load vector and the nodal displacement, respectively, which are defined in the LCS. The subscript j indicates the component number ($j = I$ or J) and i the coordinate ($i = n, t, \tau$), as shown in Figure 5.22. By equilibrium of the contact element, $\vec{f}_{In} + \vec{f}_{It} + \vec{f}_{I\tau} + \vec{f}_{Jn} + \vec{f}_{Jt} + \vec{f}_{J\tau} = 0$. F_i ($i = n, t, \tau$) is the external nodal load in the i direction. $\{R\} = \displaystyle\sum_{x=1}^{n} \begin{pmatrix} F_n \\ F_t \\ F_\tau \end{pmatrix}_x$, where x is the node number of body I or body J. The displacement and force must satisfy the equilibrium equations in the three contact conditions (note $\{n, t, \tau\}$ is the LCS).

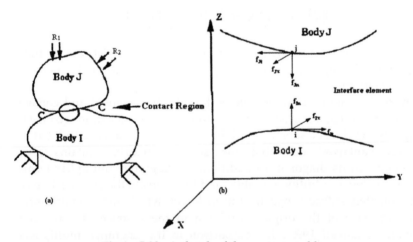

Figure 5.22. A sketch of the contact problem.

Open Condition. When the normal nodal force F_n is positive (tension), the contact is broken and no force is transmitted. The displacement change in normal and tangential directions, denoted respectively by Δu_i $(i = n, t, \tau)$, is then

$$\Delta u_n = u_{Jn} - u_{In} + \delta_n, \quad f_{Ji} = f_{Ii} = 0 \quad (i = n, t, \tau) \tag{5.17}$$

where u_{Jn} and u_{In} are the current displacements of node J and node I in normal direction, respectively. For each structural contact element, stiffness and forces are updated based upon current displacement values in order to predict new displacements and contact forces. δ_n is the gap between a pair of the potential contact points. In each increment of load, the gap status and the stiffness values are iteratively changed until convergence. As the load is increased, δ_n will change and hence should be adjusted as $\delta_n = \delta^0{}_n - \delta^T{}_n$, where $\delta^0{}_n$ is the initial gap before any deformation and $\delta^T{}_n$ is the gap change caused by the total combined normal movement at the pair of points.

Stick Condition. The force in the tangential direction (F_s), which is the composition of the nodal force in t and τ directions (F_t and F_τ), is defined only when $F_n < 0$. When the absolute value of F_s is less than $\mu | F_n |$, where μ is the Coulomb dynamic friction coefficient, there is no slide motion in the interface and the contact element responds like a linear spring. The stick condition exists if $\mu | F_n | > \| (u_{Jt} k_t + u_{J\tau} k_\tau) - (u_{It} k_t + u_{I\tau} k_\tau) \|$. That is,

$$f_{Ii} = -f_{Ji}, \quad u_{Jn} - u_{In} + \delta_n = 0, \quad u_{Ji} - u_{Ii} = 0, \quad (i = t \square \mathcal{C}), \tag{5.18}$$

where k_t and k_τ are the tangential contact stiffness in t and τ directions, respectively. In the analysis of fixture unite stiffness, set $k_t = k_\tau$.

Sliding Condition. The slide motion will occur when the absolute value of F_s is more than $\mu | F_n |$. The slide motion may occur in both element t and τ directions. That is, if $\mu | F_n | < \| (u_{Jt} k_t + u_{J\tau} k_\tau) - (u_{It} k_t + u_{I\tau} k_\tau) \|$, then

$$f_{It} = -f_{Jt} = (\pm \mu F_n)_t$$

$$f_{It} = -f_{Jt} = (\pm \mu F_n)_t$$
$$f_{In} = -f_{Jn}$$
$$u_{In} - u_{Jn} + \delta_n = 0 \tag{5.19}$$

where $(\pm \mu F_n)_t$ and $(\pm \mu F_n)_\tau$ mean the maximum friction force in t and τ directions.

5.2.3.3   Solution Procedure

The model presented in the previous section can be implemented to determine the fixture unit stiffness in clamping and machining. Because the model involves high nonlinearity, the Newton-Raphson approach is used to solve the problem. In this approach, the load is subdivided into a series of load increments, which can be applied over several load steps. At each load step, several iterations are necessary to find the solution with acceptable accuracy. The Newton-Raphson method is used first to evaluate the initial out-of-balance load vector at the beginning of the iteration at each load step. When the out-of-balance load is non-zero, the program performs a linear solution (using the initial out-of-balance loads) and checks for convergence. If the convergence criteria are not satisfied, the out-of-balance load vector is reevaluated, the stiffness matrix is updated, and a new solution is obtained. This iterative procedure continues until the solution converges. A flowchart of the analysis procedure is outlined in Figure 5.23.

5.2.4   Modeling Validation

To determine its effectiveness, the model was used to analyze two simple cases. The first case is the static contact problem of two identical beams, each having a dimension of $10 \times 10 \times 50$ inches. The left side of the first beam is fixed and its right side is connected to the second beam by contact elements (at the position y = 0 in. and z = 10 in.). A distributed compressive load (Q) is horizontally applied to the nodes on the right side of the second beam and a concentrated load (F) is vertically applied to the lowest node at the right side of the second beam, as shown in Figure 5.24. Contact conditions of the two beams are specified by the friction coefficient (μ = 0.2) and the normal contact stiffness (KN = 1.75 $\times$ 10^7 lb/in) and the tangential contact stiffness, KS = 1.75 $\times$ 10^7 lb/in.

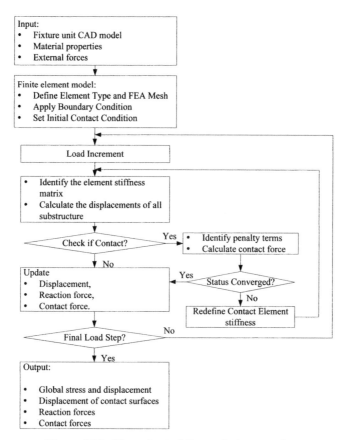

Figure 5.23. Flow chart of the analysis procedure.

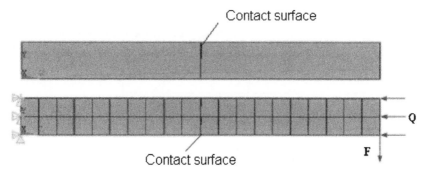

Figure 5.24. Contact problem between two beams subjected to end loads.

In the contact problem of the two beams, the contact elements are used to model the interface between two beams. When two beams are in contact, the contact element itself can be treated as three independent linear springs having stiffnesses k_t, k_π and k_n oriented in tangential and normal directions, respectively. The contact stiffnesses k_t and k_n may vary with respect to different contact conditions. When the contact stiffness becomes infinitely large, the structure should become a single continuous beam, as shown in Figure 5.25, whose analytic solution can be easily obtained.

Figure 5.25. An equivalent single continuous beam.

Figure 5.26 shows a comparison of the deflection curves of the contacted beam with different contact stiffness, as well as the analytical solution of the corresponding single continuous beam. In Figure 5.26, when the contact stiffness is sufficiently large, the deflection curve is very close to the analytical solution and the difference between the two solutions becomes invisible. When the contact stiffness is smaller, the deflection of the right beam becomes larger. The results from FEA validate the contact model and show great effects of the contact stiffness on the deformation of the contact beam. Determination of the contact stiffness is under investigation.

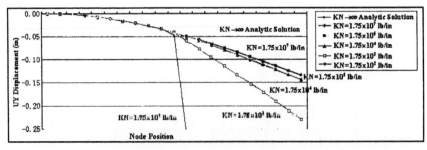

Figure 5.26. Comparison of the deflection of a beam having a contact stiffness that varies with the corresponding single continuous beam.

The other case is to analyze the typical fixture unit, which includes two deformable components (a 500 × 500 × 100 mm fixture base and a 100 × 100 × 300 mm support), as shown in Figure 5.27.

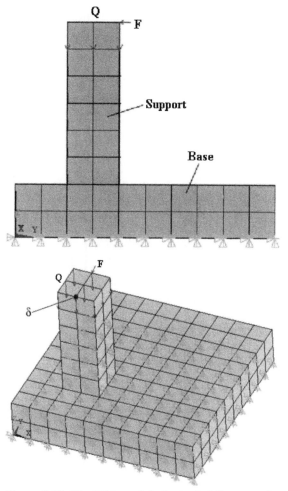

Figure 5.27. The FEA model of a typical fixture unit.

The bottom of the fixture base is fixed. An evenly distributed load, Q, is applied to the nodes on the top of the support, in simulation of the fastening force in the fixture. A concentrated load F parallel to the fixture base is applied to the node on the top of support, in simulation of the external (clamping and/or machining) force passed through the workpiece. The fixture-unit deflection is measured as δ (at the position x

= 150 mm, y = 100 mm, and z = 400 mm) in the Y direction at the top of the support, as shown in Figure 5.27. Contact conditions similar to those in Figure 5.28 are considered in this case. The contact stiffness and friction condition are considered between the fixture base and support. The numerical results of FEA simulation on fixture unit deflection under different combinations of the fastening force Q and the external force F are outlined in Table 5.15.

Table 5.15. Deflection of the fixture unit for different load combinations.

Case 1		Case 2		Case 3	
Q = 3,555 N (800 lbs)		Q = 5,337 N (1,200 lbs)		Q = 7,110 N (1,600 lbs)	
F (N)	δ (mm)	F (N)	δ (mm)	F (N)	δ (mm)
300	0.00958	300	9.62E-03	300	9.66E-03
400	0.01442	400	1.28E-02	500	1.60E-02
500	2.47E-02	600	2.16E-02	700	2.25E-02
550	3.01E-02	700	3.18E-02	900	3.90E-02
580	3.33E-02	800	4.24E-02	1,000	4.94E-02
590	3.44E-02	850	4.77E-02	1,100	6.00E-02
593.5	3.49E-02	891.5	5.25E-02	1,189	7.00E-02

Case 4		Case 5		Case 6	
Q = 8,892 N (2,000 lbs)		Q = 10,674 N (2,400 lbs)		Q = 12,456 N (2,800 lbs)	
F (N)	δ (mm)	F (N)	δ (mm)	F (N)	δ (mm)
300	9.70E-03	300	9.73E-03	300	9.77E-03
500	1.60E-02	500	1.61E-02	500	1.61E-02
700	2.24E-02	700	2.24E-02	700	2.24E-02
900	2.90E-02	900	2.87E-02	900	2.88E-02
1,100	4.62E-02	1,100	3.55E-02	1,100	3.52E-02
1,300	6.71E-02	1,300	5.34E-02	1,300	4.20E-02
1,485	8.74E-02	1,500	7.41E-02	1,500	6.05E-02
		1,700	9.54E-02	1,700	8.11E-02
		1,782	0.10479	1,900	0.10242

Note that different ranges of force F are selected for a given Q in Table 5.15. Physically, a given fastening force Q can only hold a working external force to a limit. Numerically, an extremely large value of F for a given Q may cause a solution not to converge.

A typical curve of fixture deflection versus the external force of FEA results is shown in Figure 5.28. The curve can be divided into three stages, the first linear stage (I), the second nonlinear stage (II), and the third linear stage (III), which is consistent with previous experimental results in Zhu (1993). In the first stage, for small external force F the deflection of fixture components is basically contributed to the elastic deformation. The nonlinearity of the deflection curve in the second stage is mainly caused by the interface between the fixture base and support, which dominates the overall deflection. In this stage, the support begins to separate with the fixture base, which causes a decrease of actual contact area and a rapid increase of the deflection. When the external force continuously increases, the separation becomes stabilized and the deflection tends to be linear again in the third section.

Under different fastening forces, the division of the three stages may be different. When the fastening force Q is small the contact stiffness between fixture components is also small, as well as the overall fixture unit stiffness. Figure 5.29 summarizes the deflection curves of the typical fixture unit where the fastening force is fixed in each case and external force increases. Figure 5.30 shows the experimental results of the deflection curves under different fastening forces (Zhu 1993). It is obvious that the FEA results match the experimental results in trend. The difference between experimental results and FEA results is caused by the simplification of the FEA model.

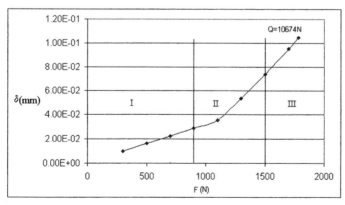

Figure 5.28. A typical deflection curve of fixture units from FEA.

It can be seen that the increase of the fastening force will enhance the fixture unit stiffness and decrease the total deformation. However, large fastening forces may cause other problems such as the wear of fixture components, especially in the case of using modular fixtures.

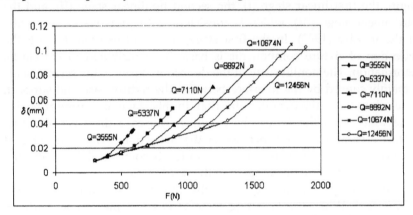

Figure 5.29. Deflection curves under different fastening forces from FEA.

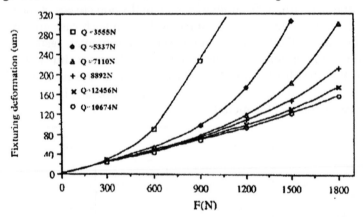

Figure 5.30. Deflection curves from previous experiments (Zhu 1993).

5.2.5   Conclusions

In this section, an FEA model of fixture unit stiffness was developed. A contact element is utilized for solving the contact problems frequently encountered in study of fixture unit stiffness. The FEA model and the analysis procedure were validated by two examples: a simple beam analysis and a simple fixture unit. The results are compared with the corresponding analytical solution and some experimental results in the literature. The agreements between those results demonstrate the great potential of the proposed model for future study of stiffness of fixture units in general configurations, such as a fixture with multiple units and components. In the analysis of the first case, it shows that contact stiffness has a great effect on the accuracy of the results. Thus the contact stiffness of fixture components is one of the key parameters in the analysis of fixture stiffness, which is assumed known in this section.

5.3 CONTACT STIFFNESS IDENTIFICATION

5.3.1   Introduction

In fixture design, a thoughtful and economic fixture-workpiece system maintains uniform maximum fixture stiffness throughout the machining process while also providing the fewest fixture components, open cutting access, and shortest setup and unloading cycles. CAFD with predictable fixture stiffness may provide an adequate design solution for fixture performance improvement. A general fixture structure is decomposed into functional units with fixture components and functional surfaces (Rong 1999). In a fixture unit, all components are connected to each other where only one is in contact directly with the fixture base and one or more in contact with the workpiece serving as the locator, clamp, or support. A sketch of fixture units in a fixture design was shown in Figure 5.19. When a workpiece is located and clamped in the fixture, the fixture units are subjected to external loads that pass through the workpiece. If the external load is known and acting on a fixture unit, and the fixture unit stiffness is also known, the displacement of the fixture unit at the contact position can be estimated based on the FEA model presented in the previous section. The fixture stiffness may contribute to production quality and machining dynamics, particularly in the tolerance-sensitive direction. The key issue in the analysis of fixture stiffness is the modeling of contact problems between fixture components of the fixture unit. The problem encountered in the modeling process is the lack of accurate system parameters, such as contact stiffness for FE formulation.

When the fixture stiffness is studied with FEA methods, it shows that the contact stiffness has a great effect on the accuracy of the results. The contact stiffness is defined as the amount of force per unit displacement required to compress an elastic contact in a particular direction, such as the normal contact stiffness and tangential contact stiffness depicted in Figure 5.31 (Yang 1998). The contact stiffness can vary with respect to its contact condition. When the stiffness becomes infinitely large, the structure system becomes the equivalent rigid body.

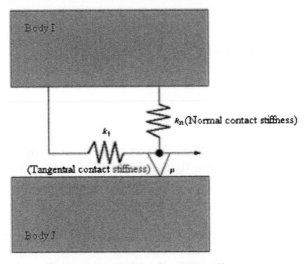

Figure 5.31. Sketch of contact stiffness.

The existing work is very preliminary in the identification of contact stiffness between fixture components, by either simply applying the Hertzian contact model or considering the effective contact area (Yeh 1999; Li 1999). In the study of joints, model parameters obtained from modal testing are widely used in joint identification. A substructure synthesis method was used to identify two types of bolted joints on an overlapped beam (Tsai 1988). The work was extended to using diagonal matrices for joint modeling and an algorithm to reduce noise (Wang 1991). The essential algorithm for identifying joint parameters transforms the assembled system into several single DOF systems using selected eigenvectors. These eigenvectors should be sensitive to the parameters to be identified. It is obvious that this method relies on the availability and accuracy of the mode shapes of the assembled structure. To extract joint parameters without interference with complicated dynamic characteristics of the substructure, a method based on rigid-body dynamics and frequency-response function measurement was developed in which rigid bodies, instead of elastic substructures, were connected to isolate the joints (Becker 1999). The substructure synthesis methods were evaluated and a generalized receptance-coupling method was proposed for multiple connections (Ren 1995). The effect of rotational stiffness was considered in the modeling of joints (Yang 2003).

Because most contact conditions of the fixture–workpiece system involve finite stiffness and the contact stiffness is a critical parameter in modeling the static and dynamic response of the system correctly, it is very important to quantify how the stiffness relates to the contact conditions, such as external load, material property, and so on. The objective of this research initiative is to develop an experimental method to determine the contact stiffness. Although static measurement of the displacement relating to the normal contact stiffness can be conducted with careful design of the experiments, it is very difficult to measure the tangential displacement where the friction plays a crucial role. Therefore, it is very difficult, if not impossible, to estimate the contact stiffness in tangential directions. In this study, a dynamic response method is used to identify the contact stiffness in both normal and tangential directions.

5.3.2 Theoretical Formulation of Normal Contact Stiffness

The idea of identification of the normal contact stiffness is that we can use the impact testing to obtain natural frequencies, along with a theoretical model, to infer normal contact stiffness. The assumptions involved are (1) the contact interfaces were modeled as discrete linear springs and (2) when preload is changed the spring stiffness will change.

When the body I is in contact with ground, the dynamic model of the entire structure can be represented in Figure 5.32. The contact surface is then modeled as a set of linear springs. The objective of this work is to identify the stiffness of these springs. When different preloads are applied, the corresponding values of spring stiffness can be obtained by experiment. In the 1D contact model, m is the mass of body I, k is the contact spring constant, p is the preload, $f(t)$ is impulse excitation, and $u(x,t)$ is the system response.

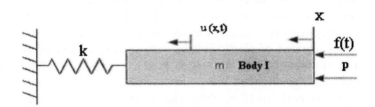

Figure 5.32. 1D contact model for normal stiffness analysis.

Consider the bar shown in Figure 5.32. The governing equation of the longitudinal vibration of the bar is

$$\rho A \frac{\partial^2 u(x,t)}{\partial t^2} = \frac{\partial}{\partial x}\left(EA \frac{\partial u(x,t)}{\partial x} \right),$$ (5.20)

where E is the Young's modulus of the bar, ρ is the mass density per unit volume, and A is the cross section area of the beam. The boundary conditions of the bar are

At x = 0: $EA \dfrac{\partial u(0,t)}{\partial x} = 0$ (5.21)

At x = 1: $EA \dfrac{\partial u(l,t)}{\partial x} = -ku$. (5.22)

Initially, the system starts from rest, from the static equilibrium position of the bar, such that when t = 0 and $u(x,0) = 0$ the initial conditions are

$$\frac{\partial u(x,0)}{\partial t} = \frac{1}{m} .$$ (5.23)

Because the natural modes are orthogonal (the proof is listed in Section 5.4.1), the modal summation solution can be set to

$$u(x,t) = X(x)q(t),$$ (5.24)

where $X(x)$ is the modal shape function and $q(t)$ is the modal response function. The modal shape function can be obtained by substituting Eq. 5.24 into Eq. 5.20, which leads to

$$X(x)\frac{d^2q(t)}{dt^2} = c^2 \frac{d^2X(x)}{dx^2} q(t)$$

$$\frac{1}{X(x)}\frac{d^2X(x)}{dx^2} = \frac{1}{c^2 q(t)}\frac{d^2 q(t)}{dt^2} = -\lambda^2,$$

(5.25)

where $c = \sqrt{\dfrac{E}{\rho}}$ and λ is a constant. The mode shapes $X_i(x)$ satisfy

$$\frac{d^2X(x)}{dx^2} + \lambda^2 X(x) = 0,$$

(5.26)

with the solution,

$$X_i(x) = C_1 \sin \lambda_i x + C_2 \cos \lambda_i x.$$

(5.27)

where C_1 and C_2 are constants; λ_i ($i = 1,2,\ldots,n$) is the system eighenvalues associated with vibration modes. According to Eqs. 5.21 and 5.22, the corresponding modal boundary conditions are

$$EA\frac{dX_i(0)}{dx} = 0$$

(5.28)

$$EA\frac{dX_i(l)}{dx} = -kX_i(l).$$

(5.29)

Substituting Eq. 5.27 into Eqs. 5.28 and 5.29, we have

$$C_1 = 0$$

(5.30)

and $X_i(x) = C_2 \cos \lambda_i x.$

(5.31)

According to the boundary condition, $EA\dfrac{dX_i(l)}{dx} = -kX_i(l)$, we have

$$\tan \lambda_i l = \frac{k}{EA\lambda_i}.$$

(5.32)

Let $k^* = \dfrac{EA}{l}$ and $\beta = \dfrac{k}{k^*}$, we have

$$\tan \lambda_i l = \frac{\dfrac{k}{k^*}}{\lambda_i l} = \frac{\beta}{\lambda_i l} \, . \tag{5.33}$$

This transcendental equation has an infinite number of solution λ_i corresponding to the modes of vibration. When β changes, the solution of λ_i will change. For example, when β changes from 0.1 to 10, then $\lambda_i l$ changes as shown in Figure 5.33.

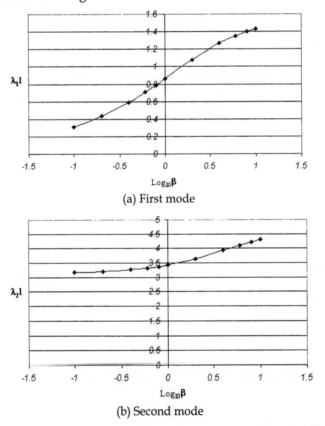

(a) First mode

(b) Second mode

Figure 5.33. Relationships between the non-dimensional natural frequencies, $\lambda_i l$, and the stiffness ratio, β, in the first four modes.

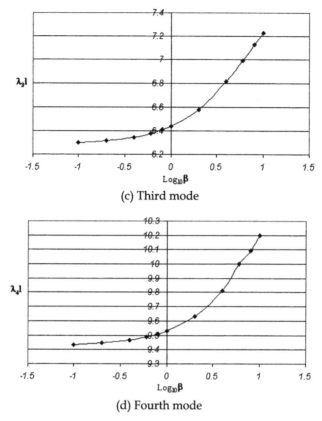

(c) Third mode

(d) Fourth mode

Figure 5.33. Relationships between the non-dimensional natural frequencies, $\lambda_4 l$, and the stiffness ratio, β, in the first four modes (Continued).

When λ_i is known, the corresponding natural frequencies can be obtained as

$$\varpi_i = \lambda_i c = \lambda_i \sqrt{\frac{E}{\rho}}. \tag{5.34}$$

If the natural frequencies can be measured in an experimental test, because the natural frequencies are related to the system characteristics, λ_i becomes known. Then β can be determined from Eq. 5.33 or from the table/curves. Finally, the contact stiffness k can be estimated based on the definition of β. According to the assumption that the contact stiffness is a function of the normal compression load, the natural frequencies can be determined in experiments under different external loads. Then the change of contact stiffness can be identified based on the change of the load, through the measurement of the natural frequency deviation. It should be noted that although any mode of the natural frequency can be used to estimate the contact stiffness some modes may be more sensitive than others to the change of the normal compressive load.

5.3.3   Theoretical Formulation of Tangential Contact Stiffness

Two fixture components are in contact at a certain number of asperities due to the inherent roughness of the surface. When they are subjected to tangential forces, the components are mutually constrained through frictional contacts. A friction model is shown in Figure 5.34. It is based on the Coulomb friction theory and valid for small displacement only. The static tangential contact stiffness results from the elasticity of asperities of the contact surfaces, and the total resulting stiffness of these contact surfaces depends on their statistical topographical parameters.

Consider a bar (fixture component I) placed into contact with the flat surface of the support under a uniform normal pressure and subjected to a force that could be a steady-state harmonic forced excitation, as shown in Figure 5.35.

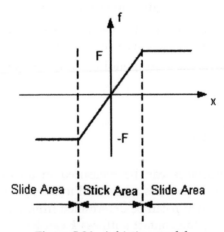

Figure 5.34. A friction model.

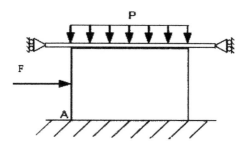

Figure 5.35. Fixture component I on the support.

It is assumed that the tangential contact stiffness increases with the normal load and is not changed when the normal force is fixed. The friction at each contact point is governed by the Coulomb Law of friction. The bar will start to deform at the contact area when the tangential force is applied. The deformation region will gradually extend inward with tangential force increased until the tangential force reaches the value μP, at which point slip will occur. When the overall bar is kept stick, the microslip will generate a Coulomb-type friction/damping force. It is assumed that the damping ratio will be fixed when the magnitude of the harmonic excitation force is not changed. The tangential contact model is shown in Figure 5.36. The friction force is given by Eq. 5.35.

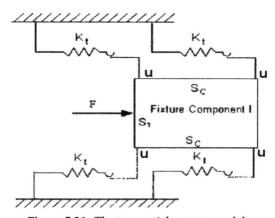

Figure 5.36. The tangential contact model.

$$f = \begin{cases} = K_t u & |u| \le \mu P / K_t \\ \mu p & otherwise \end{cases} \tag{5.35}$$

The idea of identification of the tangential contact stiffness is that we use the steady-state harmonic testing to obtain natural frequencies (along with an FEA model) to infer the tangential contact stiffness, which is modeled as a linear spring when the preload is fixed.

Consider the bar shown in Figure 5.36. From the principle of virtual displacements the system energy and work of external forces can be expressed as

$$\delta\Pi = \int_{t_1}^{t_2}\left(\int_V \frac{1}{2}\{\delta\varepsilon\}^T[D]\{\varepsilon\}dV - \int_V \frac{1}{2}\rho\{\delta\ddot{u}\}^T\{\ddot{u}\}dV + \int_{S_1}\{\delta u\}^T FdS + \int_{S_C}\{\delta u\}^T\{R_c\}dS\right)dt' \quad (5.36)$$

where $\{\delta u\}$ is a virtual displacement vector, $\{\varepsilon\}$ is the strain vector of the block, $[D]$ is the material stiffness matrix of the bar, v is the volume of the block, ρ is the mass density, $\{F\}$ is the external force vector, S_1 is the area with prescribed external forces, S_c is the contact area, and $\{R_c\}$ is the reaction force vector on the contact surface. According to the principle of virtual discplacement, Eq. 5.36 can be written as

$$\delta\Pi = \int_{t_1}^{t_2}\left(\int_V \frac{1}{2}\{\delta\varepsilon\}^T[D]\{\varepsilon\}dV - \int_V \frac{1}{2}\rho\{\delta\ddot{u}\}^T\{\ddot{u}\}dV + \int_{S_1}\{\delta u\}^T FdS + \int_{S_C}\{\delta u\}^T\{R_c\}dS\right)dt = 0. \quad (5.37)$$

The second term can be rewritten as

$$\int_{t_1}^{t_2}\int_V \rho\{\delta\ddot{u}\}^T\{\ddot{u}\}dVdt = \int_{t_1}^{t_2}\rho\frac{d\{\delta\ddot{u}\}^T}{dt}\{\ddot{u}\}dt$$

$$= \left[\rho\{\delta\ddot{u}\}^T\Big|_{t_1}^{t_2}\{\ddot{u}\}\right] - \int_{t_1}^{t_2}\rho\{\delta u\}^T\frac{d^2\{u\}}{dt^2}dt = -\int_{t_1}^{t_2}\rho\{\delta u\}^T\frac{d^2\{u\}}{dt^2}dt. \quad (5.38)$$

The displacement $\{u\}$ is a function of both space and time and can be decomposed into

$$\{u\} = [N(x)]\{d(t)\}, \quad (5.39)$$

where $[N]$ is the function of space only and $\{d\}$ is the function of time only. The combination of Eqs. 5.37 and 5.39 yields

$$\int_V [B]^T [D][B] dV \{d\} + \int_V \rho[N]^T [N] dV \{\ddot{d}\} - \int_{S_1} [N]^T \{F\} dS - \int_{S_C} [N]^T \{R_C\} dS = 0 , \quad (5.40)$$

where $[B]$ is the strain–displacement matrix. Eq. 5.40 can be written as

$$[M]\{\ddot{d}\} + \{[K] - [K_t]\}\{d\} = 0 , \quad (5.41)$$

where $[M] = \int_V \rho[N]^T [N] dV$ is the mass matrix, $\int_V [B]^T [D][B] dV \{d\} = [K]\{d\}$ is the internal force in normal direction, $[K] = \int_V [B]^T [D][B] dV \{d\}$ is the normal stiffness matrix, $\int_{S_C} [N]^T \{k_t\} dS \{d\} = [K_t]\{d\}$ is the internal force in tangential direction, $[K_t] = \int_{S_C} [N]^T \{k_t\} dS$ is the tangential stiffness matrix, and $\int_{S_1} [N]^T \{F\} dS$ is the external force vector. The integral of external force becomes 0 when an impact force is used to excite the system for obtaining the natural frequencies. Then the system natural frequency can be obtained from

$$\left([M]\omega^2 + \{[K] - [K_t]\}\right)\{d\} = 0 . \quad (5.42)$$

To identify the tangential contact stiffness, we compare two sets of the system natural frequencies: one set is identified from the measures impulse response under a lateral impulse loading and the other set is calculated from the FEA model of the system. When the contact stiffness changes with the normal load, the system natural frequencies will change (in the FEA model, the penalty function for the numerical displacement penetration reflects the change of contact stiffness). Then if the natural frequencies are measured in experiments under different normal load, the contact stiffness can be calculated inversely. For example, when the object is a 5 × 3 × 0.7 block, the FEA model is shown as in Figure 5.37. The contact stiffness is modeled as springs on the top and bottom surfaces of the specimen. The springs constrain all DOF. The impulse force was applied at the side of the specimen. The response was obtained at M point at the other side of the specimen.

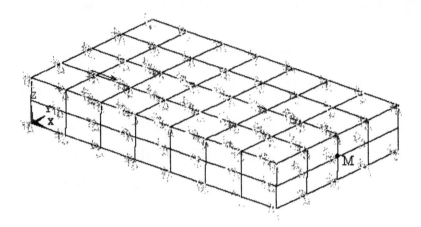

Figure 5.37. FEA model for tangential stiffness identification.

Figure 5.38 shows the relationships of the tangential stiffness versus the natural frequencies of the first two vibration modes. The results are obtained through numerical simulation.

From experiments, the frequency response is measured under different normal load. The contact stiffness can be determined based on the relationships shown in Figure 5.38.

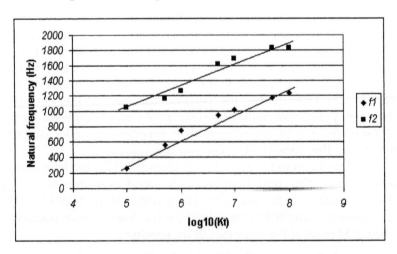

Figure 5.38. The relationship of tangential stiffness versus the lowest two natural frequencies.

5.3.4 Experiment Identification of Contact Stiffness

To verify the method of identifying contact stiffness in both normal and tangential directions, experiments were conducted. The experiment model for identification of normal and tangential contact stiffnesses is shown in Figures 5.39 and 5.40.

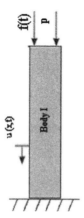

Figure 5.39. Experiment model of the normal contact stiffness identification.

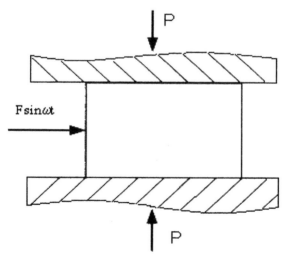

Figure 5.40. Experiment model of the tangential contact stiffness identification.

5.3.4.1 Identification of Normal Contact Stiffness

The measurement instrumentation of normal contact stiffness is relatively simple. It includes vibration sensors, impact excitation hammer with load cell, and a fast Fourier transformation (FFT) analyzer, as shown in Figure 5.41. The experiment procedure is as follows:

1. Frequency response function (FRF) of the bar is measured by using a hammer to excite the system. The first four natural frequencies of the bar are obtained.

2. According to the natural-frequency equation $\varpi_i = \lambda_i c = \lambda_i \sqrt{\dfrac{E}{\rho}}$, λ_i is calculated.

3. Based on the relationship between $\lambda_i l$ and β of the first four modes shown in Figure 5.33, β is found and then the normal contact stiffness can be obtained from the equation $\beta = \dfrac{k}{k^*}$.

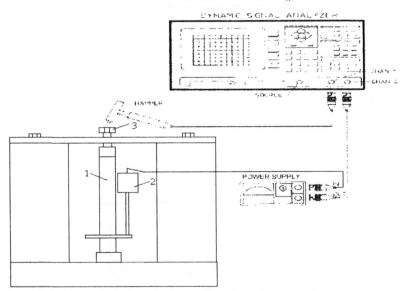

Figure 5.41. Experimental setup for identification of the normal contact stiffness.

When the natural frequencies are obtained from the experiments, along with the curves of the relationships between $\lambda_i l$ and β, the contact stiffness can be determined from each mode of vibration. However, because of the experiment errors the contact stiffness obtained from different modes may be different from each other. Arithmetic mean is used to reduce the effect of experimental errors. When the normal load changes, the natural frequency change in different modes is different. The natural frequencies corresponding to the structural stiffness may not necessarily change significantly with the change of the normal load for certain modes. The contact stiffness should be identified from the mode most sensitive to the normal load.

Figure 5.42 shows the FRF of the test system under different normal loads. Figure 5.43 shows the relationships of the natural frequencies and the normal load. The natural frequencies of the fourth mode, f_4, is most sensitive to the change of the normal load.

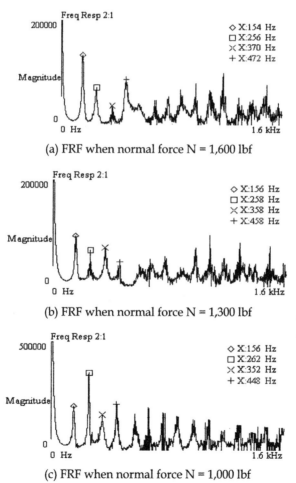

(a) FRF when normal force N = 1,600 lbf

(b) FRF when normal force N = 1,300 lbf

(c) FRF when normal force N = 1,000 lbf

Figure 5.42. Frequency response functions (FRF) of the test system.

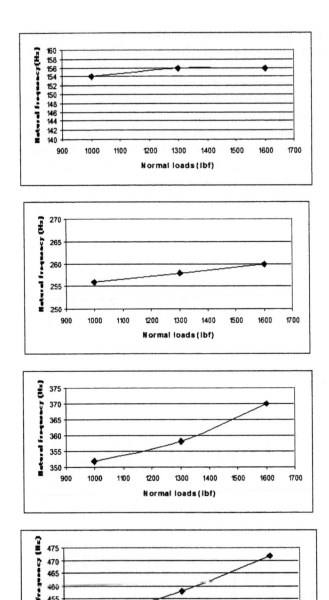

Figure 5.43. Natural frequencies versus normal loads.

Once the natural frequency is obtained from the test, the contact stiffness can be estimated by calculating $\lambda_i l$, β, and k_n. To verify the results, static measurement of the contact stiffness was conducted (see Section 5.4.2). Under the same experimental condition (i.e., the same experimental device and normal load) the contact stiffness is obtained and used in the calculation of natural frequencies and the contact stiffness and then compared with the results of dynamic tests, as shown in Figures 5.44 and 5.45.

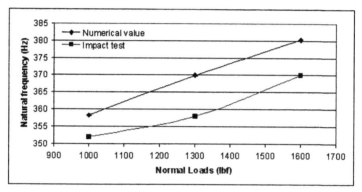

(a) Natural frequency versus normal loads

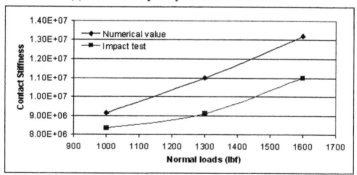

(b) Normal contact stiffness versus normal loads

Figure 5.44. Experimental result comparison.

It can be seen that the results from the dynamic tests are consistent with the numerical calculation results based on the static test results. When the dynamic test results are verified by the static test results, the dynamic test method can be used in tangential contact stiffness identification, where the static tests are too difficult to achieve.

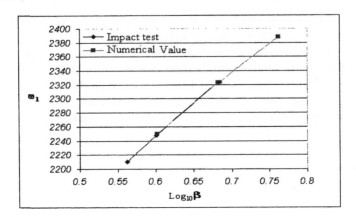

Figure 5.45. Comparison of the results in static and dynamic tests.

5.3.4.2 Identification of Tangential Contact Stiffness

The experimental setup for tangential contact stiffness identification is shown in Figure 5.46. It includes a proximity meter, an impact excitation hammer, and an FFT analyzer.

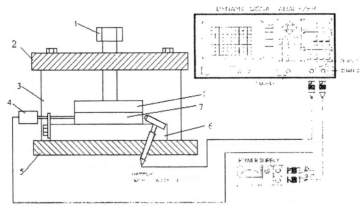

1. Preload bolt. 2. Upper plate. 3. Riser blocks. 4. Proximity meter. 5. Base plate. 6. Support plate. 7. Specimen. 8. Indenter

Figure 5.46. Experimental setup for tangential contact stiffness identification.

Figure 5.47 shows the FRF under different normal loads and Figure 5.48 shows the natural frequencies under different normal loads. According to the relationship of natural frequencies and the tangential contact stiffness shown in Figure 5.38, the tangential contact stiffness can be estimated (as shown in Figure 5.49) where an arithmetic mean was used to reduce the experimental error effect and the structure stiffness effect.

In this way, the contact stiffness can be identified through experiments. The results can be used in the fixture stiffness model presented in Section 5.2, and further used in the CAFD.

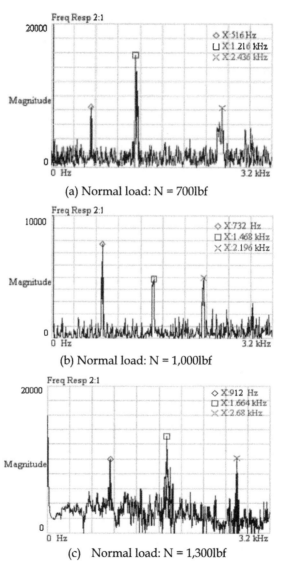

(a) Normal load: N = 700lbf

(b) Normal load: N = 1,000lbf

(c)   Normal load: N = 1,300lbf

Figure 5.47. Frequency response function under different normal loads.

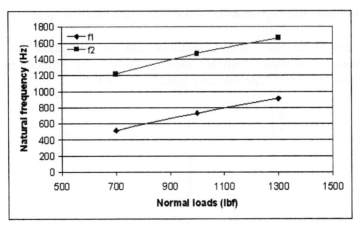

Figure 5.48. Natural frequencies of the first two modes versus the normal load.

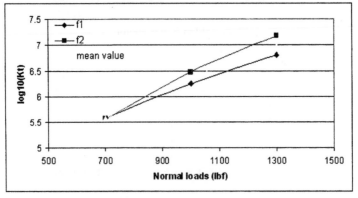

Figure 5.49. Tangential contact stiffness under different normal loads.

5.4 ADDITIONAL PROOF AND VERIFICATION

5.4.1   A Proof of Orthogonality of the Natural Modes

The mode shape functions, Eq. 5.31, for the beam in Figure 5.32 satisfies the following orthogonal relationship:

$$\int_0^l X_i(x)X_j(x)dx = 0, \quad \forall i \neq j.$$

A general proof is given as follows. Assume $X_i(x)$, $X_j(x)$ are two natural modes i and j, and λ_i, λ_j are the corresponding eigenvalues, i.e.,

$$\frac{d^2 X_i(x)}{dx^2} + \lambda_i^2 X_i(x) = 0 \text{ for mode } i \qquad (5.43)$$

$$\frac{d^2 X_j(x)}{dx^2} + \lambda_j^2 X_j(x) = 0 \text{ for mode } j. \qquad (5.44)$$

Multiply Eq. 5.43 by $X_j(x)$ and Eq. 5.44 by $X_i(x)$, subtract the second result from first, and then integrate with respect to x from x = 0 to l via

$$\int_0^l [X_j \frac{d^2 X_i}{dx^2} - X_i \frac{d^2 X_j}{dx^2}]dx + (\lambda_i^2 - \lambda_j^2)\int_0^l X_i X_j dx = 0' \qquad (5.45)$$

where

$$\int_0^l [X_j \frac{d^2 X_i}{dx^2} - X_i \frac{d^2 X_j}{dx^2}]dx$$

$$= X_j \frac{dX_i}{dx}\Big|_l^0 - \int_0^l \frac{dX_i}{dx}\frac{dX_j}{dx}dx - \left(X_i \frac{dX_j}{dx}\Big|_l^0 - \int_0^l \frac{dX_i}{dx}\frac{dX_j}{dx}dx \right) \qquad (5.46)$$

$$= X_j \frac{dX_i}{dx}\Big|_l^0 - X_i \frac{dX_j}{dx}\Big|_l^0 .$$

Applying the modal boundary condition in Eqs. 5.28 and 5.29, we have

$$\int_0^l [X_j \frac{d^2 X_i}{dx^2} - X_i \frac{d^2 X_j}{dx^2}]dx = 0. \qquad (5.47)$$

Then, substituting Eq. 5.47 into Eq. 5.45, we have

$$\int_0^l X_i(x)X_j(x)dx = 0 \qquad i \neq j$$

Note that $\int_0^l X^2{}_i(x) = \int_0^l \cos\lambda_i x \cos\lambda_i x = \int_0^l \frac{1}{2}[\cos 2\lambda_i x - 1] = \frac{1}{2}[\frac{1}{2\lambda_i}\sin(2\lambda_i l) - l] \neq 0.$

The above orthogonal relationship plays a crucial role in solving vibration problems by the modal analysis.

5.4.2   Static Measurement of Normal Contact Stiffness

Contact stiffness is a critical parameter in modeling the static and dynamic response of a fixture system. The contact stiffness is defined as the amount of force per unit displacement required to compress an elastic object in the contact region in a particular direction. It is important to quantify the relationships between the contact stiffness and the contact conditions, such as external loading, geometry, material properties and so on. Contact stiffness in the normal direction can be estimated through the measurement of static deformation of the fixture system.

5.4.2.1   Experimental Method

When a specimen is in contact with the fixture, the contact occurs at a few asperities, as shown in Figure 5.50. When the normal load increases, the contact deformation in the interface will increase nonlinearly and, in turn, the contact stiffness will change. To measure the normal contact stiffness, the total displacements can be measured under the different normal loads. In addition, the structure displacement can be theoretically calculated and subtracted from the results. Therefore, the contact deformation can be obtained and the equivalent contact stiffness K_n can be estimated.

An experimental apparatus was designed to study the normal contact stiffness to external normal loads. The study was conducted by employing cylindrical specimens of steel (AISI 4150) with different contact areas.

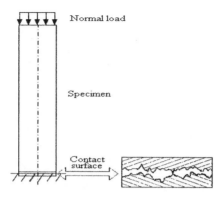

Figure 5.50. Experiment model of normal contact.

The experimental apparatus consisted of multimeters, power supply, fixture and proximity (PX032-1), as shown in Figure 5.51a. The CAD model of the experiment apparatus is shown in Figure 5.51b. The specimen was mounted into the fixture. To apply normal loads, an upper plate, two blocks, two riser plates, and the baseplate were chosen as a fixed reference frame. The preload bolt and the indenter were used to apply the external load. A torque wrench was used to control the load. Two proximities with resolution 10^{-5} inches were used for non-contact measurement of the relative displacement of the specimens. The proximities were connected to the multimeters respectively. The system output voltage signals were proportional to the distance between the transducer tip and the target.

(a) Experimental setup

Figure 5.51. Experimental setup for normal contact stiffness measurement.

(b) Contact surface

Figure 5.51. Experimental setup for normal contact stiffness measurement
(Continued).

The experiments were performed by displacing the indenter against the specimens with a specified force magnitude controlled by using the torque wrench. The normal load caused by the bolt preload can be computed as

$$P = \frac{T}{(K \times D)},$$ (5.48)

where P is the normal load, T is the wrench torque, K = 0-2 is a torque coefficient (Shigley 2001), and D is the nominal diameter of the bolts.

The resulting displacement was recorded for different normal loads. Three specimens of differencing cross-section areas were studied in the experiments. The tests with each specimen involved different magnitudes of the normal loads. The load is converted to surface pressure (i.e., the normal load per unit area). The surface pressure is calculated by the total normal load divided by the cross-section area, i.e.,

$$U = P / (\pi r^2),$$ (5.49)

where r = 0.5 inch. The output voltage was recorded and then converted to the displacement. Figure 5.52 shows the measurement results.

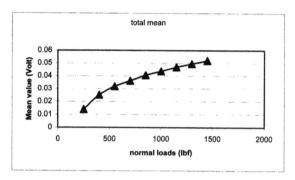

Figure 5.52. Measured voltage results versus the normal load.

To investigate effects of the testing environment on the experiment results, two series of experiments were conducted at two different dates. The output voltages for day 1 and day 2 were shown in Figure 5.53, which shows no significant difference in the results.

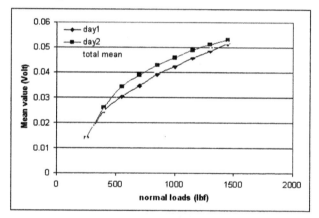

Figure 5.53. Measurement results on different days.

To obtain contact stiffness, the contact displacement and surface pressure need to be known. The contact displacement can be derived from the total displacement, which was measured with the proximities. The total displacement can be calculated from the relative output voltage based on the calibration data.

The total displacement can be decomposed into two individual displacements of specimen and connections (i.e., the displacement caused by structural deformation and the displacement due to contact dformation between the specimens). The structure displacement includes two parts: the specimen structure displacement and the fixture structure displacement. Because the structure displacements can be theoretically calculated, the contact displacement can be separated from the total displacement, as shown in Figure 5.54. Based on the arrangement of the experiment device, the structure displacement is measured as the displacement between the measuring position and the contact position. According to the definition of stiffness, the displacements were converted to the stiffness, as shown in Figure 5.55. It is clear that the non-linearilty is caused by the contact displacement and the contact stiffness is roughly linear to the normal load.

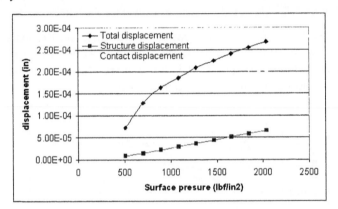

Figure 5.54. Structure and contact displacements versus surface pressure.

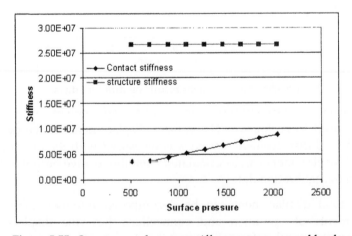

Figure 5.55. Structure and contact stiffness versus normal loads.

A Wilks-Shapiro test for normality was performed to validate the experimental results (Cook 1999; Sen 1990). The results show that the variation of the experimental data was from the random source and without significant bias to the experimental condition (Zheng 2005).

To study the effect of contact area, two different diameters of the specimens were used in the experiment. The diameters of the specimens were 1.0 and 0.75 inches, respectively. Figure 5.56 shows the related comparison of the contact stiffness measures. When the contact stiffness is defined as the ratio of normal load in unit area over contact displacement, the contact stiffness is independent of the contact area.

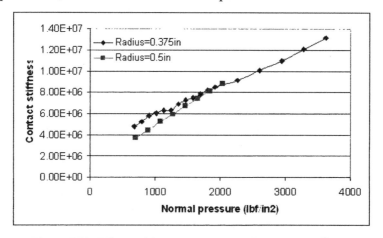

Figure 5.56. Effects of the contact area on the contact stiffness.

Once the contact stiffness data are obtained from the measurement results, a mathematical expression can be derived through statistic regression. It was proved by an adequate test that the relationship of normal pressure and contact stiffness is linear (Zheng 2005). The model is obtained in the case of steel-steel contact studied in this research, with the residual square summation (RSS) via

$$K_n = 2627844 + 2943.43708*P \text{ and RSS} = 0.9712.$$

This information can be used in fixture stiffness analysis for CAFD.

When two surfaces are in contact, the actual contact area is only a fraction of the nominal contact area due to the existence of surface irregularities. The contact stiffness may be different when the surface finish is different. Therefore, it is important to know the effect of surface finish. Three specimens with different surface finish were used in the tests of identifying contact stiffness. The surface finish profiles, R_a, were measured with a contact profilometer. The respective R_a of the specimens are 5.55 μm, 3.08 μm, and 1.44 μm. The contact stiffness data are obtained from the measurement results. Figure 5.57 shows the contact stiffness comparison of three specimens with different surface finish profiles.

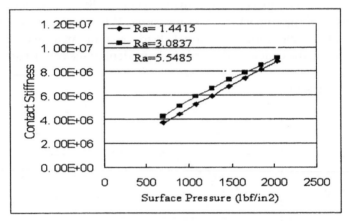

Figure 5.57. Surface finish effect on contact stiffness.

The research presented the results of experimental investigation of the contact stiffness in normal direction and tangential direction. Using the static experimental method, normal contact stiffness is estimated. It has been verified that under different load, the contact stiffness increases linearly with the load. The effect of contact surface was taken into account. The contact stiffness is independent of the contact area and the smoother surface finish leads to a higher contact stiffness. Although the normal contact stiffness can be obtained by the static method, the tangential contact stiffness is difficult to estimate. Therefore the dynamic experimental approach is developed and applied to identify the contact stiffness in both normal and tangential directions. The identification of normal contact stiffness using dynamic method was discussed and the experimental results are compared with the results of static test. It can be seen that the results from the dynamic tests are good in agreement with the numerical calculation results based on the static test results. It indicates that the developed methodology is applicable to the real situation. Hence the dynamic test is used to estimate the tangential contact stiffness. For the estimation of tangential contact stiffness, the results cannot be fully validated. But in comparison with experimental results, the estimation of tangential contact stiffness in different modes is in good agreement with the theoretical model. The results can be used CAFD.

References

Aliabadi, M. H., and C. A. Brebbia. *Computational Methods in Contact Mechanics.* Boston, MA: Computational Mechanics Publications, 1993.

Amaral, N. "Development of a Finite Element Analysis Tool for Fixture Design Integrity Verification and Optimization," M.S. thesis, Worcester Polytechnic Institute, Worcester, MA, 2001.

An, Z., S. Huang, J. Li, Y. Rong, and S. Jayaram. "Development of Automated Fixture Design Systems with Predefined Fixture Component Types: Part 1, Basic Design," *International Journal of Flexible Automation and Integrated Manufacturing* 7:3/4, pp. 321–341, 1999.

ANSYS 6.0. *ANSYS reference manuals, ANSYS,* Inc, Canonsburg, PA, 2001.

Asada, H., and A. By. "Kinematics Analysis of Workpart Fixturing for Flexible Assembly with Automatically Reconfigurable Fixtures," in *Proceedings of the IEEE International Conference on Robotics and Automation* RA-1:2, pp. 86–93, 1985.

Beards, C. F. "The Damping of Structural Vibration by Controlled Interfacial Slip-in Joints," *Journal of Vibration, Acoustic., Stress, Reliability Design* 105, pp.369–373, 1983.

Boyle, I., Y. Rong, and D. Brown. "CAFixD: A Case-based Reasoning Fixture Design Method, Framework, and Indexing Mechanism," in *Proceedings of the 24th Computers and Information in Engineering (CIE) Conference,* Salt Lake City, 28 Sept.–3 Oct., New York: ASME, 2004.

Brost, R. C., and K. Y. Goldberg. "A Complete Algorithm for Synthesizing Modular Fixtures for Polygonal Parts," *IEEE Transactions on Robots and Automation* 12:1, pp. 31–46, 1996.

Cai, W. S., J. Hu, and J. X. Yuan. "Deformable Sheet Metal Fixturing: Principles, Algorithms, and Simulations," *Journal of Engineering for Industry* 118, pp. 318–324, 1996.

Carr Lane Manufacturing Company. Online catalog at *www.carrlane.com,* 2001.

Chou, Y. C. "Automated Fixture Design for Concurrent Manufacturing Planning," *Concurrent Engineering: Research and Applications* 1, pp. 219–229, 1993.

Chou, Y. C., V. Chandru, and M. M. Barash. "A Mathematical Approach to Automatic Configuration of Machining Fixtures: Analysis and Synthesis," *Journal of Engineering for Industry* 111, pp. 299–306.

Cook, R. D., D. S. Malkus, and M. E. Plesha. *Concepts and Applications of Finite Element Analysis* (3d ed.). New York: Wiley, 1989.

Cook, R. D., and S. Weisberg. *Applied Regression Including Computing and Graphics.* New York: Wiley, 1999.

Fang, B., R. E. DeVor, and S. G. Kapoor. "Influence of Friction Damping on Workpiece-Fixture System Dynamics and Machining Stability," *Journal of Manufacturing Science and Engineering* 124, pp. 226–233, 2002.

Fuh, J.Y.H., and A.Y.C. Nee. "Verification and Optimization of Workholding Schemes for Fixture Design," *Journal of Design and Manufacturing* 4, pp. 307–318, 1994.

Grippo, P. M. M., V. Grandhi, and B. S. Thompson. "The Computer-aided Design of Modular Fixturing Systems," *International Journal of Advanced Manufacturing Technology* 2:2, pp. 75–88, 1987.

Hurtado, J. H., and S. N. Meltke. "Modeling and Analysis of the Effect of Fixture-Workpiece Conformability on Static Stability," *Transactions of the ASME Journal of Manufacturing Science and Engineering* 124, pp. 234–241, 2002.

Kang, Y., Y. Rong, and J.-C. Yang. "Computer-aided Fixture Design Verification: Part 1, The Framework and Modeling; Part 2, Tolerance Analysis; Part 3, Stability Analysis," *International Journal of Advanced Manufacturing Technology* 21, pp. 827–849, 2003.

Kashyap, S., and W. R. DeVries. "Finite Element Analysis and Optimization in Fixture Design," in *Structural Optimization*, pp. 193–201, Berlin: Springer-Verlag, 1999.

Kow, T. S., A. S. Kumar, and J.Y.H. Fuh. "An Integrated Computer-aided Modular Fixture Design System for Interference Free Design," MED-8, Manufacturing Science and Engineering, pp. 909–916, ASME IMECE, Anaheim, CA, Nov. 15–20, 1998.

Kumar, A. S., and A.Y.C. Nee. "A Framework for a Variant Fixture Design System Using Case-based Reasoning Technique," in *Computer-aided Tooling*, MED-2:1, pp. 763–775, ASME WAM, 1995.

Lee, J. D., and L. S. Haynes. "Finite Element Analysis of Flexible Fixturing Systems," *Journal of Engineering for Industry* 109, pp. 134–139, 1987.

Li, B., and S. N. Melkote. "An Elastic Contact Model for Prediction of Workpiece-Fixture Contact Forces in Clamping," *Journal of Manufacturing Science and Engineering* 121, pp. 485–493, 1999.

Liao, Y. G., and S. J. Hu. "An Integrated Model of a Fixture-Workpiece System for Surface Quality Prediction," *International Journal of Advanced Manufacturing Technology* 17, pp. 810–818, 2001.

Ma, W., Z. Lei, and Y. Rong. "FIX-DES: A Computer-aided Modular Fixture Configuration Design System," *International Journal of Advanced Manufacturing Technology* 14, pp. 21–32, 1998.

Ma, W., J. Li, and Y. Rong. "Development of Automated Fixture Planning Systems," *International Journal of Advanced Manufacturing Technology* 15, pp. 171–181, 1999.

Marin, R. A., and P. M. Ferreira. "Optimal Placement of Fixture Clamps: Part 1, Maintaining Form Closure and Independent Regions of Form Closure; Part 2, Minimizing the Maximum Clamping Forces," *Journal of Manufacturing Science and Engineering* 124, pp. 676–694, 2002.

Markus, A. "Strategies for the Automated Generation of Modular Fixtures," *Proceedings of Manufacturing Internal*, Atlanta, GA: ASME, pp. 97–103, 1988.

Mazurkiewicz, M., and W. Ostachowicz. "Theory of Finite Element Method for Elastic Contact Problems of Solid Bodies," *Computer Structure* 17, pp. 51–59, 1983.

Menassa, R., and W. R. DeVries. "Optimization Methods Applied to Selecting Support Positions in Fixture Design," *Journal of Engineering for Industry* 113, pp. 412–418, 1991.

Nee, A.Y.C., and A. S. Kumar. "A Framework for an Object/Rule-based Automated Fixture Design System," *Annals of the CIRP* 40:1, pp. 147–151, 1991.

Nnaji, B. O., S. Alladin, and P. Lyu. "Rules for an Expert Fixturing System on a CAD Screen Using Flexible Fixtures," *Journal of Intelligent Manufacturing* 1, pp. 31–48, 1990.

Patricia, J., W. Becker, and R. H. Wynn. "Using Rigid-Body Dynamics to Measure Joint Stiffness," *Mechanical Systems and Signal Processing* 13:5, pp. 789–801, 1999.

Pham, D. T., and A. de Sam Lazaro. "AUTOFIX: An Expert CAD system for Jigs and Fixtures," *International Journal of Machine Tools and Manufacture* 30:3, pp. 403–411, 1990.

Pong, P. C., R. R. Barton, and P. H. Cohen. "Optimum Fixture Design," in *Proceedings of the Second Industrial Engineering Research Conference*, May 26–28, Los Angeles, CA, pp. 6–10, Norcross, GA: IIE, 1993.

Ren, Y., and C. F. Beards. "On Substructure Synthesis with FRF Data," *Journal of Sound and Vibration*, pp. 845–866, 1995.

Rong, Y., and Y. Bai. "Automated Generation of Modular Fixture Configuration Design" *Journal of Manufacturing Science and Engineering* 119, pp. 208–219, 1997.

Rong, Y., and X. Han. "Computer-aided Reconfigurable Fixture Design," *CIRP Second International Conference on Reconfigurable Manufacturing*, Ann Arbor, MI, Aug. 20–22, 2003.

Rong, Y., and Y. Zhu. "An Application of Group Technology in Computer-aided Fixture Design," *International Journal of Systems Automation: Research and Applications* 2:4, pp. 395–405, 1992.

Rong, Y., and Y. Zhu. *Computer-aided Fixture Design*. New York: Marcel Dekker, 1999.

Roy, U., and J. Liao. "Fixturing Analysis for Stability Consideration in an Automated Fixture Design System," *Journal of Manufacturing Science and Engineering* 124, pp. 98–104, 2002.

Sen, A., and M. Srivastava. *Regression Analysis: Theory, Methods, and Applications*. New York: Springer-Verlag, 1990.

Shigley, J. E., and C. R. Mischke. *Mechanical Engineering Design*. New York: McGraw-Hill, 1989.

Sun, S. H., and J. L. Chen. "A Modular Fixture Design System Based on Case-based Reasoning," *International Journal of Advanced Manufacturing Technology* 10, pp. 389–395, 1995.

Trappey, A.J.C., C. S. Su, and J. L. Hou. "Computer-aided Fixture Analysis Using Finite Element Analysis and Mathematical Optimization Modeling," MED-2:1, *Manufacturing Science and Engineering*, pp. 777–787, ASME IMECE, 1995.

Tsai, J. S., and Y. F. Chou. "The Identification of Dynamic Characteristics of a Single Bolt Joint," *Journal of Sound and Vibration* 125:3, pp. 487–502, 1988.

Wang, J., and C. M. Liou. "Identification of Parameters of Structural Joints by Use of Noise-Contaminated FRFS," *Journal of Sound and Vibration* Vol. 141, pp. 261–277, 1990.

Wang, J. H., and M. J. Yang. "Problems and Solutions in the Parameters of Mechanical Joints," in the 3rd *International Conference in Inverse Problem in Engineering: Theory and Practice*, June 13–18, 1999, Port Ludlow, WA, U.S.A., ASME Paper No. ME 03.

Wardak, K. R., U. Tasch, and P. G. Charalambides. "Optimal Fixture Design for Drilling Through Deformable Plate Workpieces: Part 1, Model Formulation," *Journal of Manufacturing Systems* 20:1, pp. 23–31, 2001.

Whitney, D. E., R. Mantripragada, J. D. Adams, and S. J. Rhee. "Designing Assemblies," *Research in Engineering Design* 11, pp. 229–253, 1999.

Wu, S., Y. Rong, and T. Chu. "Automated Generation of Dedicated Fixture Configuration," *International Journal of Computer Applications in Technology* 10:3/4, pp. 213–235, 1997.

Wu, Y., Y. Rong, W. Ma, and S. LeClair. "Automated Modular Fixture Design: Part 1, Geometric Analysis; Part 2, Accuracy, Clamping, and Accessibility Analysis," *Robotics and Computer-integrated Manufacturing* 14, pp. 1–26, 1998.

Xiong, C., and Y. Xiong. "Stability Index and Contact Configuration Planning for Multifingered Grasp," *Journal of Robotic Systems* 15:4, pp. 183–190, 1998.

Yang, B. D., M. L. Chu, and C. H. Menq. "Stick-Slip-Separation Analysis and Non-Linear Stiffness and Damping Characterization of Friction Contacts Having Variable Normal Load," *Journal of Sound and Vibration* 210:4, pp. 461–481.

Yang, T., S.-H. Fan, and C.-S. Lin. "Joint Stiffness Identification Using FRF Measurements," *Computers and Structures* 81, pp. 2459–2566, 2003.

Yeh, J. H., and F. W. Liou. "Contact Condition Modeling for Machining Fixture Setup Processes," *International Journal of Machine Tools and Manufacture* 39, pp. 787–803, 1999.

Zheng, Y. "FEA Modeling and Contact Parameter Identification of Fixture Stiffness for Computer-aided Fixture Design," Ph.D. dissertation, Worcester Polytechnic Institute, Worcester, MA, 2005.

Zhu, Y., S. Zhang, and Y. Rong. "Experimental Study on Fixturing Stiffness of T-slot Based Modular Fixtures," in *NAMRI Transactions XXI*, pp. 231–235, Stillwater, OK: NAMRC, 1993.

CHAPTER 6

Fixture Modeling and Analysis

6.1 FIXTURE MODELING

It is one of the fundamental issues in design automation of fixtures to evaluate and control the geometric tolerances of locators in relationship to locating errors of the workpiece. This section presents a mapping model of the error space of locators and the workpiece locating error space. Given the tolerance specification of the workpiece, the geometry design requirements can be determined for all locators using the model. On the other hand, given the geometric tolerances of locators the calculating methods of locating errors of the workpiece are developed for deterministic locating, over-constrained locating, and under-constrained locating cases by using the mapping model. In the analysis of the locator and clamp configuration characteristics, the free motion cone (used to judge whether the workpiece is accessible to the fixture as well as detachable from the fixture) is defined. According to the duality theory in convex analysis, the polar of the free motion cone (namely, the constrained cone) is derived. By using the constrained cone, the positions of clamps and the feasible clamping domain are determined, where the workpiece is fully constrained. The fixturing analysis models presented are verified by several examples.

6.1.1 Fundamental Problems of Fixturing

Fixtures are used to locate and hold workpieces with locators and clamps respectively, so that the desired positions and orientations of the workpieces can be maintained during machining or manufacturing processes. Fixture design involves setup planning, fixture planning, fixture structural design, and design verification (Rong 1999). The fundamental problems to be solved in automated fixture planning include the following.

1. How to determine the locator configurations that satisfy the accessibility to the fixture and detachability from the fixture

2. How to plan an optimal locator configuration that minimizes the position and orientation errors of the workpiece

3. How to determine the clamping positions that make the workpiece fully constrained

4. How to plan a fixturing (locating and clamping) configuration that makes the fixturing stability assured with appropriately distributing the locators and clamps on the surfaces of the workpiece

5. How to plan a fixturing configuration that makes the fixturing system have the properties of minimizing the position and orientation errors of the workpiece and maximizing the stability of the fixturing system

6. How to determine the clamping forces that make the fixturing system withstand the time- and position-variant external wrench force (such as cutting forces) without sliding between the workpiece and fixels (locators and clamps)

To provide fundamental solutions to these problems, the following further modeling issues need to be studied,

7. The mapping between locator errors and the position and orientation errors of the workpiece

8. The forward problem in fixture verification: given the locator tolerances and their configurations, how to determine the position and orientation precision of the workpiece

9. The inverse problem in fixture verification: given the position and orientation variation requirement of the workpiece, how to determine the tolerances of locators

The goal of Section 6.2 is to develop a general method to solve problems 1, 3, and 7 through 9. First, a new fixturing error model is derived, which is applicable for the deterministic locating, under-constrained locating, and over-constrained locating. The characteristics of locator configurations are analyzed. The accessibility and detachability conditions are given. On the basis of form closure, the feasible clamping domain is determined. We believe that such a general fixturing error analysis model, especially applicable for all the deterministic locating, under-constrained locating, and over-constrained locating, and the method of determining the feasible clamping domain are presented for the first time.

The planning of fixturing configuration is presented in Section 6.4, the goal of which is to solve problems 2, 4, and 5. Moreover, the solution to problem 6 will be further studied.

The remainder of this section is organized as follows. A background overview is presented on previous research in fixture modeling and analysis, as well as related fields. A general fixturing error analysis model is derived, with examples to verify the model. Finally, locator and clamp configuration characteristics are investigated. On the basis of characteristic analysis, a method of determining a locator configuration that satisfies loading and unloading conditions is presented, and the feasible form-closure clamping domain is determined.

6.1.2   Related Work

The geometrical accuracy of a machined feature on a workpiece depends on, partially, the machining fixture's ability to precisely locate the workpiece, which is in fact related to locator configurations and positional accuracy of each locator. The positions of clamps affect directly the form closure of fixturing. Consequently, it is a key for fixture design automation to investigate the fundamentals of locating and clamping.

Traditionally, the design of a fixturing system has been regarded as a manual process relying on human skills and experiences. Toward fixture design becoming a science rather than an art, many researchers have been studying issues related to fixturing.

The kinematics of workpiece fixturing (Bausch 1990; Mishra 1991; Xiong 1993) are similar to those of object grasping. The goals of both fixturing and grasping are to immobilize an object kinematically by means of a suitable set of contacts. The analysis (Krishnakumar 2000; Rimon 2000) of the motion and force constraints during grasping and manipulation of rigid bodies can be extended to fixtures. Research in grasp/fixture inevitably involves closure analysis (Bicchi 1995; Ponce 1997), which can be dated back to 1885 when Reuleaux studied the form-closure mechanism for 2D and 3D objects. Form-closure is a set of mechanical constraints placed around a rigid body so that the motion of the rigid body is not permitted in any direction. For almost a hundred years, it has been known that four and seven unilateral point contacts are the minimum numbers needed for 2D and 3D form-closure respectively, and the related proofs can be found in Lakshminarayana (1978), Mishra (1991), Xiong (1993), and Zhang (2001). The analysis of form-closure shows that six locators, namely, the 3-2-1 locating principle as shown in Cai (1997), Huang (1994), Krishnakumar (2000), and Martin (2001a) and one clamp are needed to fully constrain a workpiece in a fixture. Necessary and sufficient conditions for the deterministic location of 3-2-1 locator schemes were derived (Martin 2001b). However, Czyzowicz (1991) showed that generic 2D and 3D polygonal objects can be immobilized by three and four frictionless contacts, respectively. More recently, Rimon (2000) extended their results to a much larger class of 3D objects, and stated that only four frictionless fingers or fixels are required to immobilize generic 3D objects when second-order geometrical effects (namely, curvature effects) are taken into account. To reach form closure and force closure, modular vise algorithms for designing planar fixtures (Brost 1996) and 3D modular grippers (Brown 1999) have been developed. In the algorithms, it is assumed that a part has contact with only vertical surfaces on the fingers (Brown 1999). Therefore, using vise algorithms we cannot always obtain the modular fixturing solutions for an arbitrary part (Zhang 2001). The selection of the suitable clamping region for planar fixturing has been examined (Wu 1998). However, the proposed method cannot be extended to 3D fixturing. An algorithm for finding clamping positions on 3D workpieces is presented in Martin (2001b), but how to determine the feasible clamping domain that makes the workpiece totally constrained is not given in the algorithm. In Ding (2001), an approach for automatically selecting an eligible set of form-closure fixturing surfaces for a polyhedral workpece is proposed. However, fixture accessibility and fixture detachability for workpiece are not taken into account, which are in fact the essential differences between fixturing and grasping. The reasons for those essential differences are that all fingers during grasping with multifinger hands can be considered active end effectors and that all locators during fixturing are passive elements (whereas only clamps are active). When all fingers during grasping are active, the desired position and orientation of the object to be grasped can be achieved by actively controlling the multifinger hands. Thus, robotic grasping mainly concerns holding feasibility (Rimon 2000), compliance (Lin 2000) and stability (Brost 1999; Xiong 1998, 1999). In contrast, because the position and orientation precision of the workpiece to be fixtured depends on the passive locators' tolerances and configuration fixturing for machining emphasizes more on accurate locating of the workpiece. To meet the accuracy requirement, one might not fixture a workpiece on some surfaces (Huang 1994), although they are feasible from a holding point of view.

Asada (1985) analyzed the problem of automatically locating fixture elements using robot manipulators. The kinematic problems for deterministic locating was characterized by analyzing the functional constraints posed by the fixtures on the surface of a rigid workpiece. Desirable fixture configuration characteristics are obtained for loading and unloading the workpiece successfully despite errors in workpiece manipulation. In the study of manufacturing processes, error sources were investigated for precision machining (Huang 1994; Zhang 2001), and locating error analysis models for deterministic locating were proposed (Cai 1997; Choudhuri 1999, Huang 1994; Li 1999; Rong 1999; Wang 2001). The error sensitivity equation was formed for deterministic locating (Cai 1997; Kang 2001). The impact of a locator tolerance scheme was modeled and analyzed on the potential datum-related geometric errors of linear machined features for deterministic locating (Choudhuri 1999; Rong 1999). However, over-constrained and under-constrained locating are not considered in the proposed locating error analysis models. In addition, these models neglect the sliding errors between the workpiece and locators, which means that developing a general fixturing model is necessary.

6.2 MODELING OF LOCATING DEVIATION

The geometrical accuracy of a machined feature on a workpiece depends on, partially, the machining fixture's ability to precisely locate the workpiece, which is in fact related to the locators' configuration and the position accuracy of each locator. The positions of clamps affect directly the form closure of fixturing.

Consider a general workpiece, as shown in Figure 6.1. Choose reference frame {W} fixed to the workpiece. Let {G} and {L_i} be the global frame and the ith locator frame fixed relative to it. The position of the ith contact point between the workpiece and the ith locator can be described as

$$F_i\left(\mathbf{X}_w, \quad \Theta_w, \quad \mathbf{r}_{w_i}\right) = \mathbf{X}_w + {}_w^g\mathbf{R}\mathbf{r}_{w_i} \tag{6.1}$$

and $\quad f_i\left(\mathbf{X}_{l_i}, \quad \Theta_{l_i}, \quad \mathbf{r}_{l_i}\right) = \mathbf{X}_{l_i} + {}_{l_i}^g\mathbf{R}\mathbf{r}_{l_i}, \tag{6.2}$

where $\mathbf{X}_w \in \mathfrak{R}^{3\times1}$ and $\Theta_w \in \mathfrak{R}^{3\times1}$ ($\mathbf{X}_{l_i} \in \mathfrak{R}^{3\times1}$ and $\Theta_{l_i} \in \mathfrak{R}^{3\times1}$) are the position and orientation of the workpiece (the ith locator) in the global frame {G}, $\mathbf{r}_{w_i} \in \mathfrak{R}^{3\times1}$ ($\mathbf{r}_{l_i} \in \mathfrak{R}^{3\times1}$) is the position of the ith contact point between the workpiece and the ith locator in the workpiece frame {W} (the ith locator frame {L_i}), and ${}_w^g\mathbf{R} \in \mathbf{SO}(3)$ (${}_{l_i}^g\mathbf{R} \in \mathbf{so}(3)$) is the orientation matrix of the workpiece frame {W} (the ith locator frame {L_i}) with respect to the global frame {G}. Thus we have the equation

$$F_i\left(\mathbf{X}_w, \quad \Theta_w, \quad \mathbf{r}_{w_i}\right) = f_i\left(\mathbf{X}_{l_i}, \quad \Theta_{l_i}, \quad \mathbf{r}_{l_i}\right). \tag{6.3}$$

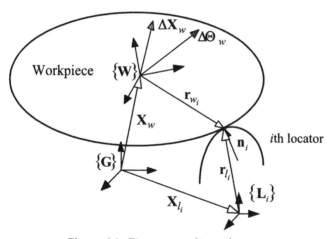

Figure 6.1. Fixture coordinate frames.

Because there may exist position error $\Delta\mathbf{X}_{l_i}$ for the ith locator and $\Delta\mathbf{X}_w$ for the workpiece, the contact between the workpiece and the ith locator will depart from its nominal position. Assume that $\Delta\mathbf{X}_w \in \mathfrak{R}^{3\times1}$ ($\Delta\Theta_w \in \mathfrak{R}^{3\times1}$), and $\Delta\mathbf{r}_{w_i} \in \mathfrak{R}^{3\times1}$ are the deviations of the position $\mathbf{X}_w \in \mathfrak{R}^{3\times1}$ (orientation $\Theta_w \in \mathfrak{R}^{3\times1}$) of the workpiece and the position of the ith contact point $\mathbf{r}_{w_i} \in \mathfrak{R}^{3\times1}$, respectively. Let $F_i\left(\mathbf{X}_w + \Delta\mathbf{X}_w, \quad \Theta_w + \Delta\Theta_w, \quad \mathbf{r}_{w_i} + \Delta\mathbf{r}_{w_i}\right)$ be the actual contact on the workpiece. Using Taylor expansion, we then obtain (neglecting the higher-order terms of errors)

$$F_i\left(\mathbf{X}_w + \Delta\mathbf{X}_w, \ \Theta_w + \Delta\Theta_w, \ \mathbf{r}_{w_i} + \Delta\mathbf{r}_{w_i}\right) = F_i\left(\mathbf{X}_w, \ \Theta_w, \ \mathbf{r}_{w_i}\right) + \frac{\partial F_i}{\partial \mathbf{X}_w} \cdot \Delta\mathbf{X}_w + \frac{\partial F_i}{\partial \Theta_w} \cdot \Delta\Theta_w + \frac{\partial F_i}{\partial \mathbf{r}_{w_i}} \cdot \Delta\mathbf{r}_{w_i},$$

(6.4)

where the second term in the right side of Eq. 6.4 is the position error of the ith contact point resulting from the position error $\Delta\mathbf{X}_w$ of the workpiece, the third term is the position error of the ith contact point resulting from the orientation error $\Delta\Theta_w$ of the workpiece, and the fourth term is the position error of the ith contact point resulting from its workpiece geometric variation $\Delta\mathbf{r}_{w_i}$ on the workpiece.

Similarly, assume that $\Delta\mathbf{X}_{l_i} \in \mathfrak{R}^{3\times1}$ ($\Delta\Theta_{l_i} \in \mathfrak{R}^{3\times1}$) and $\Delta\mathbf{r}_{l_i} \in \mathfrak{R}^{3\times1}$ are the deviations of the position $\mathbf{X}_{l_i} \in \mathfrak{R}^{3\times1}$ (orientation $\Theta_{l_i} \in \mathfrak{R}^{3\times1}$) of the ith locator and the position of the ith contact point $\mathbf{r}_{l_i} \in \mathfrak{R}^{3\times1}$, respectively. Let $f_i\left(\mathbf{X}_{l_i} + \Delta\mathbf{X}_{l_i}, \ \Theta_{l_i} + \Delta\Theta_{l_i}, \ \mathbf{r}_{l_i} + \Delta\mathbf{r}_{l_i}\right)$ be the contact on the ith locator. Using Taylor expansion, we then obtain (neglecting the higher-order terms of errors)

$$f_i\left(\mathbf{X}_{l_i} + \Delta\mathbf{X}_{l_i}, \ \Theta_{l_i} + \Delta\Theta_{l_i}, \ \mathbf{r}_{l_i} + \Delta\mathbf{r}_{l_i}\right) = f_i\left(\mathbf{X}_{l_i}, \ \Theta_{l_i}, \ \mathbf{r}_{l_i}\right) + \frac{\partial f_i}{\partial \mathbf{X}_{l_i}} \cdot \Delta\mathbf{X}_{l_i} + \frac{\partial f_i}{\partial \Theta_{l_i}} \cdot \Delta\Theta_{l_i} + \frac{\partial f_i}{\partial \mathbf{r}_{l_i}} \cdot \Delta\mathbf{r}_{l_i},$$

(6.5)

where the second term in the right side of Eq. 6.5 is the position error of the ith contact point resulting from the position error $\Delta\mathbf{X}_{l_i}$ of the ith locator, the third term is the position error of the ith contact point resulting from the orientation error $\Delta\Theta_{l_i}$ of the ith locator, and the fourth term is the position error of the ith contact point resulting from its geometric variation $\Delta\mathbf{r}_{l_i}$ on the ith locator. Although there are geometric errors during fixturing, the contact between the workpiece and locators must be maintained, which means we have the equation

$$F_i\left(\mathbf{X}_w + \Delta\mathbf{X}_w, \ \Theta_w + \Delta\Theta_w, \ \mathbf{r}_{w_i} + \Delta\mathbf{r}_{w_i}\right) = f_i\left(\mathbf{X}_{l_i} + \Delta\mathbf{X}_{l_i}, \ \Theta_{l_i} + \Delta\Theta_{l_i}, \ \mathbf{r}_{l_i} + \Delta\mathbf{r}_{l_i}\right),$$

(6.6)

i.e.,
$$F_i\left(\mathbf{X}_w, \ \Theta_w, \ \mathbf{r}_{w_i}\right) + \frac{\partial F_i}{\partial \mathbf{X}_w} \cdot \Delta\mathbf{X}_w + \frac{\partial F_i}{\partial \Theta_w} \cdot \Delta\Theta_w + \frac{\partial F_i}{\partial \mathbf{r}_{w_i}} \cdot \Delta\mathbf{r}_{w_i}$$

(6.7)

$$= f_i\left(\mathbf{X}_{l_i}, \ \Theta_{l_i}, \ \mathbf{r}_{l_i}\right) + \frac{\partial f_i}{\partial \mathbf{X}_{l_i}} \cdot \Delta\mathbf{X}_{l_i} + \frac{\partial f_i}{\partial \Theta_{l_i}} \cdot \Delta\Theta_{l_i} + \frac{\partial f_i}{\partial \mathbf{r}_{l_i}} \cdot \Delta\mathbf{r}_{l_i}.$$

Using Eq. 6.3 and only considering the influence of the geometric error ΔX_{l_i} of the ith locator on fixturing (it means that the orientation of the locator is invariable), we can represent Eq. 6.7 as

$$\left(\mathbf{I}_{3\times 3} \quad \vdots \quad -{}^g_w\mathbf{R}\mathbf{r}_{w_i} \otimes\right)\cdot\begin{pmatrix}\Delta\mathbf{X}_w \\ \Delta\Theta_w\end{pmatrix} + \Delta\mathbf{r}_{s_i} = \Delta\mathbf{X}_{l_i}, \qquad\qquad 6.8)$$

where $\Delta\mathbf{r}_{s_i} = {}^g_w\mathbf{R}\Delta\mathbf{r}_{w_i} - {}^g_{l_i}\mathbf{R}\Delta\mathbf{r}_{l_i}$ is the sliding error between the workpiece's surface and the ith locator's surface with respect to the global frame $\{\mathbf{G}\}$, $\mathbf{I}_{3\times 3} \in \Re^{3\times 3}$ is the identity matrix, and the operator $\otimes$ means

$$\mathbf{r}_{w_i} \otimes = \begin{pmatrix}r_x \\ r_y \\ r_z\end{pmatrix} \otimes = \begin{pmatrix}0 & -r_z & r_y \\ r_z & 0 & -r_x \\ -r_y & r_x & 0\end{pmatrix} \in \mathbf{so}(3) \quad (\mathbf{so}(3) \text{ is the Lie algebra of the special}$$

orthogonal group $\mathbf{SO}(3)$). The geometric meaning of the parameters in Eq. 6.8 can be found in Figure 6.2. In general, the sliding error $\Delta\mathbf{r}_{s_i}$ is usually small and can be neglected, except for some worst cases of unreasonable locator configurations. Without loss of generality, we assume that the sliding error $\Delta\mathbf{r}_{s_i}$ can be neglected. Thus Eq. 6.8 becomes

$$\left(\mathbf{I}_{3\times 3} \quad \vdots \quad -\mathbf{r}^g_{w_i} \otimes\right)\cdot\begin{pmatrix}\Delta\mathbf{X}_w \\ \Delta\Theta_w\end{pmatrix} = \Delta\mathbf{X}_{l_i}, \qquad\qquad (6.9)$$

where $\mathbf{r}^g_{w_i} = {}^g_w\mathbf{R}\mathbf{r}_{w_i} \in \Re^{3\times 1}$ is the position vector described in the global coordinate frame $\{\mathbf{G}\}$ for the ith contact point on the workpiece.

Assuming that there exists only position error Δr_{n_i} in the normal direction $\mathbf{n}_i$ for each locator, and the Z-axis direction of the coordinate frame $\{\mathbf{L}_i\}$ coincides with the normal direction $\mathbf{n}_i$, i.e., $\Delta\mathbf{X}_{l_i} = \Delta r_{n_i} \cdot \mathbf{n}_i$, then Eq. 6.9 can be described as

$$\left(\mathbf{I}_{3\times 3} \quad \vdots \quad -\mathbf{r}^g_{w_i} \otimes\right)\cdot\begin{pmatrix}\Delta\mathbf{X}_w \\ \Delta\Theta_w\end{pmatrix} = \Delta r_{n_i} \cdot \mathbf{n}_i. \qquad\qquad (6.10)$$

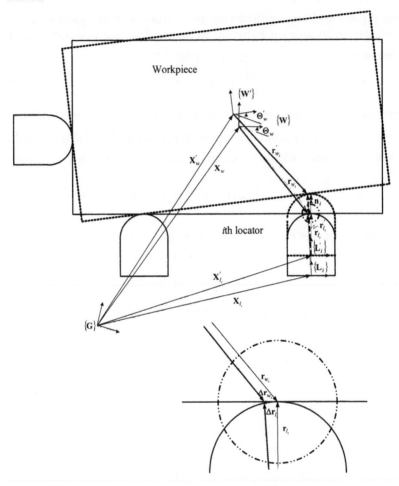

Figure 6.2. The influence of geometric error $\Delta\mathbf{X}_{l_i}$ of the ith locator on fixturing.

For a locating system of m locators, we can represent the m equations in matrix form as

$$\mathbf{G}_L^T \Delta\mathbf{X} = \mathbf{N}\Delta\mathbf{r} ,\tag{6.11}$$

where $\mathbf{G}_L = \begin{bmatrix} \mathbf{I}_{3\times3} & \cdots & \mathbf{I}_{3\times3} \\ \mathbf{r}_{w_1}^g \otimes & \cdots & \mathbf{r}_{w_m}^g \otimes \end{bmatrix} \in \Re^{6\times3m}$; $\Delta\mathbf{X} = \begin{pmatrix} \Delta\mathbf{X}_w \\ \Delta\Theta_w \end{pmatrix} \in \Re^{6\times1}$;

$\mathbf{N} = diag\begin{pmatrix} \mathbf{n}_1 & \cdots & \mathbf{n}_m \end{pmatrix} \in \Re^{3m\times m}$; and $\Delta\mathbf{r} = \begin{pmatrix} \Delta r_{n_1} & \cdots & \Delta r_{n_m} \end{pmatrix} \in \Re^{m\times1}$

Eq. 6.11 can be rewritten as

$$\mathbf{W}_L \Delta \mathbf{X} = \Delta \mathbf{r} ,\qquad\qquad (6.12)$$

where $\mathbf{W}_L = \mathbf{N}^T \mathbf{G}_L^T \in \Re^{m \times 6}$ is referred to as the locating matrix.

6.3 LOCATING CHARACTERISTICS ANALYSIS

Given tolerance specifications of the workpiece features to be machined, which can be converted into the allowed workpiece position and orientation deviation range, the geometric design requirements of all locators can be determined using Eq. 6.12. On the other hand, given the geometric tolerances of locator positions the position and orientation variation of the workpiece can be calculated as well. However, the different calculating methods need to be considered for the following locating conditions.

- Well-constraint (deterministic): The workpiece is mated at a unique position when six locators are made to contact the workpiece surface.
- Under-constraint: The six DOF of a workpiece are not fully constrained.
- Over-constraint: The six DOF of a workpiece are constrained by more than six locators.

Deterministic locating requires six locators that provide full rank of locating matrix W_L. If the rank of locating matrix is less that six, the workpiece is under-constrained (i.e., there exists at least one free motion of the workpiece that is not constrained by locators). If the locating matrix has full rank but the locating scheme has more than six locators, the workpiece is over-constrained, which indicates there exists at least one locator such that it can be removed without affecting the rank of locating matrix and the performance of locating scheme. Table 6.1 compares the rank of locating matrix and number of locators for each status.

Table 6.1. Locating completeness status.

Rank	Number of Locators	Status
< 6	-	Under-constrained
= 6	=6	Well-constrained
= 6	>6	Over-constrained

6.3.1   Locating Status Analysis

In 1985, a full rank Jacobian matrix of constraint equations was proposed
as the criterion (Asada 1985), which has made seminal contributions and
formed the basis of analytical investigations for the deterministic
locating that followed. They assumed frictionless and point contact
between fixturing elements and workpiece. The desired location is q^*, at
which a workpiece is to be positioned and piecewisely differentiable
surface function g_i (as illustrated in Figure 6.3).

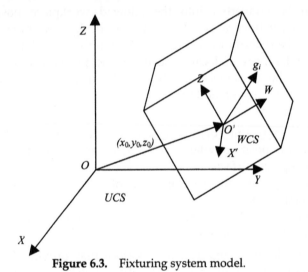

Figure 6.3.   Fixturing system model.

The surface function is defined as $g_i(q^*) = 0$. To be deterministic,
there should be a unique solution for the following equation set for all
locators:

$$g_i(q) = 0 \text{, } i=1, 2, \ldots, n, \tag{6.13}$$

where n is the number of locators and $q = [x_0, y_0, z_0, \theta_0, \phi_0, \psi_0]$
represents the position and orientation of the workpiece.
Considering just the vicinity of desired location q^*, where $q = q^* + \Delta q$,
Asada and By showed that

$$g_i(q) = g_i(q^*) + h_i \Delta q, \tag{6.14}$$

where h_i is the Jacobian matrix of geometry functions, as shown by the matrix in Eq. 6.15. The deterministic locating requirement can be satisfied if the Jacobian matrix has full rank, which leaves Eq. 6.14 with just one solution $q = q*$:

$$rank \left\{ \begin{bmatrix} \dfrac{\partial g_1}{\partial x_0} & \dfrac{\partial g_1}{\partial y_0} & \dfrac{\partial g_1}{\partial z_0} & \dfrac{\partial g_1}{\partial \theta_0} & \dfrac{\partial g_1}{\partial \phi_0} & \dfrac{\partial g_1}{\partial \psi_0} \\ \dfrac{\partial \dot{g_i}}{\partial x_0} & \dfrac{\partial \dot{g_i}}{\partial y_0} & \dfrac{\partial \dot{g_i}}{\partial z_0} & \dfrac{\partial \dot{g_i}}{\partial \theta_0} & \dfrac{\partial \dot{g_i}}{\partial \phi_0} & \dfrac{\partial \dot{g_i}}{\partial \psi_0} \\ \dfrac{\partial \dot{g_n}}{\partial x_0} & \dfrac{\partial \dot{g_n}}{\partial y_0} & \dfrac{\partial \dot{g_n}}{\partial z_0} & \dfrac{\partial \dot{g_n}}{\partial \theta_0} & \dfrac{\partial \dot{g_n}}{\partial \phi_0} & \dfrac{\partial \dot{g_n}}{\partial \psi_0} \end{bmatrix} \right\} = 6. \tag{6.15}$$

Given a locating scheme, we simply need to check the rank of the Jacobian matrix for constraint equations (or rank of locating matrix) to know whether the locating scheme provide deterministic location. Kang (2002) followed this method and implemented it to develop a geometry constraint analysis module in an automated computer-aided fixture design verification system. Their CAFDV system can calculate the Jacobian matrix and its rank to determine locating completeness. It can also analyze workpiece displacement and sensitivity to locating error.

Chou (1989) formulated the deterministic locating problem using screw theory. Let W be a matrix of a $6 \times n$ normalized locating wrench, F an $n \times 1$ intensity vector of the wrenches, and w_p the locating wrench. The equilibrium equation

$$[W][F] = -w_P \tag{6.16}$$

should have non-negative solutions for F. It is concluded that the matrix for the locating wrenches needs to be full rank to achieve deterministic location. This method has been adopted by numerous studies.

Xiong (1998) presented similar criteria to check the rank of locating matrix W_L. The study also introduced *right generalized inverse of the locating matrix* $W^r = W^T \left(W_L W_L^T \right)^{-1}$ to analyze the geometric errors of under-constrained workpieces. It has been shown that the minimum norm solution of the position and orientation errors ΔX_v of the workpiece and the geometric errors Δr of locators are related as follows:

$$\Delta X_v = W^r \Delta r. \tag{6.17}$$

Wang (2003) considered locator-workpiece contact area effects instead of applying point contact. They introduced a contact matrix and pointed out that two contact bodies should not have equal but opposite curvature at a contacting point. Carlson (2001) suggested that a linear approximation may not be sufficient for some applications, such as non-prismatic surfaces or non-small relative errors. He proposed a second-order Taylor expansion that takes locator error interaction into account. Marin (2001a) applied Chou's formulation on 3-2-1 location and formulated several easy-to-follow planning rules.

Despite the numerous analytical studies on deterministic location, less attention was paid to the analysis of non-deterministic location. There is no systematic study on how to deal with a fixture design that fails to provide deterministic location.

If deterministic location is not achieved by a fixturing system, it is as important for the designer to know what the constraint status is and how to improve the design. If the fixturing system is over-constrained, information about unnecessary locators is desired. Although under-constraint occurs, data on unconstrained motion of a workpiece can guide a designer in selecting additional locators and/or revising locating schemes more efficiently. Figure 6.4 depicts a general strategy for characterizing the geometry constraint status of a locating scheme.

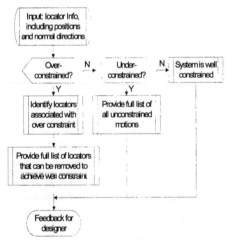

Figure 6.4. Geometry constraint status characterization.

6.3.1.1   Algorithm Development

As shown in Figure 6.5, given locator number n, locating normal vector $[a_i, b_i, c_i]$ and locating position $[x_i, y_i, z_i]$ for each locator, $i = 1, 2, ..., n$ (the $n \times 6$ locating matrix) can be determined as follows:

$$W_L = \begin{bmatrix} a_1 & b_1 & c_1 & c_1 y_1 - b_1 z_1 & a_1 z_1 - c_1 x_1 & b_1 x_1 - a_1 y_1 \\ \cdot & \cdot & \cdot & \cdot & \cdot & \cdot \\ a_i & b_i & c_i & c_i y_i - b_i z_i & a_i z_i - c_i x_i & b_i x_i - a_i y_i \\ \cdot & \cdot & \cdot & \cdot & \cdot & \cdot \\ a_n & b_n & c_n & c_n y_n - b_n z_n & a_n z_n - c_n x_n & b_n x_n - a_n y_n \end{bmatrix} \tag{6.18}$$

When $rank(W_L) = 6$ and n = 6, the workpiece is well-constrained.

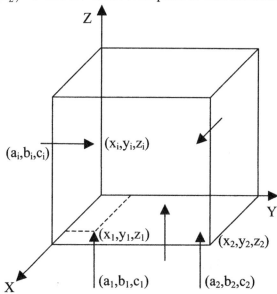

Figure 6.5. A simplified locating scheme.

When $rank(W_L) = 6$ and n > 6, the workpiece is over-constrained. This means there are n - 6 unnecessary locators in the locating scheme. The workpiece will be well-constrained without the presence of n - 6 locators. The mathematical representation for this status is that there are n - 6 row vectors in the locating matrix that can be expressed as linear combinations of the other six row vectors. The locators corresponding to that six row vectors consist of one locating scheme that provides deterministic location. The developed algorithm uses the following approach to determine the unnecessary locators

1. Find all combination of n - 6 locators.
2. For each combination, remove n - 6 locators from the locating scheme.
3. Recalculate the rank of the locating matrix for the lefthand six locators.
4. If the rank remains unchanged, the removed n - 6 locators are responsible for over-constraint.

This method may yield multiple solutions and require the designer to determine which one will achieve the best locating result. When $rank(W_L) < 6$, the workpiece is under-constrained. Assume there are n locators. Eq. 6.17 can be expressed as

$$\Delta X_v = \begin{bmatrix} \Delta x \\ \Delta y \\ \Delta z \\ \alpha_x \\ \alpha_y \\ \alpha_z \end{bmatrix} = \begin{bmatrix} w_{11} & \cdot & w_{1i} & \cdot & w_{1n} \\ w_{21} & \cdot & w_{2i} & \cdot & w_{2n} \\ w_{31} & \cdot & w_{3i} & \cdot & w_{3n} \\ w_{41} & \cdot & w_{4i} & \cdot & w_{4n} \\ w_{51} & \cdot & w_{5i} & \cdot & w_{5n} \\ w_{61} & \cdot & w_{6i} & \cdot & w_{6n} \end{bmatrix} \cdot \begin{bmatrix} \Delta r_1 \\ \cdot \\ \Delta r_i \\ \cdot \\ \Delta r_n \end{bmatrix}, \tag{6.19}$$

where $\Delta x, \Delta y, \Delta z, \alpha_x, \alpha_y, \alpha_z$ are displacement and rotation about the X, Y, and Z axes. Δr_i is the geometric error of the ith locator. w_{ij} is defined as the right-generalized inverse of the locating matrix W^r. To identify all unconstrained motions of the workpiece, $V_i = [dx_i, dy_i, dz_i, d\alpha_{xi}, d\alpha_{yi}, d\alpha_{zi}]$ is introduced such that

$$VX_v = 0. \tag{6.20}$$

When $rank(\Delta X_v) < 6$, there must exist non-zero V that satisfies Eq. 6.20. Each non-zero solution of V_i represents an unconstrained motion. Each term of V_i represents a component of that motion. For example, $[0,0,0,3,0,0]$ says that the rotation about the X axis is not constrained.

[0,1,1,0,0,0] means that the workpiece can move along the direction given by vector $[0,1,1]$. There could be infinite solutions. The solution space, however, can be constructed by $6 - rank(W_L)$ basic solutions. The following determines the basic solutions. From Eq. 6.19 and Eq. 6.20

$$VX_v = dx \cdot \Delta x + dy \cdot \Delta y + dz \cdot \Delta z + d\alpha_x \cdot \Delta\alpha_x + d\alpha_y \cdot \Delta\alpha_y + d\alpha_z \cdot \Delta\alpha_z$$

$$= dx \cdot \sum_{i=1}^{n} w_{1i} \cdot \Delta r_i + dy \cdot \sum_{i=1}^{n} w_{2i} \cdot \Delta r_i + dz \cdot \sum_{i=1}^{n} w_{3i} \cdot \Delta r_i$$

$$+ d\alpha_x \cdot \sum_{i=1}^{n} w_{4i} \cdot \Delta r_i + d\alpha_y \cdot \sum_{i=1}^{n} w_{5i} \cdot \Delta r_i + d\alpha_z \cdot \sum_{i=1}^{n} w_{6i} \cdot \Delta r_i$$

$$= \sum_{i=1}^{n} V \cdot \begin{bmatrix} w_{1i} & w_{2i} & w_{3i} & w_{4i} & w_{5i} & w_{6i} \end{bmatrix}^T \cdot \Delta r_i$$

$$= 0.$$ (6.21)

Eq. 6.21 holds for $\forall \Delta r_i$ if and only if Eq. 6.22 is true for $\forall i \ (1 \leq i \leq n)$, and thus

$$V \cdot \begin{bmatrix} w_{1i} & w_{2i} & w_{3i} & w_{4i} & w_{5i} & w_{6i} \end{bmatrix}^T = 0.$$ (6.22)

In special cases (say, all w_{1j} equal to zero) V has an obvious solution [1, 0, 0, 0, 0, 0], indicating the displacement along the X axis is not constrained. This is easy to understand because $\Delta x = 0$ in this case, implying that the corresponding position/orientation error of the workpiece is not dependent on any locator errors. Hence, the associated motion is not constrained by locators. In addition, a combined motion is not constrained if one of the elements in ΔX_v can be expressed as linear combination of other elements; for example, $\exists w_{1j} \neq 0, \ w_{2j} \neq 0$, $w_{1j} = -w_{2j}$ for $\forall j$. In this scenario, the workpiece cannot move along the X or Y axis. However, it can move along the diagonal line between the X and Y axes defined by vector [1, 1, 0]. To find solutions for general cases, the following strategy was developed.

1. Eliminate dependent row(s) from the locating matrix. Let $r = rank(W_L)$ and n = number of locators. If $r < n$, create a vector in $(n-r)$ dimension space $U = \lfloor u_1 \ . \ u_j \ . \ u_{n-r} \rfloor$ $(1 \leq j \leq n-r, \ 1 \leq u_j \leq n)$. Select u_j such that $rank(W_L) = r$ still holds after setting all terms of all u_jth row(s) equal to zero. Set $r \times 6$ to the modified locating matrix

$$W_L = \begin{bmatrix} a_1 & b_1 & c_1 & c_1 y_1 - b_1 z_1 & a_1 z_1 - c_1 x_1 & b_1 x_1 - a_1 y_1 \\ . & . & . & . & . & . \\ a_i & b_i & c_i & c_i y_i - b_i z_i & a_i z_i - c_i x_i & b_i x_i - a_i y_i \\ . & . & . & . & . & . \\ a_n & b_n & c_n & c_n y_n - b_n z_n & a_n z_n - c_n x_n & b_n x_n - a_n y_n \end{bmatrix}_{r \times 6},$$

where $i = 1, 2, \ldots, n \; (i \neq u_j)$.

2. Compute the 6 x n right-generalized inverse of the modified locating matrix

$$W^r = W^T \left(W_L W_L^T \right)^{-1} = \begin{bmatrix} w_{11} & . & w_{1i} & . & w_{1r} \\ w_{21} & . & w_{2i} & . & w_{2r} \\ w_{31} & . & w_{3i} & . & w_{3r} \\ w_{41} & . & w_{4i} & . & w_{4r} \\ w_{51} & . & w_{5i} & . & w_{5r} \\ w_{61} & . & w_{6i} & . & w_{6r} \end{bmatrix}_{6 \times r} .$$

3. Trim w^r down to a $r \times r$ full rank matrix. Let $r = rank(W_L) < 6$. Construct a $(6-r)$ dimension vector $Q = \lfloor q_1 \; . \; q_j \; . \; q_{6-r} \rfloor$ $(1 \leq j \leq 6-r, \; 1 \leq q_j \leq n)$. Select q_j such that $rank(w^r) = r$ still holds after setting all terms of all $q_j{}^{th}$ row(s) equal to zero. Set $r \times r$ to the modified inverse matrix

$$w^{rm} = \begin{bmatrix} w_{11} & . & w_{1i} & . & w_{1r} \\ . & . & . & . & . \\ w_{l1} & . & w_{li} & . & w_{lr} \\ . & . & . & . & . \\ w_{61} & . & w_{6i} & . & w_{6r} \end{bmatrix}_{6 \times 6},$$

where $l = 1, 2, \ldots, 6 \; (l \neq q_j)$.

4. Normalize the free motion space. Assume $V = [V_1 \; V_2 \; V_3 \; V_4 \; V_5 \; V_6]$ is one of the basic solutions of Eq. 6.20,

with all six terms undetermined. Select a term q_k from vector Q $(1 \leq k \leq 6-r)$. Set

$$\begin{cases} V_{q_k} = -1 \\ V_{q_j} = 0 (j = 1,2,...,6-r, j \neq k). \end{cases}$$

5. Calculate undetermined terms of V. V is also a solution of Eq. 6.22. The r undetermined terms can be found by

$$\begin{bmatrix} v_1 \\ \cdot \\ v_s \\ \cdot \\ v_6 \end{bmatrix} = \begin{bmatrix} w_{q_k 1} \\ \cdot \\ w_{q_k i} \\ \cdot \\ w_{q_k r} \end{bmatrix} \cdot \begin{bmatrix} w_{11} & \cdot & w_{1i} & \cdot & w_{1r} \\ \cdot & \cdot & \cdot & \cdot & \cdot \\ w_{l1} & \cdot & w_{li} & \cdot & w_{lr} \\ \cdot & \cdot & \cdot & \cdot & \cdot \\ w_{61} & \cdot & w_{6i} & \cdot & w_{6r} \end{bmatrix}^{-1},$$

where $s = 1,2,...,6$ $(s \neq q_j, s \neq q_k)$, and $l = 1,2,...,6$ $(l \neq q_j)$.

6. Repeat step 4 (select another term from Q) and step 5 until all $(6-r)$ basic solutions have been determined.

6.3.1.2 Case Study

Based on the algorithm, a C++ program was developed to identify the under-constraint status and unconstrained motions.

Example 1. In a surface grinding operation, a workpiece is located on a fixture system, as shown in Figure 6.6. The normal vector and position of each locator are as follows:

L₁: [0, 0, 1]′, [1, 3, 0]′
L₂: [0, 0, 1]′, [3, 3, 0]′
L₃: [0, 0, 1]′, [2, 1, 0]′
L₄: [0, 1, 0]′, [3, 0, 2]′
L₅: [0, 1, 0]′, [1, 0, 2]′.

Consequently, the locating matrix is determined as

$$W_L = \begin{bmatrix} 0 & 0 & 1 & 3 & -1 & 0 \\ 0 & 0 & 1 & 3 & -3 & 0 \\ 0 & 0 & 1 & 1 & -2 & 0 \\ 0 & 1 & 0 & -2 & 0 & 3 \\ 0 & 1 & 0 & -2 & 0 & 1 \end{bmatrix}.$$

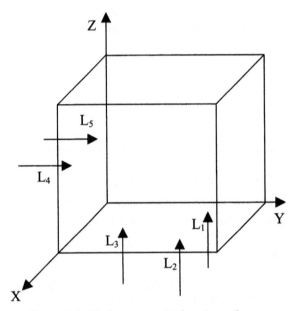

Figure 6.6. Under-constraint locating scheme.

This locating system provides under-constraint locating in that $rank(W_L) = 5 < 6$. The program then calculates the right-generalized inverse of the locating matrix via

$$W^r = \begin{bmatrix} 0 & 0 & 0 & 0 & 0 \\ 0.5 & 0.5 & -1 & -0.5 & 1.5 \\ 0.75 & -1.25 & 1.5 & 0 & 0 \\ 0.25 & 0.25 & -0.5 & 0 & 0 \\ 0.5 & -0.5 & 0 & 0 & 0 \\ 0 & 0 & 0 & 0.5 & -0.5 \end{bmatrix}.$$

The first row is recognized as a dependent row because removal of this row does not affect rank of the matrix. The other five rows are independent rows. A linear combination of the independent rows is found according to the requirement in step 5 for under-constraint status. The solution for this special case is obvious, in that all the coefficients are zero. Hence, the unconstrained motion of the workpiece can be determined as

$$V_i = [\,-1, 0, 0, 0, 0, 0\,].$$

This indicates that the workpiece can move in the X direction. Based on this result, an additional locator should be employed to constrain displacement of workpiece along the X axis.

Example 2. Figure 6.7 shows a knuckle with 3-2-1 locating. The normal vector and position of each locator in this initial design are as follows:

Figure 6.7. Knuckle under 3-2-1 locating.

L_1: [0, 1, 0]', [896, -877, -515]'
L_2: [0, 1, 0]', [1060, -875, -378]'
L_3: [0, 1, 0]', [1010, -959, -612]'
L_4: [0.9955, -0.0349, 0.088]', [977, -902, -624]'

L_5: [0.9955, -0.0349, 0.088]', [977, -866, -624]'
L_6: [0.088, 0.017, -0.996]', [1034, -864, -359]'

The locating matrix of this scheme is

$$
W_L = \begin{bmatrix}
0 & 1 & 0 & 515.0 & 0 & 0.8960 \\
0 & 1 & 0 & 378.0 & 0 & 1.0600 \\
0 & 1 & 0 & 612.0 & 0 & 1.0100 \\
0.9955 & -0.0349 & 0.0880 & -101.2445 & -707.2664 & 0.8638 \\
0.9955 & -0.0349 & 0.0880 & -98.0728 & -707.2664 & 0.8280 \\
0.0880 & 0.0170 & -0.9960 & 866.6257 & 998.2466 & 0.0936
\end{bmatrix}
$$

where $rank(W_L) = 5 < 6$ reveals that the workpiece is under-constrained. One of the first five rows can be removed. Assume the first row (i.e., locator L_1) is removed from W_L. The modified locating matrix thus becomes

$$
\begin{bmatrix}
0 & 1 & 0 & 378.0 & 0 & 1.0600 \\
0 & 1 & 0 & 612.0 & 0 & 1.0100 \\
0.9955 & -0.0349 & 0.0880 & -101.2445 & -707.2664 & 0.8638 \\
0.9955 & -0.0349 & 0.0880 & -98.0728 & -707.2664 & 0.8280 \\
0.0880 & 0.0170 & -0.996 & 866.6257 & 998.2466 & 0.0936
\end{bmatrix}
$$

The right-generalized inverse of the modified locating matrix is

$$
W^r = \begin{bmatrix}
1.8768 & -1.8607 & -20.6665 & 21.3716 & 0.4995 \\
3.0551 & -2.0551 & -32.4448 & 32.4448 & 0 \\
-1.0956 & 1.0862 & 12.0648 & -12.4764 & -0.2916 \\
-0.0044 & 0.0044 & 0.0061 & -0.0061 & 0 \\
0.0025 & -0.0025 & 0.0065 & -0.0069 & 0.0007 \\
-0.0004 & 0.0004 & 0.0284 & -0.0284 & 0
\end{bmatrix}
$$

The program checked the dependent row and found that every row is dependent on the other five rows. Without loosing generality, the first row is regarded as a dependent row. A linear combination of the other five rows to match the first row is then calculated. The computation yields the unconstrained motion

$$
V_i = [-1, 0, -1.713, -0.0432, -0.0706, 0.04].
$$

This motion is displacement in the [-1, 0, -1.713] direction combined with a rotation defined by [-0.0432, -0.0706, 0.04]. To revise this locating scheme, another locator should be added to constrain at least one component of the unconstrained motion of the workpiece.

6.3.2   Locating Error Analysis

Once the locating status has been identified, the algorithms discussed in the following can be implemented to analyze locating errors.

6.3.2.1 Well-constrained

With deterministic locating, we can obtain the exact position and orientation errors ΔX of the workpiece from Eq. 6.12 as follows, given the geometric errors Δr of locators:

$$\Delta X = W_L^{-1} \Delta r , \tag{6.23}$$

where W_L^{-1} is the inverse of the locating matrix W_L.

6.3.2.2 Over-constrained

In this situation, given the geometric errors Δr of locators, the exact position and orientation errors ΔX of the workpiece cannot be obtained from Eq. 6.12, but the least square solution can be calculated as follows:

$$\Delta X = W_L^{+l} \Delta r , \tag{6.24}$$

where $W_L^{+l} = \left(W_L^T W_L \right)^{-1} W_L^T$ is the left-generalized inverse of the locating matrix W_L.

6.3.2.3 Under-constrained

When $rank(W_L) = m' < 6$, the fixture is referred to as under-constrained locating. Given the geometric errors Δr of locators, the position and orientation errors ΔX of the workpiece are represented as follows:

$$\Delta X = \Delta X_v + \Delta X_n , \tag{6.25}$$

where $\Delta\mathbf{X}_v = \mathbf{W}_L^{+r}\Delta\mathbf{r} \in \mathbf{V}$ is referred to as the minimum norm solution of the position and orientation errors $\Delta\mathbf{X}$ of the workpiece, $\mathbf{W}_L^{+r} = \mathbf{W}_L^T\left(\mathbf{W}_L\mathbf{W}_L^T\right)^{-1}$ is the right-generalized inverse of the locating matrix $\mathbf{W}_L$, $\mathbf{V}$ is a subspace of Euclidean 6D vector space $\mathbf{E}^6$, $\Delta\mathbf{X}_n = \left(\mathbf{I}_{6\times6} - \mathbf{W}_L^{+r}\mathbf{W}_L\right)\lambda \in \mathbf{N}(\mathbf{W}_L)$ is referred to as the null solution of the position and orientation errors $\Delta\mathbf{X}$ of the workpiece, the null space $\mathbf{N}(\mathbf{W}_L)$ of the locating matrix $\mathbf{W}_L$ is a special subspace of $\mathbf{E}^6$ [the dimension of the null space $\mathbf{N}(\mathbf{W}_L)$ is $6-m'$], $\mathbf{N}(\mathbf{W}_L)$ and $\mathbf{V}$ are orthogonal complements of each other, we denote this by $\mathbf{V}^\perp = \mathbf{N}(\mathbf{W}_L)$, or equivalently, $\mathbf{N}^\perp(\mathbf{W}_L) = \mathbf{V}$, $\mathbf{E}^6 = \mathbf{V} \oplus \mathbf{N}(\mathbf{W}_L)$, $\mathbf{I}_{6\times6}$ is an identity matrix, and $\lambda = \left(\lambda_1, \cdots, \lambda_6\right)^T \in \mathfrak{R}^{6\times1}$ is an arbitrary vector. The following example explains the geometric meaning of Eq. 6.25.

Example 3. In a surface grinding operation, a workpiece is located under-constrained on a machine table. The coordinate frame $\{\mathbf{W}\}$ is used to describe the position and orientation of the machining surface of the workpiece. To simplify the calculation, without loss of generality, the machine table coordinate frame $\{\mathbf{G}\}$ is defined as shown in Figure 6.8. The positions, errors in the normal direction, and unit normal vectors of three locators (equivalent) are as follows:

$$\mathbf{r}_{w_1}^g = \begin{pmatrix} a_1 \\ b_1 \\ 0 \end{pmatrix}, \quad \mathbf{r}_{w_2}^g = \begin{pmatrix} a_2 \\ b_2 \\ 0 \end{pmatrix}, \quad \mathbf{r}_{w_3}^g = \begin{pmatrix} a_3 \\ b_3 \\ 0 \end{pmatrix}, \quad \Delta\mathbf{r} = \begin{pmatrix} \Delta r_{n_1} \\ \Delta r_{n_2} \\ \Delta r_{n_3} \end{pmatrix},$$

$$\mathbf{n}_1 = \begin{pmatrix} 0 \\ 0 \\ 1 \end{pmatrix}, \quad \mathbf{n}_2 = \begin{pmatrix} 0 \\ 0 \\ 1 \end{pmatrix}, \quad \mathbf{n}_3 = \begin{pmatrix} 0 \\ 0 \\ 1 \end{pmatrix},$$

where $a_1 = 0$, $a_2 = a$, $a_3 = -a$, $b_1 = -b$, $b_2 = b$, $b_3 = b$, a and b are nonzero constants.

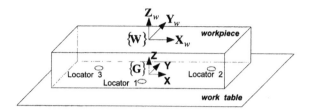

Figure 6.8. Under-constrained locating.

We obtain the locating matrix $\mathbf{W}_L$ as follows:

$$\mathbf{W}_L = \begin{bmatrix} 0 & 0 & 1 & b_1 & -a_1 & 0 \\ 0 & 0 & 1 & b_2 & -a_2 & 0 \\ 0 & 0 & 1 & b_3 & -a_3 & 0 \end{bmatrix}. \tag{6.26}$$

It is clear that $rank(\mathbf{W}_L) = m = 3$ as long as the three locators are not on the same straight line. In this situation, the right-generalized reverse $\mathbf{W}_L^{+r}$ exists, and

$$\mathbf{W}_L^{+r} = \begin{bmatrix} 0 & 0 & 0 \\ 0 & 0 & 0 \\ \dfrac{1}{2} & \dfrac{1}{4} & \dfrac{1}{4} \\ -\dfrac{1}{2b} & \dfrac{1}{4b} & \dfrac{1}{4b} \\ 0 & -\dfrac{1}{2a} & \dfrac{1}{2a} \\ 0 & 0 & 0 \end{bmatrix}. \tag{6.27}$$

Further, using Eqs. 6.25 and 6.27 we obtain the locating errors of the workpiece as follows

$$\Delta \mathbf{X} = \Delta \mathbf{X}_v + \Delta \mathbf{X}_n = \begin{pmatrix} \lambda_1 \\ \lambda_2 \\ \dfrac{1}{4}\left(2\Delta r_{n_1} + \Delta r_{n_2} + \Delta r_{n_3}\right) \\ \dfrac{1}{4b}\left(-2\Delta r_{n_1} + \Delta r_{n_2} + \Delta r_{n_3}\right) \\ \dfrac{1}{2a}\left(-\Delta r_{n_2} + \Delta r_{n_3}\right) \\ \lambda_6 \end{pmatrix}, \tag{6.28}$$

where

$$\Delta \mathbf{X}_v = \mathbf{W}_L^{+r} \Delta \mathbf{r} = \begin{pmatrix} 0 \\ 0 \\ \dfrac{1}{4}\left(2\Delta r_{n_1} + \Delta r_{n_2} + \Delta r_{n_3}\right) \\ \dfrac{1}{4b}\left(-2\Delta r_{n_1} + \Delta r_{n_2} + \Delta r_{n_3}\right) \\ \dfrac{1}{2a}\left(-\Delta r_{n_2} + \Delta r_{n_3}\right) \\ 0 \end{pmatrix} \tag{6.29}$$

$$\Delta \mathbf{X}_n = \left(\mathbf{I}_{6\times6} - \mathbf{W}_L^{+r}\mathbf{W}_L\right)\lambda = \begin{pmatrix} \lambda_1 \\ \lambda_1 \\ 0 \\ 0 \\ 0 \\ \lambda_6 \end{pmatrix}. \tag{6.30}$$

Eq. 6.29 gives the translational error in the $\mathbf{Z}$-axis direction and the rotational errors around $\mathbf{X}$- and $\mathbf{Y}$-axis directions for the workpiece. The three error elements can thus be determined. However, Eq. 6.30 shows the translational errors in $\mathbf{X}$- and $\mathbf{Y}$-axis directions and the rotational error around $\mathbf{Z}$-axis direction for the workpiece. The three error elements thus cannot be determined. Moreover, we can find that $\Delta \mathbf{X}_v$ and $\Delta \mathbf{X}_n$ are orthogonal to each other via Eqs. 6.29 and 6.30. All of the results are the characteristics of under-constrained locating.

Once the position and orientation precision of the workpiece is determined, the position errors of a set of critical points on the workpiece can be calculated to evaluate the variation of a feature. Assuming that the position of the jth critical point is represented as $\mathbf{p}_j$, then the mapping between the position errors $\Delta \mathbf{p}_j$ of the jth critical evaluating point and the position and orientation errors $\Delta \mathbf{X}$ of the workpiece can be described as

$$\Delta \mathbf{p}_j = \left(\mathbf{I}_{3\times3} \; \vdots \; -\mathbf{p}_j \otimes\right)\Delta \mathbf{X}. \tag{6.31}$$

In the quality evaluation of manufacturing, the position variations of critical points are often examined in certain directions. For example, the position variations of a set of vertices in the normal direction of the workpiece surface may be considered when its flatness needs to be evaluated. Using Eq. 6.31, the mapping between the position variation v_{j_map} of the jth critical point along the direction $\mathbf{Q}_k$ and the position and orientation errors $\Delta \mathbf{X}$ of the workpiece are described as follows:

$$v_{j\_map} = \left(\mathbf{Q}_k^T \;\vdots\; -\mathbf{Q}_k^T \mathbf{p}_j \otimes \right) \Delta \mathbf{X} . \tag{6.32}$$

Eq. 6.32 can be rewritten as

$$v_{j\_map} = \mathbf{U} \cdot \Delta \mathbf{X} = \begin{pmatrix} u_1 & u_2 & u_3 & u_4 & u_5 & u_6 \end{pmatrix} \begin{pmatrix} \Delta x \\ \Delta y \\ \Delta z \\ \Delta \theta_x \\ \Delta \theta_y \\ \Delta \theta_z \end{pmatrix}, \tag{6.33}$$

where $\mathbf{U} = \left(\mathbf{Q}_k^T \;\vdots\; -\mathbf{Q}_k^T \mathbf{p}_j \otimes \right) = \begin{pmatrix} u_1 & u_2 & u_3 & u_4 & u_5 & u_6 \end{pmatrix}$,

$\Delta \mathbf{X} = \left(\Delta \mathbf{X}_w^T, \; \Delta \mathbf{\Theta}_w^T \right)^T = \begin{pmatrix} \Delta x & \Delta y & \Delta z & \Delta \theta_x & \Delta \theta_y & \Delta \theta_z \end{pmatrix}^T .$

In general, the position and orientation errors are independent and random thus, the position variation magnitude v_j of the jth critical point along the direction $\mathbf{Q}_k$ can be described as

$$v_j = \sqrt{\left(u_1 \cdot \Delta x \right)^2 + \left(u_2 \cdot \Delta y \right)^2 + \left(u_3 \cdot \Delta z \right)^2 + \left(u_4 \cdot \Delta \theta_x \right)^2 + \left(u_5 \cdot \Delta \theta_y \right)^2 + \left(u_6 \cdot \Delta \theta_z \right)^2 } . \tag{6.34}$$

6.3.3   Conclusions

There inherently exists position error for every locator. The position errors of locators will affect the position and orientation precision of the workpiece. How to evaluate and control such error influence is one of the fundamental questions in fixture automation design and planning.

This section derives a general fixturing error mapping model, which is applicable for deterministic, under-constrained, and over-constrained locating. An algorithm for checking the locating status has been proposed and implemented. This algorithm can identify an under-constrained status and unconstrained motions. It can also recognize an over-constrained status and redundant locators. Then, using an appropriate method the position and orientation variation of the workpiece can be determined for a given locator configuration and position tolerances. On the other hand, given the tolerance of the position and orientation variation of the workpiece the tolerances of locators can be designed. Moreover, the position variations of the critical evaluation points on the workpiece can be calculated using the model, which is important in the quality evaluation of manufacturing.

6.4 LOCATOR AND CLAMP CONFIGURATION CHARACTERISTICS

6.4.1   Locator Configuration Characteristics

The locating variation of workpieces is related to the positions and errors of locators. When the tolerance specifications of machining surfaces on a workpiece are given, the design requirements of locators can be determined. In reverse, given the geometric errors of locators the position and orientation errors of workpieces can be calculated. However, some locator configuration may affect workpiece accessibility to the fixture, as well as detachability (Asada 1985) from the fixture.

Consider a workpiece as shown in Figure 6.9. Assuming that the instantaneous velocity (twist) of the workpiece can be written as $\Delta \mathbf{D} = \left(\Delta \mathbf{D}_w^T \quad \Delta \Phi_w^T \right)^T \in \mathfrak{R}^{6 \times 1}$, where $\Delta \mathbf{D}_w \in \mathfrak{R}^{3 \times 1}$ and $\Delta \Phi_w \in \mathfrak{R}^{3 \times 1}$ are called the instantaneous linear and angular velocity of the workpiece, respectively, and there exists only one point of contact between each locator and the workpiece that is frictionless, the motion constraint of the workpiece by the ith locator can be represented as

$$\mathbf{n}_i^T \Delta \mathbf{d}_i \geq 0, \quad i = 1, \cdots, m \tag{6.35}$$

where $\Delta \mathbf{d}_i \in \mathfrak{R}^{3 \times 1}$ is the infinitesimal motion of the workpiece at the ith point of contact between the workpiece and the ith locator.

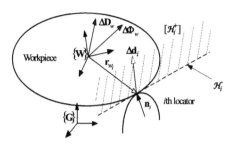

Figure 6.9. Closed-half space and hyperplane.

Eq. 6.35 implies that the ith contact point on the workpiece can only move in the closed half-space $\left[\mathcal{H}_i^+\right] = \left\{\Delta\mathbf{d}_i \in \mathfrak{R}^3 \mid \mathbf{n}_i^T \Delta\mathbf{d}_i \geq 0\right\}$, which is divided by the plane $\mathcal{H}_i = \left\{\Delta\mathbf{d}_i \in \mathfrak{R}^3 \mid \mathbf{n}_i^T \Delta\mathbf{d}_i = 0\right\}$ from the entire space $\mathfrak{R}^3$, as shown in Figure 6.4. Because

$$\Delta\mathbf{d}_i = \Delta\mathbf{D}_w + \Delta\Phi_w \times \mathbf{r}_{w_i}^g, \tag{6.36}$$

Eq. 6.35 can be rewritten as

$$\mathbf{n}_i^T \begin{pmatrix} \mathbf{I}_{3\times 3} \\ \mathbf{r}_{w_i}^g \otimes \end{pmatrix}^T \Delta\mathbf{D} \geq 0. \tag{6.37}$$

Eq. 6.37 implies that the workpiece can only move in the closed half-space $\left[\mathcal{H}_i^+\right] = \left\{\Delta\mathbf{D} \in \mathfrak{R}^6 \mid \mathbf{n}_i^T \begin{pmatrix} \mathbf{I}_{3\times 3} \\ \mathbf{r}_{w_i}^g \otimes \end{pmatrix}^T \Delta\mathbf{D} \geq 0\right\}$, which is divided by the hyperplane

$\mathcal{H}_i = \left\{\Delta\mathbf{D} \in \mathfrak{R}^6 \mid \mathbf{n}_i^T \begin{pmatrix} \mathbf{I}_{3\times 3} \\ \mathbf{r}_{w_i}^g \otimes \end{pmatrix}^T \Delta\mathbf{D} = 0\right\}$ from the entire space $\mathfrak{R}^6$. Then the motion

constraint of the workpiece with m points of contact can be represented as

$$\mathbf{N}^T \mathbf{G}_L^T \Delta\mathbf{D} \geq 0; \tag{6.38}$$

that is, $\mathbf{W}_L \Delta\mathbf{D} \geq 0$. $\tag{6.39}$

Eq. 6.39 means that the workpiece can only move in a convex polyhedral cone in the space $\mathfrak{R}^6$ as follows:

$$\mathcal{K} = \left\{ \Delta \mathbf{D} \in \mathfrak{R}^6 \,\middle|\, \mathbf{W}_L \Delta \mathbf{D} \geq \mathbf{0} \right\} \qquad (6.40)$$

The convex polyhedral cone $\mathcal{K}$ is referred to as the free motion cone, which is generated by intersecting m closed half-spaces

$$\left[\mathcal{H}_i^+\right] = \left\{ \Delta \mathbf{D} \in \mathfrak{R}^6 \,\middle|\, \mathbf{n}_i^T \begin{pmatrix} \mathbf{I}_{3\times3} \\ \mathbf{r}_{w_i}^g \otimes \end{pmatrix}^T \Delta \mathbf{D} \geq 0 \right\} \quad (i = 1, \; \cdots, \; m).$$

If the free motion cone $\mathcal{K}$ does not contain any other elements than 0, that is,

$$\mathcal{K} = \left\{ \mathbf{0} \right\}, \qquad (6.41)$$

then the locating of the fixture is referred to as form-closure. From the standpoint of motion, form closure means that all motion degrees of freedom of the workpiece are eliminated under the geometric constraints of fixels (fixture elements). If the configuration of locators satisfies Eq. 6.41 for a workpiece, the workpiece is neither accessible to the fixture nor detachable from the fixture. In contrast, if the free motion cone $\mathcal{K}$ contains a nonzero element, the workpiece will be able to move in one or more related directions, which means the workpiece is accessible to the fixture as well as detachable from the fixture. Thus, Eq. 6.41 can be used as a qualitative measure for judging accessibility and detachability.

6.4.2   Clamp Configuration Characteristics: Accessibility and Detachability

According to the duality theory in convex analysis (Rockafellar 1970), the polar $\mathcal{K}^\circ$ of the free motion cone $\mathcal{K}$ can be described as

$$\mathcal{K}^\circ = \left\{ \mathbf{f}_{cl} \in \mathfrak{R}^6 \,\middle|\, \begin{array}{l} \langle \mathbf{f}_{cl}, \; \Delta \mathbf{D} \rangle = \mathbf{f}_{cl}^T \Delta \mathbf{D} \leq 0 \\ \text{for all } \Delta \mathbf{D} \in \mathcal{K} \end{array} \right\}, \qquad (6.42)$$

where $\mathbf{f}_{cl} \in \mathfrak{R}^h$ is the clamping force (wrench) ($h = 3$ for 2D fixturing, $h = 6$ for 3D fixturing).

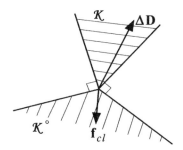

Figure 6.10. Convex polyhedral cone and its polar.

Figure 6.10 shows the 2D free motion cone $\mathcal{K}$ and its polar $\mathcal{K}^\circ$. It is clear that the free motion cone $\mathcal{K}$ is a subspace of $\mathfrak{R}^6$ because of m locator constraints, and $\mathcal{K}^\circ$ is the orthogonally complementary subspace. In general, for any non-empty closed convex $\mathcal{K}$, $\mathcal{K}^\circ$ consists of all the vectors normal to $\mathcal{K}$ at $\mathbf{0}$, whereas $\mathcal{K}$ consists of all vectors normal to $\mathcal{K}^\circ$ at $\mathbf{0}$, and $\left(\mathcal{K}^\circ\right)^\circ = \mathcal{K}$. We call the polar $\mathcal{K}^\circ$ the constrained cone.

The function of clamping is to balance the force system, which consists of the weight of the workpiece, cutting forces, and support forces of locators, so that the desired position and orientation of the workpiece can be maintained. To obtain the desired position and orientation, the clamping force $\mathbf{f}_{cl}$ should be inside the constrained cone $\mathcal{K}^\circ$. In other words, if there exist feasible motion directions of the workpiece after it is located, or if the workpiece is accessible to the fixture as well as detachable from the fixture, the directions of the clamping force $\mathbf{f}_{cl}$ can be determined using Eq. 6.42. Furthermore, from Eq. 6.42, it can be found that the clamp configuration is not a point but a feasible domain, which satisfies the constraints given by Eq. 6.42.

6.4.3   Judgment of Form Closure Fixturing

Let f_{n_i} be the normal force of the ith fixel exerted on the workpiece. The wrench generated by the force f_{n_i} can then be represented as

$$\begin{pmatrix} \mathbf{f}_i \\ \mathbf{\tau}_i \end{pmatrix} = \begin{pmatrix} \mathbf{n}_i \\ \mathbf{r}^g_{w_i} \times \mathbf{n}_i \end{pmatrix} f_{n_i} . \tag{6.43}$$

Thus, the wrench $\mathbf{F} = \begin{pmatrix} \mathbf{f}^T & \mathbf{\tau}^T \end{pmatrix}^T \in \mathfrak{R}^6$ generated by q contact forces can be described as

$$\mathbf{F} = \begin{pmatrix} \mathbf{f} \\ \mathbf{\tau} \end{pmatrix} = \mathbf{GNf}_c , \tag{6.44}$$

where $G = \begin{bmatrix} \mathbf{I}_{3\times3} & \cdots & \mathbf{I}_{3\times3} \\ \mathbf{r}^g_{w_1} \otimes & \cdots & \mathbf{r}^g_{w_q} \otimes \end{bmatrix} \in \mathfrak{R}^{6\times3q}$ is called the fixturing matrix,

$\mathbf{f}_c = \begin{pmatrix} f_{n_1}, & \cdots, & f_{n_q} \end{pmatrix}^T \in \mathfrak{R}^{q\times1}$ is referred to as the fixturing force, $\mathbf{f} = \sum_{i=1}^{q} \mathbf{f}_i$,

$\mathbf{\tau} = \sum_{i=1}^{q} \mathbf{\tau}_i$.

If the force $\mathbf{F}$ represents the external wrench exerted on the workpiece, then Eq. 6.44 describes the equilibrium constraint of the fixturing forces. In addition, because the fixels can only push, not pull the workpiece, the unilateral constraint of the contact force f_{n_i} must be satisfied. This is represented as

$$f_{n_i} \geq 0, \quad i = 1, \cdots, q . \tag{6.45}$$

In the viewpoint of force, form closure means that we can always find positive contact force f_{n_i} ($i = 1, \cdots, q$) to balance any external wrench exerted on the workpiece under the geometric constraints of locators and clamps, which can be judged using the following principle (Lakshminaryana 1978):

If and only if the constraint matrix **GN** *is full column rank, there exists a vector* $\mathbf{0} < \mathbf{y} \in \mathfrak{R}^{q\times1}$ *such that* $\mathbf{GNy} = \mathbf{0}$, *then the fixturing with q frictionless contact points is form-closure.*

Example 4. Consider the planar locating of a polygonal workpieces as shown in Figure 6.11a. Here, three locators are used to locate the planar workpiece. The position and normal vectors of each locator are as follows:

$$\mathbf{r}_1 = \begin{pmatrix} a \\ 0 \end{pmatrix}, \quad \mathbf{r}_2 = \begin{pmatrix} 0 \\ b \end{pmatrix}, \quad \mathbf{r}_3 = \begin{pmatrix} c \\ d \end{pmatrix},$$

$$\mathbf{n}_1 = \begin{pmatrix} 0 \\ 1 \end{pmatrix}, \quad \mathbf{n}_2 = \begin{pmatrix} 1 \\ 0 \end{pmatrix}, \text{ and } \mathbf{n}_3 = \begin{pmatrix} -\dfrac{\sqrt{2}}{2} \\ -\dfrac{\sqrt{2}}{2} \end{pmatrix}$$

Using the position and normal vectors, we can describe the motion constraint (Eq. 6.39) of the workpiece with three locators as

$$\begin{bmatrix} 0 & 1 & a \\ 1 & 0 & -b \\ -\dfrac{\sqrt{2}}{2} & -\dfrac{\sqrt{2}}{2} & \dfrac{\sqrt{2}}{2}(d-c) \end{bmatrix} \begin{pmatrix} \varDelta D_x \\ \varDelta D_y \\ \varDelta \varphi_z \end{pmatrix} \geq 0, \tag{6.46}$$

where a, b, c, and d are positive constants.

From Eq. 6.46, it can be found that the free motion cone $\mathcal{K}$ does not contain any other elements than 0. Thus the corresponding locating configuration is neither accessible to the fixture nor detachable from the fixture when the workpiece trajectories are limited to planar motion only.

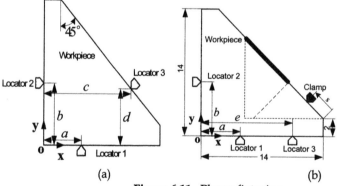

(a)   (b)

Figure 6.11. Planar fixturing.

Now we change the locating configuration from Figure 6.11a to Figure 6.11b. The position and normal vectors of each locator are as follows:

$$\mathbf{r}_1 = \begin{pmatrix} a \\ 0 \end{pmatrix}, \quad \mathbf{r}_2 = \begin{pmatrix} 0 \\ b \end{pmatrix}, \quad \mathbf{r}_3 = \begin{pmatrix} e \\ 0 \end{pmatrix},$$

$$\mathbf{n}_1 = \begin{pmatrix} 0 \\ 1 \end{pmatrix}, \quad \mathbf{n}_2 = \begin{pmatrix} 1 \\ 0 \end{pmatrix}, \text{ and } \mathbf{n}_3 = \begin{pmatrix} 0 \\ 1 \end{pmatrix}.$$

Similarly, we can obtain the motion constraints of locating corresponding to Figure 6.6b via

$$\begin{bmatrix} 0 & 1 & a \\ 1 & 0 & -b \\ 0 & 1 & e \end{bmatrix} \begin{pmatrix} \Delta D_x \\ \Delta D_y \\ \Delta \varphi_z \end{pmatrix} \geq 0, \tag{6.47}$$

where a, b, and e are positive constants.

From Eq. 6.47, we can find that the free motion cone $\mathcal{K}$ contains the following motion elements:

$$\Delta \mathbf{D} = \begin{pmatrix} \Delta D_x \\ \Delta D_y \\ \Delta \varphi_z \end{pmatrix} = \begin{pmatrix} 1 \\ 0 \\ 0 \end{pmatrix} \text{ or } \Delta \mathbf{D} = \begin{pmatrix} 0 \\ 1 \\ 0 \end{pmatrix} \text{ or } \Delta \mathbf{D} = \begin{pmatrix} u_x \\ u_y \\ 0 \end{pmatrix}$$

$$(u_x \geq 0, u_y \geq 0, \text{ and } u_x^2 + u_y^2 = 1).$$

The function of clamping is to resist the motions along these directions. Using Eq. 6.42, we obtain the constrained cone $\mathcal{K}^\circ$ via

$$\mathcal{K}^\circ = \left\{ \mathbf{f}_{cl} = \begin{pmatrix} f_x \\ f_y \\ m_z \end{pmatrix} \in \mathfrak{R}^3 \left| \begin{array}{c} \left\{ \begin{array}{c} f_x \leq 0 \\ f_y \leq 0 \\ u_x f_x + u_y f_y \leq 0 \end{array} \right\} \\ \text{for all } \Delta \mathbf{D} = \begin{pmatrix} u_x \\ u_y \\ 0 \end{pmatrix} \in \mathcal{K} \end{array} \right. \right\}, \tag{6.48}$$

where f_x, f_y, and m_z are the elements of the clamping force. If the clamp is located on the slanting edge, as shown in Figure 6.11b, the clamping force direction can be written as

$$\hat{\mathbf{f}}_{cl} = \begin{pmatrix} -\frac{\sqrt{2}}{2} \\ -\frac{\sqrt{2}}{2} \\ 0 \end{pmatrix}. \tag{6.49}$$

It is clear that the clamping force described by Eq. 6.49 is inside the constrained cone $\mathcal{K}^{\circ}$, which implies that the clamping configuration is a feasible domain. That is, the position of clamping can be chosen at any place on the slanting edge (Figure 6.11b).

   Assume that $a = 2$, $b = 8$, $e = 12$, and the position of a clamp is represented by s for the fixturing configuration shown in Figure 6.6b. The constraint matrix $\mathbf{GN}$ is then written as

$$\mathbf{GN} = \begin{bmatrix} 0 & 1 & 0 & -\frac{\sqrt{2}}{2} \\ 1 & 0 & 1 & -\frac{\sqrt{2}}{2} \\ 2 & -8 & 12 & -\frac{\sqrt{2}}{2}\left(12 - \sqrt{2} \cdot s\right) \end{bmatrix}. \tag{6.50}$$

According to the principle of form-closure fixturing, assuming there exists a vector $\mathbf{y} = \left(y_1, \ y_2, \ y_3, \ y_4\right)^T > 0$ such that $\mathbf{GNy} = \mathbf{0}$, using Eq. 6.50 we yield the position constraint of the clamp is as

$$4\sqrt{2} < s < 9\sqrt{2}. \tag{6.51}$$

This means that the position of the clamp can be chosen only in the bold solid line areas shown in Figure 6.6b to totally constrain the workpiece.

Example 5. Consider a 3D case of a polyhedral workpiece located with six locators, as shown in Figure 6.12a. The workpiece is a cubic rigid body with one corner cut out. The position and normal vectors of each locator are as follows:

$$\mathbf{r}_1 = \begin{pmatrix} 0.8a \\ 0.5a \\ 0 \end{pmatrix}, \ \mathbf{r}_2 = \begin{pmatrix} 0.2a \\ 0.5a \\ 0 \end{pmatrix}, \ \mathbf{r}_3 = \begin{pmatrix} 0.8a \\ 0 \\ 0.5a \end{pmatrix},$$

$$\mathbf{r}_4 = \begin{pmatrix} 0.2a \\ 0 \\ 0.5a \end{pmatrix}, \ \mathbf{r}_5 = \begin{pmatrix} 0 \\ 0.5a \\ 0.5a \end{pmatrix}, \ \mathbf{r}_6 = \begin{pmatrix} 0.8a \\ 0.8a \\ 0.4a \end{pmatrix}, \ a > 0$$

$$\mathbf{n}_1 = \begin{pmatrix} 0 \\ 0 \\ 1 \end{pmatrix}, \ \mathbf{n}_2 = \begin{pmatrix} 0 \\ 0 \\ 1 \end{pmatrix}, \ \mathbf{n}_3 = \begin{pmatrix} 0 \\ 1 \\ 0 \end{pmatrix},$$

$$\mathbf{n}_4 = \begin{pmatrix} 0 \\ 1 \\ 0 \end{pmatrix}, \ \mathbf{n}_5 = \begin{pmatrix} 1 \\ 0 \\ 0 \end{pmatrix}, \text{ and } \mathbf{n}_6 = \begin{pmatrix} -\frac{\sqrt{3}}{3} \\ -\frac{\sqrt{3}}{3} \\ -\frac{\sqrt{3}}{3} \end{pmatrix}.$$

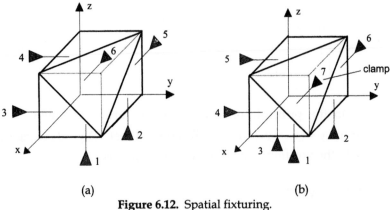

(a)   (b)

Figure 6.12. Spatial fixturing.

Using the position and normal vectors, we can describe the motion constraint (Eq. 6.39) of the workpiece with six locators as

$$\begin{bmatrix} 0 & 0 & 1 & 0.5a & -0.8a & 0 \\ 0 & 0 & 1 & 0.5a & -0.2a & 0 \\ 0 & 1 & 0 & -0.5a & 0 & 0.8a \\ 0 & 0 & 0 & -0.5a & 0 & 0.2a \\ 1 & 0 & 0 & 0 & 0.5a & -0.5a \\ -\frac{\sqrt{3}}{3} & -\frac{\sqrt{3}}{3} & -\frac{\sqrt{3}}{3} & -\frac{2\sqrt{3}}{15}a & \frac{2\sqrt{3}}{15}a & 0 \end{bmatrix} \begin{pmatrix} \Delta D_x \\ \Delta D_y \\ \Delta D_z \\ \Delta \varphi_x \\ \Delta \varphi_y \\ \Delta \varphi_z \end{pmatrix} \geq 0. \quad (6.52)$$

From Eq. 6.52, we can find that the free motion cone $\mathcal{K}$ does not contain any other elements than 0. Thus, the corresponding locating configuration is neither accessible to the fixture nor detachable from the fixture.

Now we change the locating configuration from Figure 6.12a to Figure 6.12b. The position and normal vectors of each locator are as follows

$$
\mathbf{r}_1 = \begin{pmatrix} 0.8a \\ 0.5a \\ 0 \end{pmatrix}, \quad \mathbf{r}_2 = \begin{pmatrix} 0.2a \\ 0.8a \\ 0 \end{pmatrix}, \quad \mathbf{r}_3 = \begin{pmatrix} 0.2a \\ 0.2a \\ 0 \end{pmatrix},
$$

$$
\mathbf{r}_4 = \begin{pmatrix} 0.8a \\ 0 \\ 0.5a \end{pmatrix}, \quad \mathbf{r}_5 = \begin{pmatrix} 0.2a \\ 0 \\ 0.5a \end{pmatrix}, \quad \mathbf{r}_6 = \begin{pmatrix} 0 \\ 0.5a \\ 0.5a \end{pmatrix}
$$

$$
\mathbf{n}_1 = \begin{pmatrix} 0 \\ 0 \\ 1 \end{pmatrix}, \quad \mathbf{n}_2 = \begin{pmatrix} 0 \\ 0 \\ 1 \end{pmatrix}, \quad \mathbf{n}_3 = \begin{pmatrix} 0 \\ 0 \\ 1 \end{pmatrix},
$$

$$
\mathbf{n}_4 = \begin{pmatrix} 0 \\ 1 \\ 0 \end{pmatrix}, \quad \mathbf{n}_5 = \begin{pmatrix} 0 \\ 1 \\ 0 \end{pmatrix}, \text{ and } \mathbf{n}_6 = \begin{pmatrix} 1 \\ 0 \\ 0 \end{pmatrix}.
$$

The motion constraints of locating corresponding to Figure 6.12b can be described as

$$
\begin{bmatrix}
0 & 0 & 1 & 0.5a & -0.8a & 0 \\
0 & 0 & 1 & 0.8a & -0.2a & 0 \\
0 & 0 & 1 & 0.2a & -0.2a & 0 \\
0 & 1 & 0 & -0.5a & 0 & 0.8a \\
0 & 1 & 0 & -0.5a & 0 & 0.2a \\
1 & 0 & 0 & 0 & 0.5a & -0.5a
\end{bmatrix}
\begin{pmatrix}
\Delta D_x \\
\Delta D_y \\
\Delta D_z \\
\Delta \varphi_x \\
\Delta \varphi_y \\
\Delta \varphi_z
\end{pmatrix} \geq 0.
\tag{6.53}
$$

From Eq. 6.53, we can find that the free motion cone $\mathcal{K}$ contains the following motion elements:

$$\mathbf{\Delta D} = \begin{pmatrix} \Delta D_x \\ \Delta D_y \\ \Delta D_z \\ \Delta \varphi_x \\ \Delta \varphi_y \\ \Delta \varphi_z \end{pmatrix} = \begin{pmatrix} 1 \\ 0 \\ 0 \\ 0 \\ 0 \\ 0 \end{pmatrix} \quad \text{or} \quad \mathbf{\Delta D} = \begin{pmatrix} 0 \\ 1 \\ 0 \\ 0 \\ 0 \\ 0 \end{pmatrix}$$

$$\text{or} \quad \mathbf{\Delta D} = \begin{pmatrix} 0 \\ 0 \\ 1 \\ 0 \\ 0 \\ 0 \end{pmatrix} \quad \text{or} \quad \mathbf{\Delta D} = \begin{pmatrix} u_x \\ u_y \\ u_z \\ 0 \\ 0 \\ 0 \end{pmatrix},$$

where $u_x \geq 0$, $u_y \geq 0$, $u_z \geq 0$, and $u_x^2 + u_y^2 + u_z^2 = 1$. Thus, according to Eq. 6.42, we obtain the constrained cone $\mathcal{K}^\circ$ via

$$\mathcal{K}^\circ = \left\{ \mathbf{f}_{cl} = \begin{pmatrix} f_x \\ f_y \\ f_z \\ m_x \\ m_y \\ m_z \end{pmatrix} \in \mathfrak{R}^6 \;\middle|\; \begin{array}{l} f_x \leq 0 \\ f_y \leq 0 \\ f_z \leq 0 \\ u_x f_x + u_y f_y + u_z f_z \leq 0 \\[4pt] \text{for all } \mathbf{\Delta D} = \begin{pmatrix} u_x \\ u_y \\ u_z \\ 0 \\ 0 \\ 0 \end{pmatrix} \in \mathcal{K} \end{array} \right\}, \tag{6.54}$$

where f_x, f_y, f_z, m_x, m_y and m_z are the elements of the clamping force. If the clamp is located on the slanting surface, as shown in Figure 6.12b, the clamping force direction can be written as

$$\hat{\mathbf{f}}_{cl} = \begin{pmatrix} -\frac{\sqrt{3}}{3} \\ -\frac{\sqrt{3}}{3} \\ -\frac{\sqrt{3}}{3} \\ 0 \\ 0 \\ 0 \end{pmatrix}. \tag{6.55}$$

It is clear that the clamping force described by Eq. 6.55 is inside the constrained cone $\mathcal{K}°$, which implies that the clamping configuration is a feasible domain (the slanting surface, as shown in Figure 6.12b) the position of clamping can be chosen at any place on the slanting surface.

Assume that the position of a clamp is represented by $\mathbf{r}_{cl} = (x, \ y, \ 2a - x - y)^T$ as shown in Figure 6.12b. The constraint matrix $\mathbf{GN}$ is then written as

$$\mathbf{GN} = \begin{bmatrix} 0 & 0 & 0 & 0 & 0 & 1 & -\frac{\sqrt{3}}{3} \\ 0 & 0 & 0 & 1 & 1 & 0 & -\frac{\sqrt{3}}{3} \\ 1 & 1 & 1 & 0 & 0 & 0 & -\frac{\sqrt{3}}{3} \\ 0.5a & 0.8a & 0.2a & 0.5a & -0.5a & 0 & \frac{\sqrt{3}}{3}(2a-x-2y) \\ -0.8a & -0.2a & -0.2a & 0 & 0 & 0.5a & -\frac{\sqrt{3}}{3}(2a-2x-y) \\ 0 & 0 & 0 & 0.8a & 0.2a & -0.5a & \frac{\sqrt{3}}{3}(y-x) \end{bmatrix}. \tag{6.56}$$

According to the principle of form-closure fixturing, assuming that there exists a vector $\mathbf{y} = (y_1, \ \cdots, \ y_7)^T > 0$ such that $\mathbf{GNy} = \mathbf{0}$ then using Eq. 6.56 the position constraints of the clamp are as follows:

$$\begin{cases} 2x + y - 1.7a > 0 \\ 3y - 1.7a > 0 \\ 6.3a - 4x - 5y > 0 \\ |y - x| < 0.3a \\ x + y > a. \end{cases} \tag{6.57}$$

The corresponding feasible clamping domain is shown in Figure 6.13. The position of the clamp can be chosen only in the closure areas inside the bold solid line to totally constrain the workpiece.

6.4.4   Conclusions

Based on an analysis of the free motion cone defined in this section, the accessibility of the workpiece to the fixture and the detachability of the workpiece from the fixture are developed. Using the polar of the free motion cone (namely, the constrained cone) the directions of clamping forces are found. The method of determining the feasible clamping domain in which the fixturing of the workpiece is form closure has been presented. The effectiveness of the proposed model and related method is verified by several examples.

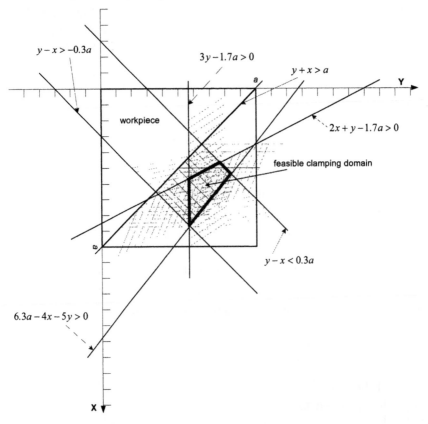

Figure 6.13. A bird's-eye view of the workpiece.

6.5 FIXTURE PLANNING INDEXES

Locating errors associated with workpieces are determined by the position errors and the configuration of locators. Clamps are employed to fix workpieces in place after they are located. In this section, three indexes are defined for fixture planning: (1) the locating robustness index, used to evaluate the configurations of locators; (2) the stability index, used to evaluate the capability to withstand any external disturbance wrench for the fixturing system; and (3) the fixturing resultant index, used to evaluate simultaneously the robustness against positional errors associated with locators and stability under external disturbance wrench. It has been proved that the stability index is invariant under linear coordinate transformation, change of torque origin, and change of dimensional unit. Three corresponding planning methods, which are in fact the nonlinear programming methods, are examined here. These determine optimal locator and clamp configuration. Constraints in the planning methods include the constraints of the degrees of freedom of the workpiece, the locating surface constraint, accessibility and detachability constraints, clamping direction constraints, and clamping domain constraints. Finally, an example is given to verify the effectiveness of the proposed planning methods, and to compare the planning results of the three methods.

6.5.1   Introduction

In Section 6.1, a mapping model between the error space of locators and the workpiece locating error space was built up, which is applicable to deterministic locating, over-constrained locating, and under-constrained locating. It was investigated how to determine the feasible locator configurations where the workpiece is accessible to the fixture (as well as detachable from the fixture) and how to determine the feasible clamping domains that make the workpiece fully constrained after it is located and clamped. How to define and find an optimal fixturing configuration in such feasible configurations is one of the fundamentals to be addressed in fixture automation design and planning.

In general, fixture locating errors always exist even though one seeks to minimize these errors. Even though locators may have the same position tolerances, different locator configurations may result in different position and orientation errors of workpieces. In some cases, because of an unreasonable locator configuration ranges in position and orientation variation of workpieces are unacceptable. Thus, in the past few years many researchers have investigated fixture planning.

A practical automated fixture planning method was developed with integration of CAD, based on predefined locating modes and operational rules (Rong 1999). A variational method for planning locator configuration has also been developed (Cai 1997). The conditions for optimizing the deterministic 3-2-1 locator scheme for low reaction forces and small locator locating errors have been analyzed, and near-optimal deterministic 3-2-1 location scheme synthesis algorithms have been developed (Martin 2001b), but the coupling effect of each locator error on the position and orientation errors of workpieces has been neglected. Furthermore, how to determine optimal clamp configuration has not been taken into account.

It is known that compliance can play a significant role in fixturing (Lin 1999). A stiffness quality measure for compliant grasps or fixtures has been defined (Lin 1999, 2000). During fixturing, fixture elements have to protect the workpiece from deflecting under the load of machining forces. Measurement methods can indicate the number and location of fixture elements that best suit the given task.

The effectiveness of four fixture layout optimization methods, namely, continuous GA, discrete GA, continuous SQP, and discrete SQP, have been investigated. Documentation on this can be found in Vallapuzha (2002b). A method was presented that minimizes workpiece location errors due to localized elastic deformation of the workpiece at the fixturing points by optimally placing the locators and clamps around the workpiece (Li 1999). The similar fixture layout optimization problem that minimizes the deformation of the machined surface due to clamping and machining forces can be solved using the genetic algorithm (Krishnakumar 2000). In fact, compared with position errors of locators, localized elastic deformation of the workpiece at fixturing points is not the predominant factor.

Robust fixture design is considered in Cai (1997) and Wang (1999). If the position of a workpiece is insensitive to locator errors, the locating of the workpiece is known as robust locating. The sensitivity equation is the foundation for robust fixture design. The linearized sensitivity equation and quadratic sensitivity equation (Carlson 2001) were developed for a deterministic locating scheme without considering the sliding of contacts between workpieces and locators. Two performance measures that includes maximization of workpiece localization accuracy and minimization of norm and dispersion of locator contact forces are presented in Wang (1999, 2001). Corresponding algorithms for planning optimal fixture layout for deterministic locating were also developed. Algorithms for computing placement of (frictionless) point fingers that put a polygonal part in form-closure (and placement of point fingers that achieve second-order immobility of a polygonal part) are presented in van der Stappenn (2000). An algorithm for computing n-finger form-closure grasps on polygonal objects is proposed by Liu (2000). Based on the duality between convex hulls and convex polytopes, the qualitative test of 3D form-closure grasps is formulated as an LP problem by Liu (1999). Algorithms for synthesis of two-fingered (Vallapuzha 2002a) and three-fingered (Ponce 2000) force-closure grasps for arbitrary polygonal objects have also been proposed. In Han (2000), the nonlinear friction cone constraints between part and fingertips are transformed into linear matrix inequalities (LMIs), and the basic grasp analysis problems are formulated as a set of convex optimization problems involving LMIs. Contact stability (Montana 1992) and the stability of grasped objects (Trinkle 1992; Howard 1996; Xiong 1998, 1999) were investigated for determining optimal contact configuration. An approach for automatically selecting an eligible set of fixturing surfaces for a polyhedral workpiece and calculating optimal fixturing points on the selected surfaces is described in Ding (2001). However, accessibility to the fixture and detachability from the fixture of the workpiece has not been taken into account in Ding's algorithms, which are in fact the essential differences between fixturing and grasping. An algorithm for finding the clamping positions on 3D workpieces with planar and cylindrical faces was presented in Martin (2001). However, how to determine the feasible clamping domain that makes the workpiece totally constrained is not given in Martin's algorithm. More recently, Rimon (2000) considered the second-order geometrical effects of workpieces, and stated that only four frictionless fingers or fixels are required to immobilize a generic 3D workpiece. However, the algorithm for planning the optimal positions of four fixels was not given.

6.5.2   Evaluation Indexes of Fixturing

6.5.2.1 Evaluation Index of Locator Configurations

A fixture is a device used in manufacturing, assembling, and inspecting to hold a workpiece. The position and orientation precision of workpieces is related to the position tolerances of locators. Moreover, given the same position tolerances of locators the locating errors of a workpiece may vary with different configurations of locators. The mapping relation between the locating errors and the errors of locators can be described as $\mathbf{W}_L : \Delta\mathbf{X} \in \mathfrak{R}^6 \mapsto \Delta\mathbf{r} \in \mathfrak{R}^m$. That is,

$$\mathbf{W}_L \Delta\mathbf{X} = \Delta\mathbf{r}, \tag{6.58}$$

where $\mathbf{W}_L \in \mathfrak{R}^{m \times 6}$ is the locating matrix (m is the number of locators), which is related to the configuration of locators; $\Delta\mathbf{X} = \left(\Delta\mathbf{X}_w^T \quad \Delta\boldsymbol{\Theta}_w^T\right)^T \in \mathfrak{R}^{6 \times 1}$ is the locating errors of the workpiece; $\Delta\mathbf{X}_w \in \mathfrak{R}^{3 \times 1}$ and $\Delta\boldsymbol{\Theta}_w \in \mathfrak{R}^{3 \times 1}$ are the position and orientation errors of the workpiece, respectively; and $\Delta\mathbf{r} \in \mathfrak{R}^{m \times 1}$ is the position errors of the m locators.

Using Eq. 6.58, locating errors can be determined for given position errors of locators and the configuration of locators. In general, the position errors of all locators are independent and random. From the standpoint of statistics, the six elements of the locating errors for deterministic locating can be described as

$$\Delta\mathbf{X}_w^x = \sqrt{\left(\mathbf{W}_L^{-1}(1,1)\cdot\Delta\mathbf{r}(1)\right)^2 + \cdots + \left(\mathbf{W}_L^{-1}(1,6)\cdot\Delta\mathbf{r}(6)\right)^2}$$
$$\Delta\mathbf{X}_w^y = \sqrt{\left(\mathbf{W}_L^{-1}(2,1)\cdot\Delta\mathbf{r}(1)\right)^2 + \cdots + \left(\mathbf{W}_L^{-1}(2,6)\cdot\Delta\mathbf{r}(6)\right)^2}$$
$$\Delta\mathbf{X}_w^z = \sqrt{\left(\mathbf{W}_L^{-1}(3,1)\cdot\Delta\mathbf{r}(1)\right)^2 + \cdots + \left(\mathbf{W}_L^{-1}(3,6)\cdot\Delta\mathbf{r}(6)\right)^2} \tag{6.59}$$
$$\Delta\boldsymbol{\Theta}_w^x = \sqrt{\left(\mathbf{W}_L^{-1}(4,1)\cdot\Delta\mathbf{r}(1)\right)^2 + \cdots + \left(\mathbf{W}_L^{-1}(4,6)\cdot\Delta\mathbf{r}(6)\right)^2}$$
$$\Delta\boldsymbol{\Theta}_w^y = \sqrt{\left(\mathbf{W}_L^{-1}(5,1)\cdot\Delta\mathbf{r}(1)\right)^2 + \cdots + \left(\mathbf{W}_L^{-1}(5,6)\cdot\Delta\mathbf{r}(6)\right)^2}$$
$$\Delta\boldsymbol{\Theta}_w^z = \sqrt{\left(\mathbf{W}_L^{-1}(6,1)\cdot\Delta\mathbf{r}(1)\right)^2 + \cdots + \left(\mathbf{W}_L^{-1}(6,6)\cdot\Delta\mathbf{r}(6)\right)^2},$$

where $\mathbf{W}_L^{-1}(i,j)$ is the ith row, jth column, element of the inverse matrix $\mathbf{W}_L^{-1}$ of the locating matrix $\mathbf{W}_L$, and $\Delta\mathbf{r}(k)$ is the position error of the kth locator. Similarly, for over-constrained locating the six elements of the locating errors can be written as

$$
\begin{aligned}
\Delta X_w^x &= \sqrt{\left(W_L^{+l}(1,1)\cdot \Delta r(1)\right)^2 + \cdots + \left(W_L^{+l}(1,m)\cdot \Delta r(m)\right)^2} \\
\Delta X_w^y &= \sqrt{\left(W_L^{+l}(2,1)\cdot \Delta r(1)\right)^2 + \cdots + \left(W_L^{+l}(2,m)\cdot \Delta r(m)\right)^2} \\
\Delta X_w^z &= \sqrt{\left(W_L^{+l}(3,1)\cdot \Delta r(1)\right)^2 + \cdots + \left(W_L^{+l}(3,m)\cdot \Delta r(m)\right)^2} \\
\Delta \Theta_w^x &= \sqrt{\left(W_L^{+l}(4,1)\cdot \Delta r(1)\right)^2 + \cdots + \left(W_L^{+l}(4,m)\cdot \Delta r(m)\right)^2} \\
\Delta \Theta_w^y &= \sqrt{\left(W_L^{+l}(5,1)\cdot \Delta r(1)\right)^2 + \cdots + \left(W_L^{+l}(5,m)\cdot \Delta r(m)\right)^2} \\
\Delta \Theta_w^z &= \sqrt{\left(W_L^{+l}(6,1)\cdot \Delta r(1)\right)^2 + \cdots + \left(W_L^{+l}(6,m)\cdot \Delta r(m)\right)^2},
\end{aligned}
\tag{6.60}
$$

where $W_L^{+l}(i,j)$ is the ith row, jth column, element of the left-generalized inverse matrix $W_L^{+l} = \left(W_L^T W_L\right)^{-1} W_L^T \in \Re^{6\times m}$ of the locating matrix W_L.

However, because the eliminated DOF of the workpiece are fewer than six for under-constrained locating, only partial elements of the locating errors can be determined corresponding to the special constrained directions. In Example 3 of Section 6.3.2, the related elements of the locating errors would be represented as

$$
\begin{aligned}
\Delta X_w^z &= \sqrt{\left(W_L^{+r}(3,1)\cdot \Delta r(1)\right)^2 + \cdots + \left(W_L^{+r}(3,m)\cdot \Delta r(m)\right)^2} \\
\Delta \Theta_w^x &= \sqrt{\left(W_L^{+r}(4,1)\cdot \Delta r(1)\right)^2 + \cdots + \left(W_L^{+r}(4,m)\cdot \Delta r(m)\right)^2} \\
\Delta \Theta_w^y &= \sqrt{\left(W_L^{+r}(5,1)\cdot \Delta r(1)\right)^2 + \cdots + \left(W_L^{+r}(5,m)\cdot \Delta r(m)\right)^2},
\end{aligned}
\tag{6.61}
$$

where $W_L^{+r}(i,j)$ is the ith row, jth column element of the right-generalized inverse matrix $W_L^{+r} = W_L^T \left(W_L W_L^T\right)^{-1} \in \Re^{6\times m}$ of the locating matrix W_L ($m = 3$ in Example 3 of Section 6.3.2). The remaining elements of the locating errors, which cannot be determined, are not related to the requirement of machining.

Eqs. 6.59, 6.60 and 6.61 show that elements of the locating errors of the workpiece depend on the position tolerances and configurations of locators. Generally, the position errors of locators cannot be fully eliminated because decreasing the position errors means increasing the cost of manufacturing.

To solve the issue of how to plan an optimal locator configuration minimizing the locating errors of workpieces, the norm of the locating error of the workpiece is defined as the evaluation index of locator configurations. That is,

$$\Omega_R = \|\Delta \mathbf{X}\|^2 .$$

(6.62)

Although there exist errors for each locator, their influence on the locating errors of the workpiece can be minimized using the index Ω_R to plan the locator configuration. The index Ω_R is called the locating robustness index of fixtures.

6.5.2.2 Stability Index

Once a workpiece is fully located and clamped in a fixture (i.e., the fixturing is form-closure) all motion DOF of the workpiece are eliminated under the geometric constraints of locators and clamps. From the standpoint of force, the form closure fixturing means that the fixture can balance any external wrench exerted on the workpiece by using a set of positive contact forces. The relation between the external wrench $\mathbf{F} \in \mathfrak{R}^{6 \times 1}$ and the positive contact forces $\mathbf{f}_c \in \mathfrak{R}^{q \times 1}$ (q is the number of contacts) described in Section 6.3 is

$$\mathbf{F} = \mathbf{G} \mathbf{N} \mathbf{f}_c ,$$

(6.63)

where $\mathbf{G} = \begin{bmatrix} \mathbf{I}_{3\times3} & \cdots & \mathbf{I}_{3\times3} \\ \mathbf{r}_{w_1}^g \otimes & \cdots & \mathbf{r}_{w_q}^g \otimes \end{bmatrix} \in \mathfrak{R}^{6\times3q}$ is the fixturing matrix and $\mathbf{N} = diag(\mathbf{n}_1 \quad \cdots \quad \mathbf{n}_q) \in \mathfrak{R}^{3q \times q}$ is the normal vector matrix consisting of q normal vectors $\mathbf{n}_j$ ($j = 1, \cdots, q$) at contacts.

Note that the capability of withstanding the external wrench varies for different contact configurations. Obviously, to obtain a more stable fixturing, it is very important to find the optimal fixel contact configuration. Using Eq. 6.63, the fixturing stability index can be defined (Xiong 1999) as

$$\Omega_S = \sqrt{det(\mathbf{G} \mathbf{N} \mathbf{N}^T \mathbf{G}^T)} .$$

(6.64)

The larger the value of index Ω_S the farther the fixel contact configuration is from a singular configuration. This implies that the fixturing has greater ability to withstand any disturbance wrench on the workpiece.

6.5.2.3 Properties of the Stability Index

It can be seen that changing the coordinate frame, the torque origin, and the dimensional unit results in a different fixturing matrix, but the stability index remains the same. It can be proved.

Theorem: Fixturing stable index Ω_S is invariant under a linear coordinate transformation and a change of torque origin. The index is similarly invariant under a change of dimensional unit.

Proof: At first, the invariance of the index is proved under linear coordinate transformations. Let the change of the coordinate frame from $oxyz$ to $o'x'y'z'$ be denoted by $\mathbf{T} = \begin{bmatrix} \mathbf{C} & \mathbf{P} \\ 0 & 1 \end{bmatrix}$, where $\mathbf{C} \in \mathbf{SO}(3)$ is the orthogonal orientation matrix with determinant $+1$, $\mathbf{P} \in \Re^{3 \times 1}$ is the position vector. The representation of the matrix $\mathbf{GN}$ is then transformed under $\mathbf{T}$ (assuming the torque origin coincides with the origin of each coordinate frame) to

$$\widetilde{\mathbf{GN}} = \begin{bmatrix} \mathbf{I}_{3\times3} & \cdots & \mathbf{I}_{3\times3} \\ \tilde{\mathbf{r}}_1 \otimes & \cdots & \tilde{\mathbf{r}}_q \otimes \end{bmatrix} diag[\tilde{\mathbf{n}}_1 \quad \cdots \quad \tilde{\mathbf{n}}_q] = \begin{bmatrix} \mathbf{Cn}_1 & \cdots & \mathbf{Cn}_q \\ \mathbf{Cr}_1 \otimes \mathbf{Cn}_1 & \cdots & \mathbf{Cr}_q \otimes \mathbf{Cn}_q \end{bmatrix}$$

$$= \begin{bmatrix} \mathbf{C} & 0 \\ 0 & \mathbf{C} \end{bmatrix} \begin{bmatrix} \mathbf{I}_{3\times3} & \cdots & \mathbf{I}_{3\times3} \\ \mathbf{r}_1 \otimes & \cdots & \mathbf{r}_q \otimes \end{bmatrix} diag[\mathbf{n}_1 \quad \cdots \quad \mathbf{n}_q] = \mathbf{T}_1 \mathbf{GN},$$

(6.65)

where $\mathbf{T}_1 = \begin{bmatrix} \mathbf{C} & 0 \\ 0 & \mathbf{C} \end{bmatrix}$ and $det(\mathbf{T}_1) = 1$. Thus we obtain

$$\widetilde{\Omega}_S = \sqrt{det(\widetilde{\mathbf{GN}}\widetilde{\mathbf{N}}^T \widetilde{\mathbf{G}}^T)} = \sqrt{det(\mathbf{T}_1 \mathbf{GNN}^T \mathbf{G}^T \mathbf{T}_1^T)} = \sqrt{det(\mathbf{GNN}^T \mathbf{G}^T)} = \Omega_S.$$

(6.66)

Then the invariance of the index is proved under the change of the torque origin. Assuming the torque origin changes from $\mathbf{o}$ to $\mathbf{s}$, and $\mathbf{p}_1, \cdots, \mathbf{p}_q$ are contact points between the workpiece and fixels, the representative of the matrix $\mathbf{GN}$ is transformed to

$$\tilde{\mathbf{G}}\tilde{\mathbf{N}} = \begin{bmatrix} \mathbf{I}_{3\times3} & \cdots & \mathbf{I}_{3\times3} \\ \tilde{\mathbf{r}}_1 \otimes & \cdots & \tilde{\mathbf{r}}_q \otimes \end{bmatrix} diag[\mathbf{n}_1 \quad \cdots \quad \mathbf{n}_q] = \begin{bmatrix} \mathbf{n}_1 & \cdots & \mathbf{n}_q \\ \overline{\mathbf{sp}_1} \otimes \mathbf{n}_1 & \cdots & \overline{\mathbf{sp}_q} \otimes \mathbf{n}_q \end{bmatrix} = \begin{bmatrix} \mathbf{n}_1 & \cdots & \mathbf{n}_q \\ (\overline{\mathbf{so}} + \mathbf{r}_1) \otimes \mathbf{n}_1 & \cdots & (\overline{\mathbf{so}} + \mathbf{r}_q) \otimes \mathbf{n}_q \end{bmatrix}$$

$$= \begin{bmatrix} \mathbf{I}_{3\times3} & \mathbf{0} \\ \overline{\mathbf{so}} \otimes & \mathbf{I}_{3\times3} \end{bmatrix} \begin{bmatrix} \mathbf{I}_{3\times3} & \cdots & \mathbf{I}_{3\times3} \\ \mathbf{r}_1 \otimes & \cdots & \mathbf{r}_q \otimes \end{bmatrix} diag[\mathbf{n}_1 \quad \cdots \quad \mathbf{n}_q] = \mathbf{T}_2\mathbf{GN},$$

$$(6.67)$$

where $\mathbf{T}_2 = \begin{bmatrix} \mathbf{I}_{3\times3} & \mathbf{0} \\ \overline{\mathbf{so}} \otimes & \mathbf{I}_{3\times3} \end{bmatrix}$ and $det(\mathbf{T}_2) = 1$. Thus we obtain

$$\tilde{\Omega}_S = \sqrt{det(\tilde{\mathbf{G}}\tilde{\mathbf{N}}\tilde{\mathbf{N}}^T\tilde{\mathbf{G}}^T)} = \sqrt{det(\mathbf{T}_2\mathbf{GNN}^T\mathbf{G}^T\mathbf{T}_2^T)} = \sqrt{det(\mathbf{GNN}^T\mathbf{G}^T)} = \Omega_S. \quad (6.68)$$

Finally, the invariance of the index is similarly proved under the change of the dimensional unit. Assuming the position vector is changed from $\mathbf{r}_i$ to $\tilde{\mathbf{r}}_i = k\mathbf{r}_i$, where k is a positive constant factor, the new representative of the matrix $\mathbf{GN}$ is transformed to

$$\tilde{\mathbf{G}}\tilde{\mathbf{N}} = \begin{bmatrix} \mathbf{I}_{3\times3} & \cdots & \mathbf{I}_{3\times3} \\ \tilde{\mathbf{r}}_1 \otimes & \cdots & \tilde{\mathbf{r}}_q \otimes \end{bmatrix} diag[\mathbf{n}_1 \quad \cdots \quad \mathbf{n}_q] = \begin{bmatrix} \mathbf{n}_1 & \cdots & \mathbf{n}_q \\ k\mathbf{r}_1 \otimes \mathbf{n}_1 & \cdots & k\mathbf{r}_q \otimes \mathbf{n}_q \end{bmatrix} \quad (6.69)$$

$$= \begin{bmatrix} \mathbf{I}_{3\times3} & \mathbf{0} \\ \mathbf{0} & k\mathbf{I}_{3\times3} \end{bmatrix} \begin{bmatrix} \mathbf{I}_{3\times3} & \cdots & \mathbf{I}_{3\times3} \\ \mathbf{r}_1 \otimes & \cdots & \mathbf{r}_q \otimes \end{bmatrix} diag[\mathbf{n}_1 \quad \cdots \quad \mathbf{n}_q] = \mathbf{T}_3\mathbf{GN},$$

where $\mathbf{T}_3 = \begin{bmatrix} \mathbf{I}_{3\times3} & \mathbf{0} \\ \mathbf{0} & k\mathbf{I}_{3\times3} \end{bmatrix}$ and $det(\mathbf{T}_3) = k^3$. Thus we obtain

$$\tilde{\Omega}_S = \sqrt{det(\tilde{\mathbf{G}}\tilde{\mathbf{N}}\tilde{\mathbf{N}}^T\tilde{\mathbf{G}}^T)} = \sqrt{det(\mathbf{T}_3\mathbf{GNN}^T\mathbf{G}^T\mathbf{T}_3^T)} = k^3\sqrt{det(\mathbf{GNN}^T\mathbf{G}^T)} = k^3\Omega_S$$

$$(6.70)$$

6.5.2.4 Fixture Planning Index

There may be more than one index for evaluating the fixturing quality of a workpiece in different viewpoints. Because the objective may differ among indexes, some trade-off measures are often required to reach the resultant fixturing quality. In the configuration planning of fixtures, two objectives can be reached simultaneously: (1) the locator configuration having the greatest robustness to the geometric errors of locators and (2) the fixturing configuration having the greatest ability to withstand any disturbance wrench on the workpiece. Obviously, the two objectives are not the same. Thus, to reach simultaneously the two objectives a corresponding index is defined as

$$\Omega = \frac{w\Omega_R}{\Omega_S}.\tag{6.71}$$

In general, the value of the index Ω_R is much smaller than the value of the stable index Ω_S. To increase comparability, a weighting factor w is used in Eq. 6.71.

6.5.3   Fixture Planning

6.5.3.1 Constraints

No matter which index is applied in the configuration planning of fixtures, the planning must satisfy a series of constraints, which can be described as follows

1. *Degrees of Freedom (DOF)*: In regard to machining, some or all DOF of the workpiece must be eliminated by properly arranging locators to position the workpiece. The DOF constraint is written as

    $$rank(\mathbf{W}_L) = r.\tag{6.72}$$

    When $r = m < 6$, the workpiece is under-constrained; when $r = m = 6$, the workpiece is well-constrained; and when $r = 6$ and $m > 6$, the workpiece is over-constrained.

2. *Locating Surfaces*: Theoretically, locators can be arranged randomly on the surfaces of the workpiece as long as the DOF constraint is satisfied. However, because of different machining requirements, the positions of locators cannot be selected at an arbitrary place on the surfaces of the workpiece. For example, the contacts of locators cannot be chosen on the machining surface. The locating surface constraint for the ith locator $\mathbf{r}_i$ is written as

    $$\mathbf{r}_i = \left\{ \begin{pmatrix} x_i, & y_i, & z_i \end{pmatrix}^T \middle| \begin{array}{l} S_j(x_i, \ y_i, \ z_i) = 0 \\ i = 1, \ \cdots, \ m; \\ j = 1, \ \cdots, \ J \end{array} \right\},\tag{6.73}$$

    where $S_j(x_i, \ y_i, \ z_i) = 0$ means the ith locator $\mathbf{r}_i$ on the jth surface of the workpiece.

3. *Accessibility and Detachability*: Because all locators are passive fixels, the locator configurations must make the workpiece accessible to the workpiece and detachable from the workpiece. The accessibility and detachability constraints are represented as

$$\mathcal{K} \neq \{\mathbf{0}\}. \tag{6.74}$$

Eq. 6.74 shows that the free motion cone $\mathcal{K}$ contains non-zero elements. That is, the workpiece can be moved in one or some related directions before clamping. Thus, the workpiece is accessible and detachable.

4. *Clamping Directions*: The goal of clamping is to oppose the remaining motion possibility of the work piece after it is located. As mentioned previously, the clamping forces must be in the polar $\mathcal{K}^{\circ}$ of the free motion cone $\mathcal{K}$. That is,

$$\mathbf{f}_{cl}^{T}\Delta\mathbf{D} \leq 0, \tag{6.75}$$

where $\mathbf{f}_{cl} \in \mathfrak{R}^{6} \subset \mathcal{K}^{\circ}$ is the clamping force and $\Delta\mathbf{D} \in \mathfrak{R}^{6} \subset \mathcal{K}$ is the remaining motion of the workpiece after it is located.

5. *Clamping Domain*: Generally, the same clamping force directions may be chosen in different clamping positions. However, in some clamping positions the workpiece is not fully constrained with locators and clamps. The feasible clamping domain should be determined using the principle of form-closure fixturing. The clamping domain constraints, which are derived from the principle of form-closure fixturing, are written as

$$\left\{ \begin{array}{l} \mathbf{r}_{h} = \left\{ (x_{h}, \quad y_{h}, \quad z_{h})^{T} \left| \begin{array}{l} SC_{v}(x_{h}, \quad y_{h}, \quad z_{h}) = 0 \\ h = m+1, \quad \cdots, \quad q; \\ v = 1, \quad \cdots, \quad V \end{array} \right. \right\} \\ g_{k}(\mathbf{r}_{1}, \quad \cdots, \quad \mathbf{r}_{q}) > 0, \quad k = 1, \quad \cdots, \quad K, \end{array} \right. \tag{6.76}$$

where $SC_{v}(x_{h}, \quad y_{h}, \quad z_{h}) = 0$ means the hth clamp $\mathbf{r}_{h}$ on the vth surface of the workpiece and k inequalities $g_{k}(\mathbf{r}_{1}, \quad \cdots, \quad \mathbf{r}_{q}) > 0$ form the feasible clamping domains.

6.5.3.2 Planning Methods

According to different objective functions, configuration planning is divided into the following three methods.

1. *Minimizing Locating Error Planning Method*: In this method, the objective function is the locating robustness index Ω_R. The corresponding planning method is described as

$$
\begin{aligned}
&\min_{\substack{\mathbf{r}_i \in \Phi \\ (t=1, \cdots, m)}} \quad \Omega_R(\mathbf{r}_1, \cdots, \mathbf{r}_m) \\
&subject \quad to\ (\Phi) \\
&\left\{
\begin{array}{l}
rank\ (\mathbf{W}_L) = r \\
\mathbf{r}_i = \left\{ (x_i, \ y_i, \ z_i)^T \middle|
\begin{array}{l}
S_j(x_i, \ y_i, \ z_i) = 0 \\
i = 1, \ \cdots, \ m\ ; \\
j = 1, \ \cdots, \ J
\end{array}
\right\} \\
\mathcal{K} \neq \{\mathbf{0}\}.
\end{array}
\right.
\end{aligned}
\tag{6.77}
$$

Thus, finding the optimal locator configuration is expressed as a constrained nonlinear programming whereby constraints include the DOF constraint, the locating surface constraints, and the accessibility and detachability conditions. Using this method, an optimal locator configuration can be found to minimize the locating errors of the workpiece under the constraints.

2. *Stability-based Planning Method*: In this method, the objective function is the stability index Ω_S. The corresponding planning method is represented as

$$
\begin{aligned}
&\max_{\substack{\mathbf{r}_i \in \Phi \\ (t=1, \cdots, q)}} \quad \Omega_S(\mathbf{r}_1, \cdots, \mathbf{r}_q) \\
&subject \quad to\ (\Phi) \\
&\left\{
\begin{array}{l}
rank\ (\mathbf{W}_L) = r \\
\mathbf{r}_i = \left\{ (x_i, \ y_i, \ z_i)^T \middle|
\begin{array}{l}
S_j(x_i, \ y_i, \ z_i) = 0 \\
i = 1, \ \cdots, \ m\ ; \\
j = 1, \ \cdots, \ J
\end{array}
\right\} \\
\mathcal{K} \neq \{\mathbf{0}\} \\
\mathbf{f}_{cl}^T \Delta \mathbf{D} \leq 0 \\
\mathbf{r}_h = \left\{ (x_h, \ y_h, \ z_h)^T \middle|
\begin{array}{l}
SC_v(x_h, \ y_h, \ z_h) = 0 \\
h = m+1, \ \cdots, \ q\ ; \\
v = 1, \ \cdots, \ V.
\end{array}
\right\} \\
g_k(\mathbf{r}_1, \ \cdots, \ \mathbf{r}_q) > 0, \ k = 1, \ \cdots, \ K
\end{array}
\right.
\end{aligned}
\tag{6.78}
$$

Here, the problem of the configuration planning is also changed into a constrained nonlinear programming whereby constraints include the clamping direction constraints, and clamping domain constraints in addition to those for nonlinear programming using Eq. 6.68. Using this method, an optimal fixturing configuration can be found with the strongest capability to withstand the external disturbance wrench exerted on the workpiece under the constraints.

3. *Fixturing Resultant Index-based Planning Method*: In this method, the objective function is the fixturing resultant evaluation index Ω. The corresponding planning method is stated as

$$\min_{\substack{r_i \in \Phi \\ (t=1, \cdots, q)}} \left(\Omega(r_1, \cdots, r_q) = \frac{w\Omega_R(r_1, \cdots, r_m)}{\Omega_S(r_1, \cdots, r_q)} \right)$$

subject to (Φ)

$$\begin{cases} rank(\mathbf{W}_L) = r \\ r_i = \left\{ (x_i, y_i, z_i)^T \middle| \begin{matrix} S_j(x_i, y_i, z_i) = 0 \\ i = 1, \cdots, m; \\ j = 1, \cdots, J \end{matrix} \right\} \\ \mathcal{K} \neq \{0\} \\ \mathbf{f}_{cl}^T \Delta \mathbf{D} \leq 0 \\ r_h = \left\{ (x_h, y_h, z_h)^T \middle| \begin{matrix} SC_v(x_h, y_h, z_h) = 0 \\ h = m+1, \cdots, q; \\ v = 1, \cdots, V \end{matrix} \right\} \\ g_k(r_1, \cdots, r_q) > 0, \ k = 1, \cdots, K. \end{cases}$$

(6.79)

Similarly, the third planning method is still a constrained nonlinear programming with the same constraints as the second planning method. By using this method, an optimal fixturing configuration can be found to minimize the locating errors of the workpiece, and to achieve the strongest capability to withstand the external disturbance wrench exerted on the workpiece under the constraints.

6.5.4 Case Study

To verify our proposed configuration planning methods, an example is given for planning the fixture configuration for a workpiece. Assume that the side length of the cubic workpiece (a cubic rigid body with one corner cut out) is 500 (units). The locator and clamp configuration is shown in Figure 6.14.

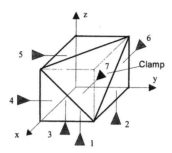

Figure 6.14. Locator and clamp layout.

First, by using the planning method described by Eq. 6.77, the optimal locator configuration is obtained with some midway configurations during the process of iteration, as outlined in Table 6.2 and shown in Figure 6.15.

Table 6.2. Optimal locator configuration and some midway configurations (using the locating robustness index Ω_R as the objective function).

Configuration No.	Locators' Coordinates	Notation
1	(300, 250, 0); (100, 300, 0); (100, 100, 0); (300, 0, 250); (100, 0, 250); (0, 250, 250)	Initial configuration
2	(317.76, 250.56, 0); (84.42, 323.70, 0); (92.92, 65.18, 0); (324.88, 0, 262.25); (63.06, 0, 222.68); (0, 234.92, 237.75)	Some midway configurations
3	(329.77, 250.95, 0); (73.89, 339.71, 0); (88.14, 41.65, 0); (341.69, 0, 270.53); (38.10, 0, 204.21); (0, 224.74, 229.47)	
4	(348.09, 251.53, 0); (57.81, 364.16, 0); (80.84, 5.74, 0); (367.35, 0, 283.16); (0, 0, 176.02); (0, 209.18, 216.84)	
5	(351.54, 251.99, 0); (55.20, 367.25, 0); (78.53, 0, 0); (370.99, 0, 284.59); (0, 0, 171.88); (0, 206.29, 214.01)	
6	(405.97, 262.94, 0); (18.27, 400.02, 0); (31.98, 0, 0); (416.32, 0, 297.09); (0, 0, 118.69); (0, 160.27, 164.16)	
7	(432.90, 268.35, 0); (0, 416.23, 0); (8.96, 0, 0); (438.75, 0, 303.27); (0, 0, 92.37); (0, 137.49, 139.49)	
8	(442.48, 270.43, 0); (0, 421.84, 0); (0, 0, 0); (446.82, 0, 305.19); (0, 0, 82.67); (0, 128.26, 130.09)	Optimal configuration

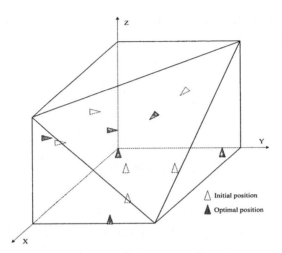

Figure 6.15. Robust locator configuration.

The changes of the position and orientation errors of the workpiece during the process of iteration are shown in Figures 6.16 and 6.17. It can be found that the position and orientation errors of the workpiece always show a decreasing trend during the iteration process, and reach their minima at the optimal locator configuration. However, we cannot determine the optimal clamp position using the locating robustness index Ω_R to plan the fixture configuration.

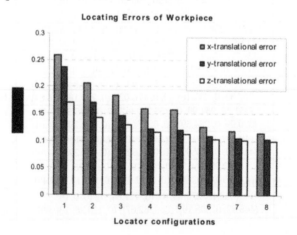

Figure 6.16. Changes of the position errors of the workpiece (using the robustness index Ω_R as the objective function).

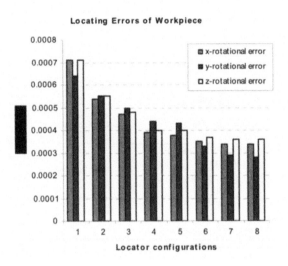

Figure 6.17. Changes of the orientation errors of the workpiece
(using the robustness index Ω_R as the objective function).

Now the fixture configuration for the workpiece is planned using the second planning method (described by Eq. 6.78 whereby the stability index Ω_S is the objective function. Under the constraints expressed in Eq. 6.78, the optimal fixture configuration is obtained with some midway configurations during the process of iteration, as shown in Table 6.3 and Figure 6.18. The changes of the position and orientation errors of the workpiece, and the stability index Ω_S during the process of iteration, are shown in Figures 6.19, 6.20, and 6.21, respectively. From Figures 6.19 and 6.20, it can be found that the changes of the position and orientation errors of the workpiece are not related to the iteration process. In other words, the position and orientation errors of the workpiece do not reach their minima at the optimal locator configuration. However, the stability index Ω_S shows an increasing trend during the iteration process from Figure 6.22 (the value of the stability index Ω_S is normalized in Figure 6.22), and reaches its maximum in the vicinity of the third fixture configuration. It then slightly decreases along with the iteration, but is still close to the maximum, which means that the position constraints of the clamp in the vicinity of the third fixture configuration are not fully satisfied although the objective function approaches its maximum. After several iterations, the optimal fixture configuration is obtained which maximizes the stability index Ω_S and satisfies all constraints described in Eq. 6.78.

Table 6.3. Optimal fixture configuration and some midway configurations (using the fixturing stability index Ω_S as the objective function)

Config-uration No.	Fixture Configurations		Notation
	Locators' Coordinates	Clamp Coordinates	
1	(300, 250, 0); (100, 300, 0); (100, 100, 0); (300, 0, 250); (100, 0, 250); (0, 250, 250)	(300, 300, 400)	Initial configuration
2	(500, 244.15, 0); (0, 500, 0); (0, 0, 0); (500, 0, 269.22); (0, 0, 256.42); (0, 333.12, 206.85)	(320.01, 228.13, 451.86)	Some midway configurations
3	(500, 109.55, 0); (0, 500, 0); (0, 0, 0); (500, 0, 238.55); (0, 0, 319.54); (0, 373.88, 146.69)	(326.48, 200.36, 473.17)	
4	(500, 53.11, 0); (0, 500, 0); (0, 0, 0); (500, 0, 225.39); (0, 0, 352.28); (0, 397.27, 122.66)	(326.69, 223.96, 449.35)	
5	(500, 46.44, 0); (0, 500, 0); (0, 0, 0); (500, 0, 225.39); (0, 0, 345.77); (0, 397.30, 122.69)	(326.67, 223.97, 449.36)	
6	(500, 50.35, 0); (0, 500, 0); (0, 0, 0); (500, 0, 225.26); (0, 0, 350.32); (0, 397.34, 122.60)	(326.69, 224.03, 449.28)	
7	(500, 173.47, 0); (0, 500, 0); (0, 0, 0); (500, 0, 220.05); (0, 0, 499.99); (0, 398.90, 118.95)	(327.39, 226.28, 446.33)	
8	(500, 156.80, 0); (0, 500, 0); (0, 0, 0); (500, 0, 218.06); (0, 0, 486.52); (0, 400.83, 118.90)	(326.76, 227.59, 445.65)	
9	(500, 0.002, 0); (0, 500, 0); (0, 0, 0); (500, 0, 199.36); (0, 0, 360.04); (0, 419.07, 118.42)	(320.84, 239.90, 439.26)	Optimal configuration

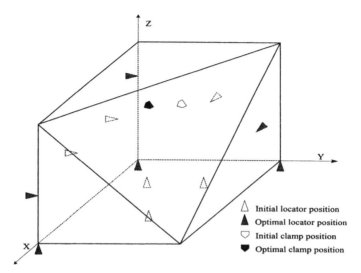

Figure 6.18. Stable fixture configuration.

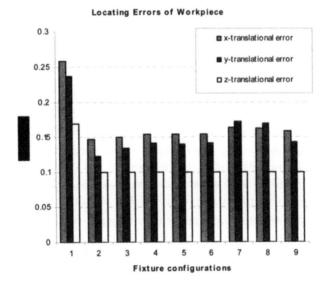

Figure 6.19. Changes of the position errors of the workpiece (using the stability index Ω_S as the objective function).

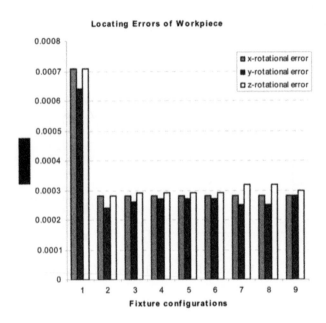

Figure 6.20. Changes of the orientation errors of the workpiece (using the stability index Ω_S as the objective function).

Finally, the fixture configuration for the workpiece is planned using the third planning method (Eq. 6.79) whereby the fixturing resultant evaluation index Ω is used as the objective function and the weighting factor is chosen as $w = 10^8$. The optimal fixture configuration and some midway configurations during the process of iteration are outlined in Table 6.4 and shown in Figure 6.22. The changes of the position and orientation errors of the workpiece, and the stability index Ω_S during the process of iteration, are shown in Figures 6.23, 6.24, and 6.25, respectively. From Figures 6.23 and 6.24, it can be seen that the changes of the position and orientation errors of the workpiece always show a decreasing trend during the iteration process, and reach their minima at the optimal fixture configuration. At the same time, the stability index Ω_S shows an increasing trend during the iteration process from Figure 6.25, and reaches its maximum at the optimal fixture configuration.

Table 6.4. Optimal fixture configuration and some midway configurations.

Config-uration No.	Fixture Configurations		Notation
	Locators' Coordinates	Clamp Coordinates	
1	(300, 250, 0); (100, 300, 0); (100, 100, 0); (300, 0, 250); (100, 0, 250); (0, 250, 250)	(300, 300, 400)	Initial configuration
2	(315.66, 250.48, 0); (87.68, 317.23, 0); (95.38, 81.05, 0); (316.62, 0, 252.00); (79.32, 0, 242.92); (247.38, 247.05)	(299.86, 296.24, 403.90)	
3	(325.76, 250.79, 0); (79.74, 328.33, 0); (92.41, 68.85, 0); (327.32, 0, 253.28); (65.99, 0, 238.36); (0, 245.69, 245.15)	(299.77, 293.82, 406.41)	
4	(340.44, 251.25, 0); (68.19, 344.47, 0); (88.08, 51.09, 0); (342.89, 0, 255.15); (46.61, 0, 231.72); (0, 243.22, 242.38)	(299.65, 290.29, 410.06)	
5	(355.65, 251.72, 0); (56.23, 361.20, 0); (83.59, 32.69, 0); (359.02, 0, 257.09); (26.53, 0, 224.85); (0, 240.68, 239.51)	(299.51, 286.64, 413.85)	Some midway configurations
6	(373.54, 252.27, 0); (42.15, 380.88, 0); (78.31, 11.05, 0); (378.00, 0, 259.36); (2.90, 0, 216.76); (0, 237.68, 236.14)	(299.36, 282.35, 418.29)	
7	(375.74, 252.34, 0); (40.42, 383.30, 0); (77.66, 8.39, 0); (380.33, 0, 259.65); (0, 0, 215.77); (0, 237.31, 235.72)	(299.34, 281.82, 418.84)	
8	(382.97, 252.82, 0); (35.05, 390.46, 0); (75.07, 0, 0); (387.44, 0, 260.22); (0, 0, 212.86); (0, 235.94, 234.09)	(299.12, 280.17, 420.71)	
9	(432.30, 257.56, 0); (0, 434.95, 0); (54.84, 0, 0); (433.31, 0, 262.25); (0, 0, 194.90); (0, 225.60, 221.52)	(296.68, 269.37, 433.95)	
10	(500, 265.13, 0); (0, 494.90, 0); (25.18, 0, 0); (496.33, 0, 263.76); (0, 0, 170.78); (0, 210.07, 203.47)	(292.01, 254.08, 453.91)	Optimal configuration

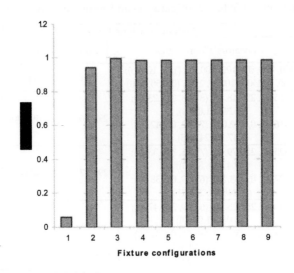

Figure 6.21. Stability index Ω_S during the process of iteration.

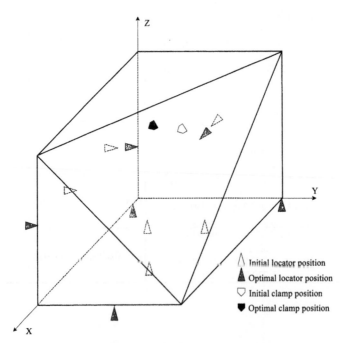

Figure 6.22. Optimal fixture configuration.

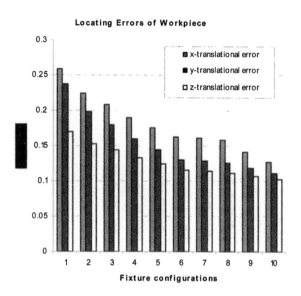

Figure 6.23. Changes of the position errors of the workpiece.

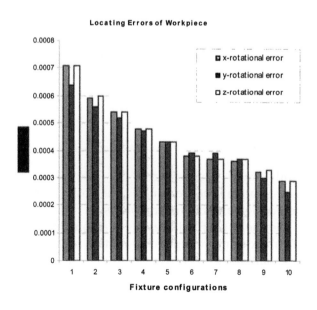

Figure 6.24. Changes of the orientation errors of the workpiece.

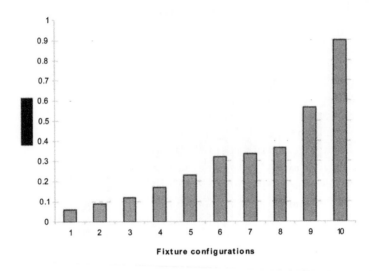

Figure 6.25. Change of the stability index Ω_S .

Comparing the three planned results shows that the nonlinear programming using Eq. 6.79 is a more effective method of finding the optimal fixture configuration that minimizes locating errors and allows the fixturing system to have the strongest capability to withstand any external disturbance wrench.

6.5.5   Conclusions

There inherently exists position error for every locator. When the configuration of locators is not reasonable, the position errors of locators will affect the position and orientation precision of the workpiece more significantly. It is one of the fundamentals of automated fixture design to plan the optimal fixturing configuration so that high locating accuracy and fixturing stability for a workpiece can be obtained.

This section first defined three performance indexes: the locating robustness index, the stability index of fixturing, and the fixturing resultant index. The locating robustness index is used to evaluate the configurations of locators. Minimizing the locating robustness index means that the position and orientation of the workpiece are most insensitive to locator errors. The stability index, which is invariant under a linear coordinate transformation and a change of torque origin (and invariant under a change of the dimensional unit) is used to evaluate the capability to withstand any external disturbance wrench for the fixturing system. The larger the value of the stability index the farther the fixel contact configuration is from a singular configuration. This implies that the fixturing has greater ability to withstand any disturbance wrench on the workpiece. The fixturing resultant index is used to evaluate simultaneously the robustness to the position errors of locators and the stability under the external disturbance wrench for the fixturing system. Minimizing the fixturing resultant index implies that the position and orientation of the workpiece are most insensitive to locator errors at the same time the fixturing system has the strongest capability to withstand any external disturbance wrench. Then the corresponding three nonlinear programming methods for planning the optimal fixturing configuration are formulated. A set of constraints, such as the constraints of DOF of the workpiece, the locating surface constraint, accessibility and detachability constraints, the clamping direction constraints, and the clamping domain constraints is taken into account in the programming methods. Finally, the effectiveness of the three programming methods is verified by an example. Comparison of the planning results of the three methods shows that the third programming method in which the objective is to minimize the fixturing resultant index is an effective and reasonable planning method. The proposed programming methods are applicable to all types of locating schemes, including deterministic locating, under-constrained locating, and over-constrained locating.

References

Asada, H., and A. B. By. "Kinematic Analysis of Workpart Fixturing for Flexible Assembly with Automatically Reconfigurable Fixtures," *IEEE Journal of Robotics and Automation* 1:2, pp. 86–93, 1985.

Bausch, J. J., and K. Y. Toumi. "Kinematic Methods for Automated Fixture Reconfiguration Planning," *Proceedings of the IEEE International Conference on Robotics and Automation*, pp. 1396–1401, 1990.

Bicchi, A. "On the Closure Properties of Robotic Grasping," *International Journal of Robotics Research* 14:4, pp. 319–334, 1995.

Brost, R. C, and K. Y. Goldberg. "A Complete Algorithm for Designing Planar Fixtures Using Modular Components," *IEEE Transactions of Robotics and Automation* 12:1, pp. 31–46, 1996.

Brost, R. C., M. Fischer, and G. Hirzinger. "A Fast and Robust Grasp Planner for Arbitrary 3D Objects," *Proceedings of the IEEE International Conference on Robotics and Automation*, pp. 1890–1896, 1999.

Brown, R. G., and R. C. Brost. "A 3-D Modular Gripper Design Tool," *IEEE Transactions of Robotics and Automation* 15:1, pp. 174–186, 1999.

Cai, W., S. J. Hu, and J. X. Yuan. "A Variational Method of Robust Fixture Configuration Design for 3-D Workpieces," *Transactions of the ASME, Journal of Manufacturing Science and Engineering* 119, pp. 593–602, 1997.

Carlson, J. S. "Quadratic Sensitivity Analysis of Fixtures and Locating Schemes for Rigid Parts," *Transactions of the ASME, Journal of Manufacturing Science and Engineering* 123, pp. 462–472, 2001.

Chou, Y. C., V. Chandru, and M. M. Barash. "A Mathematical Approach to Automatic Configuration of Machining Fixtures: Analysis and Synthesis," *Transactions of the ASME, Journal of Engineering for Industry* 111, pp. 299–306, 1989.

Choudhuri, S. A., and E. C. DeMeter. "Tolerance Analysis of Machining Fixture Locators," *Transactions of the ASME, Journal of Manufacturing Science and Engineering* 121, pp. 273–281, 1999.

Czyzowicz, J., I. Stojmenovic, and J. Urrutia. "Immobilizing a Polytope," *Proceedings of the Second Workshop on Algorithms and Data Structures (WADS91)*, *Lecture Notes in Computational Science* 519, pp. 214–227, 1991.

Ding, D., Y. H. Liu, M. Y. Wang, and S. Wang. "Automatic Selection of Fixturing Surfaces and Fixturing Points for Polyhedral Workpieces," *IEEE Transactions of Robotics and Automation* 17:6, pp. 833–841, 2001.

Han, L., J. C. Trinkle, and Z. X. Li. "Grasp Analysis as Linear Matrix Inequality Problems," *IEEE Transactions of Robotics and Automation* 16:6, pp. 663–674, 2000.

Howard, W. S., and V. Kumar. "On the Stability of Grasped Objects," *IEEE Transactions of Robotics and Automation* 12:6, pp. 904–917, 1996.

Huang, K., and P. Gu. "Tolerance Analysis in Setup and Fixture Planning for Precision Machining," *Proceedings of the Fourth International Conference on Computer Integrated Manufacturing and Automation Technology*, pp. 289–305, 1994.

Kang, Y. "Computer-Aided Fixture Design Verification," Ph.D. dissertation, Worcester Polytechnic Institute, Worcester, MA, 2001.

Kang, Y., Y. Rong, J. Yang, and W. Ma. "Computer-aided Fixture Design Verification," *Assembly Automation* 22, pp. 350–359, 2002.

Kerr, J., and B. Roth. "Analysis of Multifingered Hands," *International Journal of Robotics Research* 4:4, pp. 3–17, 1986.

Krishnakumar, K., and S. N. Melkote. "Machining Fixture Layout Optimization Using the Genetic Algorithm," *International Journal of Machine Tools and Manufacture* 40, pp. 579–598, 2000.

Lakshminarayana, K. "Mechanics of Form Closure," Paper No. 78-DET-32, pp. 2–8, American Society of Mechanical Engineering, New York, 1978.

Li, B., and S. N. Melkote. "Improved Workpiece Location Accuracy Through Fixture Layout Optimization," *International Journal of Machine Tools and Manufacture* 39, pp. 871–883, 1999.

Lin, Q., and J. W. Burdick. "A Task-Dependent Approach to Minimum-Deflection Fixtures," *Proceedings of the IEEE International Conference on Robotics and Automation*, pp. 1562–1569, 1999.

Lin, Q., J. W. Burdick, and E. Rimon. "A Stiffness-Based Quality Measure for Compliant Grasps and Fixtures," *IEEE Transactions of Robotics and Automation* 17:5, pp. 679–697, 2000.

Liu, Y. H. "Qualitative Test and Force Optimization of 3-D Frictional Form-Closure Grasps Using Linear Programming," *IEEE Transactions of Robotics and Automation* 15:1, pp. 163–173, 1999.

Liu, Y. H. "Computing n-Finger Form-Closure Grasps on Polygonal Objects," *International Journal of Robotics Research* 19:2, pp. 149–158, 2000.

Marin, R. A., and P. M. Ferreira. "Kinematic Analysis and Synthesis of Deterministic 3-2-1 Locator Schemes for Machining Fixtures," *Transactions of the ASME, Journal of Manufacturing Science and Engineering* 123, pp. 708–719, 2001a.

Marin, R. A., and P. M. Ferreira. "Optimal Placement of Fixture Clamps," *Proceedings of the IEEE/ASME International Conference on Advanced Intelligent Mechatronics*, pp. 314–319, 2001b.

Mishra, B. "Workholding Analysis and Planning," IEEE/RSI International Workshop on Intelligent Robots and Systems, IROS '91, pp. 53–57, 1991.

Montana, D. J. "Contact Stability for Two-Fingered Grasps," *IEEE Transactions of Robotics and Automation* 8:4, pp. 421–430, 1992.

Ponce, J., and B. Faverjon. "On Computing Three-Finger Force-Closure Grasps of Polygonal Objects," *IEEE Transactions of Robotics and Automation* 11:6, pp. 868–881, 1995.

Ponce, J., S. Sullivan, A. Sudsang, J. D. Boissonnat, and J. P. Merlet. "On Computing Four-Finger Equilibrium and Force-Closure Grasps of Polyhedral Objects," *International Journal of Robotics Research* 16:1, pp. 11–35, 1997.

Rimon, E. "A Curvature-Based Bound on the Number of Frictionless Fingers Required to Immobilize Three-Dimensional Objects," *IEEE Transactions of Robotics and Automation* 17:5, pp. 679–697, 2000.

Rockafellar, R. T. *Convex Analysis*. Princeton, MA: Princeton University Press, 1970.

Rong, Y., and Y. Zhu. *Computer-aided Fixture Design*. New York: Marcel Dekker, 1999.

Trinkle, J. C. "On the Stability and Instantaneous Velocity of Grasped Frictionless Objects," *IEEE Transactions of Robotics and Automation* 8:5, pp. 560–572, 1992.

Tung, C. P., and A. C. Kak. "Fast Construction of Force-Closure Grasps," *IEEE Transactions of Robotics and Automation* 12:4, pp. 615–626, 1996.

Vallapuzha, S., E. C. DeMeter, S. Choudhuri, and R. P. Khetan. "An Investigation into the Use of Spatial Coordinates of the Genetic Algorithm Based Solution of the Fixture Layout Optimization Problem," *International Journal of Machine Tools and Manufacture* 42, pp. 265–275, 2002a.

Vallapuzha, S., E. C. DeMeter, S. Choudhuri, and R. P. Khetan. "An Investigation into the Use of Spatial Coordinates of the Genetic Algorithm Based Solution of the Fixture Layout Optimization Problem," *International Journal of Machine Tools and Manufacture* 42, pp. 265–275, 2002b.

van der Stappen, A. F., C. Wentink, and M. H. Overmars. "Computing Immobilizing Grasps of Polygonal Parts," *International Journal of Robotics Research* 19:5, pp. 467–479, 2000.

Wang, M. Y. "Precision Workpiece Fixturing and Localization," International Conference on Advanced Manufacturing Technology, Xi'an, China, 1999.

Wang, M. Y., and D. M. Pelinescu. "Optimizing Fixture Layout in a Point Set Domain," *IEEE Transactions of Robotics and Automation* 17:3, pp. 312–323, 2001.

Wang, M. Y., T. Liu, and D. M. Pelinescu. "Fixture Kinematic Analysis Based on the Full Contact Model of Rigid Bodies," *Journal of Manufacturing Science and Engineering* 125, pp. 316–324, 2003.

Wu, Y., Y. Rong, W. Ma, and S. LeClair. "Automated Modular Fixture Design: Geometric Analysis," *Robotics and Computer-Integrated Manufacturing* 14:1, pp. 1–15, 1998.

Xiong, C. H., and Y. L. Xiong. "Stability Index and Contact Configuration Planning for Multifingered Grasp," *Journal of Robotic Systems* 15:4, pp. 183–190, 1998.

Xiong, C. H., Y. F. Li, H. Ding, and Y. L. Xiong. "On the Dynamic Stability," *International Journal of Robotics Research* 18:9, pp. 951–958, 1999.

Xiong, C. H., Y. F. Li, Y. Rong, and Y. L. Xiong. "Qualitative Analysis and Quantitative Evaluation of Fixturing," *Robotics and Computer Integrated Manufacturing* 18:5-6, pp. 335–342, 2002.

Zhang, Y., W. Hu, Y. Kang, Y. Rong, and D. W. Yen. "Locating Error Analysis and Tolerance Assignment for Computer–aided Fixture Design," *International Journal of Production Research* 39:15, pp. 3529–3545, 2001.

Index

Printed and bound by CPI Group (UK) Ltd, Croydon, CR0 4YY

03/10/2024

01040418-0008